Lecture Notes in Mathematics

Edited by A. Dold and B. Eckmann

1133

Krzysztof C. Kiwiel

Methods of Descent for Nondifferentiable Optimization

Springer-Verlag
Berlin Heidelberg New York Tokyo

Author

Krzysztof C. Kiwiel
Systems Research Institute, Polish Academy of Sciences
ul. Newelska 6, 01-447 Warsaw, Poland

Mathematics Subject Classification: 49-02, 49D37, 65-02, 65K05, 90-02, 90C30

ISBN 3-540-15642-9 Springer-Verlag Berlin Heidelberg New York Tokyo
ISBN 0-387-15642-9 Springer-Verlag New York Heidelberg Berlin Tokyo

Printing and binding: Beltz Offsetdruck, Hemsbach/Bergstr.
2146/3140-543210

Math Sci
Sep

PREFACE

This book is about numerical methods for problems of finding the largest or smallest values which can be attained by functions of several real variables subject to several inequality constraints. If such problems involve continuously differentiable functions, they can be solved by a variety of methods well documented in the literature. We are concerned with more general problems in which the functions are locally Lipschitz continuous,but not necessarily differentiable or convex. More succintly, this book is about numerical methods for nondifferentiable optimization.

Nondifferentiable optimization, also called nonsmooth optimization, has many actual and potential applications in industry and science. For this reason, a great deal of effort has been devoted to it during the last decade. Most research has gone into the theory of nonsmooth optimization, while surprisingly few algorithms have been proposed, these mainly by C.Lemaréchal, R.Mifflin and P.Wolfe. Frequently such algorithms are conceptual, since their storage and work per iteration grow infinitely in the course of calculations. Also their convergence properties are usually weaker than those of classical methods for smooth optimization problems.

This book gives a complete state-of-the-art in general-purpose methods of descent for nonsmooth minimization. The methods use piecewise linear approximations to the problem functions constructed from several subgradients evaluated at certain trial points. At each iteration, a search direction is found by solving a quadratic programming subproblem and then a line search produces both the next improved approximation to a solution and a new trial point so as to detect gradient discontinuities. The algorithms converge to points satisfying necessary optimality conditions. Also they are widely applicable, since they require only a weak semismoothness hypothesis on the problem functions which is likely to hold in most applications.

A unifying theme of this book is the use of subgradient selection and aggregation techniques in the construction of methods for nondifferentiable optimization. It is shown that these techniques give rise in a totally systematic manner to new implementable and globally convergent modifications and extensions of all the most promising algorithms which have been recently proposed. In effect, this book should give the reader a feeling for the way in which the subject has developed and is developing, even though it mainly reflects the author's research.

This book does not discuss methods without a monotonic descent (or ascent) property, which have been developed in the Soviet Union.

The reason is that the subject of their effective implementations is still a mystery. Moreover, these subgradient methods are well described in the monograph of Shor (1979). We refer the reader to Shor's excellent book (its English translation was published by Springer-Verlag in 1985) for an extensive discussion of specific nondifferentiable optimization problems that arise in applications. Due to space limitations, such applications will not be treated in this book.

In order to make the contents of this book accessible to as wide a range of readers as possible, our analysis of algorithms will use only a few results from nonsmooth optimization theory. These, as well as certain other results that may help the reader in applications, are briefly reviewed in the introductory chapter, which also contains a review of representative existing algorithms. The reader who has basic familiarity with nonsmooth functions may skip this chapter and start reading from Chapter 2, where methods for unconstrained convex minimization are described in detail. The basic constructions of Chapter 2 are extended to the unconstrained nonconvex case in two fundamentally different ways in Chapters 3 and 4, giving rise to competitive methods. Algorithms for constrained convex problems are treated in Chapter 5, and their extensions to the nonconvex case are described in Chapter 6. Chapter 7 presents new versions of the bundle method of Lemaréchal and its extensions to constrained and nonconvex problems. Chapter 8 contains a few numerical results.

The book should enable research workers in various branches of science and engineering to use methods for nondifferentiable optimization more efficiently. Although no computer codes are given in the text, the methods are described unambiguously, so computer programs may readily be written.

The author would like to thank Claude Lemaréchal and Dr. A.Ruszczyński for introducing him to the field of nonsmooth optimization, and Prof. K.Malanowski for suggesting the idea of the book. Without A.Ruszczyński's continuing help and encouragement this book would not have been written. Part of the results of this book were obtained when the author worked for his doctoral dissertation under the supervision of Prof. A.P.Wierzbicki at the Institute of Automatic Control of the Technical University of Warsaw. The help of Prof. R.Kulikowski and Prof. J.Hołubiec from the Systems Research Institute of the Polish Academy of Sciences, where this book was written, is gratefully acknowledged. Finally, the author wishes to thank Mrs. I.Forowicz and Mrs. E.Grudzińska for patiently typing the manuscript.

TABLE OF CONTENTS

CHAPTER 1

Fundamentals

1. Introduction

The nonlinear programming problem, also known as the mathematical programming problem, can be taken to have the form

$$P: \text{ minimize } f(x), \text{ subject to } F_i(x) \le 0 \text{ for } i=1,\dots,m,$$

where the objective function f and the constraint functions F_i are real-valued functions defined on the N-dimensional Euclidean space R^N. The value of $m \ge 0$ is finite; when $m=0$ the problem is unconstrained. Often the optimization problem P is smooth: the problem functions f and F_i are continuously differentiable, i.e. they have continuous gradients ∇f and ∇F_i, $i=1,\dots,m$. But in many applications this is not true. Nonsmooth problems are the subject of nonsmooth optimization, also called nondifferentiable optimization.

Owing to actual and potential applications in industry and science, recently much research has been conducted in the area of nonsmooth optimization both in the East (see the excellent monographs by Gupal (1979), Nurminski (1979) and Shor (1979)) and in the West (see the comprehensive bibliographies of Gwinner (1981) and Nurminski (1982)).

Nonsmooth problems that arise in applications have certain common features. They are more complex and have poorer analytical properties than standard mathematical programming problems, cf. (Bazaraa and Shetty, 1979; Pshenichny and Danilin, 1975). A single evaluation of the problem functions usually requires solutions of auxiliary optimization subproblems. In particular, it is very common to encounter a nondifferentiable function which is the pointwise supremum of a collection of functions that may themselves be differentiable - a max function.

Functions with discontinuous gradients, such as max functions, cannot be minimized by classical nonlinear programming algorithms. This observation applies both to gradient-type algorithms (the method of steepest descent, conjugate direction methods, quasi-Newton methods) and to direct search methods which do not require calculation of derivatives (the method of Nelder and Mead, the method of Powell, etc.), see (Lemarechal, 1978 and 1982; Wolfe, 1975).

This work is concerned with numerical methods for finding (approximate) solutions to problem P when the problem functions are locally Lipschitzian, i.e. Lipschitz continuous on each bounded subset of R^N, but not necessarily differentiable.

The advent of F.H.Clarke's (1975) analysis of locally Lipschitzian functions provided a unified approach to both nondifferentiable and nonconvex problems (Clarke, 1976). Clarke's subdifferential analysis, the pertinent part of which is briefly reviewed in the following section, suffices for establishing properties of a vast class of optimization problems that arise in applications (Pshenichny, 1980; Rockafellar, 1978).

2. Basic Results of Nondifferentiable Optimization Theory

In this section we describe general properties of nondifferentiable optimization problems that are the subject of this work. Basic familiarity is, however, assumed. Source material may be found in (Clarke, 1975; Clarke, 1976; Rockafellar, 1970; Rockafellar, 1978; Rockafellar, 1981).

The section is organized as follows. First, we review concepts of differentiability and elementary properties of the Clarke subdifferential. The proofs are omitted, because only simple results, such as Lemma 2.2, will be used in subsequent chapters. Other results, in particular the calculus of subgradients, should help the reader who is mainly interested in applications. Secondly, we study convex first order approximations to nondifferentiable functions. Such approximations are then used for deriving necessary conditions of optimality for nondifferentiable problems. Our approach is elementary and may appear artificial. However, it yields useful interpretations of the algorithms described in subsequent chapters.

The following notation is used. We denote by $\langle\cdot,\cdot\rangle$ and $|\cdot|$, respectively, the usual inner product and norm in finite-dimensional, real Euclidean space. R^N denotes Euclidean space of dimension $N<\infty$. We use x_i to denote the i-th component of the vector x. Thus

$\langle x,y\rangle = \sum_{i=1}^{N} x_i y_i$ and $|x| = \langle x,x\rangle^{1/2}$ for $x,y \in R^N$. Superscripts are used to denote different vectors, e.g. x^1 and x^2. All vectors are column vectors. However, for convenience a column vector in R^{N+n} is sometimes denoted by (x,y) even though x and y are column vectors in R^N and R^n, respectively. $[x,y]$ denotes the line segment joining x and y in R^N, i.e. $[x,y]=\{z\in R^N: z=\lambda x+(1-\lambda)y$ for some λ satisfying $0\le\lambda\le 1\}$.

A set $S\subset R^N$ is called <u>convex</u> if $[x,y]\subset S$ for all x and y belonging to S. A linear combination $\sum_{j=1}^{k} \lambda_j x^j$ is called a <u>convex combination</u> of points $x^1,\dots,x^k$ in R^N if each $\lambda_j \ge 0$ and $\sum_{j=1}^{k} \lambda_j = 1$. The <u>convex hull</u> of a set $S\subset R^N$, denoted conv S, is the set of all convex combinations of points in S. conv S is the smallest convex set containing S, and S is convex if and only if S=conv S. An important property of convex hulls is described in

Lemma 2.1 (Caratheodory's theorem; see Theorem 17.1 in (Rockafellar, 1970)).

If $S \subset R^N$ then $x \in \text{conv } S$ if and only if x is expressible as a convex combination of N+1 (not necessarily different) points of S.

Any nonzero vector $g \in R^N$ and number γ define a hyperplane

$$H = \{x \in R^N : \langle g,x \rangle = \gamma\},$$

which is a translation of the (N-1)-dimensional subspace $\{x \in R^N: \langle g,x\rangle=0\}$ of R^N. H divides R^N into two closed half-spaces $\{x \in R^N: \langle g,x \rangle \le \gamma\}$ and $\{x \in R^N: \langle g,x \rangle \ge \gamma\}$, respectively. We say that H is a supporting hyperplane to a set $S \subset R^N$ at $\bar{x} \in S$ if $\langle g,\bar{x} \rangle = \gamma$ and $\langle g,x \rangle \le \gamma$ for all $x \in S$. Any closed convex set S can be described as an intersection of all the closed half-spaces that contain S.

We use the set notation

$$S^1+S^2 = \{z^1+z^2 : z^1 \in S^1, z^2 \in S^2\},$$

$$\text{conv}\{S^i: i=1,2\} = \text{conv}\{z: z \in S^1 \cup S^2\}$$

for any subsets S^1 and S^2 of R^N.

A function $f:R^N \longrightarrow R$ is called convex if

$$f(\lambda x^1+(1-\lambda)x^2) \le \lambda f(x^1)+(1-\lambda)f(x^2) \quad \text{for all } \lambda \in [0,1] \text{ and } x^1,x^2 \in R.$$

This is equivalent to the epigraph of f

$$\text{epi } f = \{(x,\beta) \in R^{N+1} : \beta \ge f(x)\}$$

being a convex subset of R^{N+1}. A function $f:R^N \longrightarrow R^1$ is called concave if the function $(-f)(x)=-f(x)$ is convex. If $F_i:R^N \longrightarrow R$ is convex and $\lambda_i \ge 0$ for each $i=1,\ldots,k$, then the functions

$$\phi_1(x) = \sum_{i=1}^{k} \lambda_i f_i(x),$$
$$\phi_2(x) = \max\{f_i(x): i=1,\ldots,k\} \qquad (2.1)$$

are convex.

A function $f:R^N \longrightarrow R$ is strictly convex if $f(\lambda x^1+(1-\lambda)x^2) < \lambda f(x^1)+(1-\lambda)f(x^2)$ for all $\lambda \in (0,1)$ and $x^1 \ne x^2$. For instance, the function $|.|^2$ is strictly convex.

A function $f: R^N \longrightarrow R$ is said to be locally Lipschitzian if for each bounded subset B of R^N there exists a Lipschitz constant $L = L(B) < \infty$ such that

$$|f(x^1)-f(x^2)| \le L\,|x^1-x^2| \quad \text{for all} \quad x^1,x^2 \in B. \tag{2.2}$$

Then in particular f is continuous. Examples of locally Lipschitzian functions include continuously differentiable functions, convex functions, concave functions and any linear combination or pointwise maximum of a finite collection of such functions, cf. (2.1).

Following (Rockafellar, 1978), we shall now describe differentiability properties of locally Lipschitzian functions. Henceforth let f denote a function satisfying (2.2) and let x be an interior point of B, i.e. $x \in \text{int}\, B$.

The Clarke generalized directional derivative of f at x in a direction d

$$f^o(x;d) = \limsup_{y\to x,\, t\downarrow 0} [f(y+td)-f(y)]/t \tag{2.3}$$

is a finite, convex function of d and $f^o(x;d) \le L|d|$. The Dini upper directional derivative of f at x in a direction d

$$f^D(x;d) = \limsup_{t\downarrow 0} [f(x+td)-f(x)]/t \tag{2.4}$$

exists for each $d \in R^N$ and satisfies

$$f(x+td) \le f(x)+tf^D(x;d)+o(t), \tag{2.5}$$

where $o(t)/t\to 0$ as $t\downarrow 0$. The limit

$$f'(x;d) = \lim_{t\downarrow 0} [f(x+td)-f(x)]/t \tag{2.6}$$

is called the (one-sided) directional derivative of f at x with respect to d, if it exists. The two-sided derivative (the Gateaux derivative) corresponds to the case $f'(x;-d)=-f'(x;d)$. Clearly,

$$\begin{aligned} f^D(x;d) &\le f^o(x;d), \\ f'(x;d) &\le f^D(x;d), \end{aligned} \tag{2.7}$$

whenever $f'(x;d)$ exists.

If $f'(x;d)$ is linear in d (Gateaux differentiable at x)

$$f'(x;d) = \langle g_f, d \rangle \quad \text{for all} \quad d \in R^N, \tag{2.8}$$

then the vector g_f is called the <u>gradient</u> of f at x and denoted by $\nabla f(x)$. The components of $\nabla f(x)=(\frac{\partial f}{\partial x_1}(x),\ldots,\frac{\partial f}{\partial x_N}(x))$ are the coordinate-wise two-sided partial derivatives of f at x. The function f is (Frechet) <u>differentiable</u> at x if

$$f(x+d)=f(x)+\langle\nabla f(x),d\rangle+o(|d|) \quad \text{for all } d\in R^N, \tag{2.9}$$

where $o(t)/t\to 0$ as $t\downarrow 0$. The above relation is equivalent to

$$\lim_{d'\to d,\,t\downarrow 0} [f(x+td')-f(x)]/t=\langle\nabla f(x),d\rangle \quad \text{for all } d\in R^N. \tag{2.10}$$

If

$$\lim_{y\to x,\,t\downarrow 0} [f(y+td)-f(y)]/t = \langle\nabla f(x), d\rangle \quad \text{for all } d \text{ in } R^N, \tag{2.11}$$

then f is called <u>strictly differentiable</u> at x. In this case f is differentiable at x and the gradient $\nabla f:R^N\to R^N$ is continuous at x relative to its domain

$$\text{dom } \nabla f = \{y\in R^N\colon f \text{ is differentiable at } y\}$$

It is known that a locally Lipschitzian function $f:R^N\to R$ is differentiable at almost all points $x\in R^N$, and moreover that the gradient mapping ∇f is locally bounded on its domain. Suppose that (2.2) holds for some neighborhood B of a point $x\in R^N$. Then

$$\langle\nabla f(y),d\rangle = f'(y;d) = \lim_{t\downarrow 0}[f(y+td)-f(y)]/t \le L|d|$$

for all $y\in B\cap \text{dom } \nabla f$ and $d\in R^N$, and this implies

$$|\nabla f(y)|\le L \text{ for all } y\in B\cap \text{dom } \nabla f. \tag{2.12}$$

Since dom ∇f is dense in B, there exist sequences $\{y^j\}$ such that f is differentiable at y^j and $y^j\to x$. The corresponding sequence of gradients $\{\nabla f(y^j)\}$ is bounded and has accumulation points (each being the limit of some convergent subsequence). It follows that the set

$$M_f(x) = \{z\in R^N\colon \nabla f(y^j)\to z \text{ for some sequence } y^j\to z \text{ with f differentiable at } y^j\} \tag{2.13a}$$

is nonempty, bounded and closed. The set

$$\partial f(x) = \text{conv } M_f(x) \tag{2.13b}$$

is called the <u>subdifferential</u> of f at x (called the generalized gradient by Clarke (1975)). Each element $g_f \in \partial f(x)$ is called a <u>subgradient</u> of f at x. Thus

$$\partial f(x)=\text{conv}\{\lim \nabla f(y^j): y^j \to x, \quad f \text{ differentiable at } y^j\}. \tag{2.14}$$

In particular therefore, $\partial f(x)=-\partial(-f)(x)$. Three immediate consequences of the definition are listed in

<u>Lemma 2.2</u>. (i) $\partial f(x)$ is a nonnempty convex compact set.

(ii) The point-to-set mapping $\partial f(\cdot)$ is <u>locally bounded</u> (bounded on bounded subsets of R^N), i.e. if $B \subset R^N$ is bounded then the set $\{g_f \in \partial f(y): y \in B\}$ is bounded.

(iii) $\partial f(\cdot)$ is <u>upper semicontinuous</u>, i.e. if a sequence $\{y^j\}$ converges to x and $g_f^j \in \partial f(y^j)$ for each j then each accumulation point g_f of $\{g_f^j\}$ satisfies $g_f \in \partial f(x)$.

In general, $\partial f(x)$ does not reduce to $\nabla f(x)$ when the gradient ∇f is discontinuous at x.

<u>Lemma 2.3</u>. The following are equivalent:

(i) $\partial f(x)$ consists of a single vector;

(ii) $\nabla f(x)$ exists and ∇f is continuous at x relative to dom ∇f;

(iii) f is strictly differentiable at x.

Moreover, when these properties hold one has $\partial f(x)=\{\nabla f(x)\}$.

Frequently $\partial f(x)$ is a singleton for almost every x. A locally Lipschitzian function $f:R^N \to R$ is <u>subdifferentially regular</u> at $x \in R^N$ if for every $d \in R^N$ the ordinary directional derivative (2.6) exists and coincides with the generalized one in (2.3):

$$f'(x;d) = f^o(x;d) \quad \text{for all } d. \tag{2.15}$$

If (2.15) holds at each $x \in R^N$ then $\partial f(x)$ is actually single-valued at almost every x. Below we give two important examples of subdifferentially regular functions.

Lemma 2.4. If f is convex then f is subdifferentially regular and

$$f'(x;d) = \max\{\langle g_f, d\rangle : g_f \in \partial f(x)\} \quad \text{for all } x,d. \tag{2.16}$$

Lemma 2.5. Suppose that

$$f(x) = \max\{f_u(x) : u \in U\} \qquad \text{for all } x \in R^N, \tag{2.17}$$

where the index set U is a compact topological space (e.g. a finite set in the discrete topology), each f_u is locally Lipschitzian, uniformly for u in U, and the mappings $f_u(x)$ and $\partial f_u(x)$ are upper semicontinuous in (x,u) (e.g. each f_u is a differentiable function such that $f_u(x)$ and $\nabla f_u(x)$ depend continuously on (x,u)). Let

$$U(x) = \{u \in U : f_u(x) = f(x)\}. \tag{2.18}$$

Then f is locally Lipschitzian and

$$\partial f(x) \subset \text{conv}\,\{\partial f_u(x) : u \in U(x)\}. \tag{2.19}$$

If each f_u is subdifferentially regular at x, then so is f, equality holds in (2.19), and

$$f'(x;d) = \max\{\langle g_u, d\rangle : g_u \in \partial f_u(x), u \in U(x)\} \quad \text{for all } d. \tag{2.20}$$

Corollary 2.6. Suppose that

$$f(x) = \max\{f_i(x) : i \in I\} \qquad \text{for all } x \text{ in } R^N, \tag{2.21}$$

where the index set I is finite, and let $I(x)=\{i \in I : f_i(x)=f(x)\}$.

(i) If each f_i is continuously differentiable then

$$\begin{aligned} f'(x;d) &= \max\{\langle \nabla f_i(x), d\rangle : i \in I(x)\} \qquad \text{for all } d, \\ \partial f(x) &= \text{conv}\{\nabla f_i(x) : i \in I(x)\}. \end{aligned} \tag{2.22}$$

(ii) If each f_i is convex then

$$\begin{aligned} f'(x;d) &= \max\{\langle g_{f_i}, d\rangle : g_{f_i} \in \partial f_i(x),\ i \in I(x)\} \quad \text{for all } d, \\ \partial f(x) &= \text{conv}\{g_{f_i} \in \partial f_i(x) : i \in I(x)\}. \end{aligned} \tag{2.23}$$

When f is smooth, there exists an apparatus for computing ∇f in terms of the derivatives of other functions from which f is composed. The calculus of subgradients, which generalizes rules like $\nabla(f_1+f_2)(x) = \nabla f_1(x)+\nabla f_2(x)$, is based on the following results.

Lemma 2.7. Let $g:R^n \to R$ and $h_i:R^N \to R$, $i=1,\ldots,n$, be locally Lipschitzian. Let $h(x)=(h_1(x),\ldots,h_n(x))$ and $(goh)(x)=g(h(x))$ for all $x\in E^N$. Then goh is locally Lipschitzian and

$$\partial(goh)(x) \subset \text{conv}\left\{ \sum_{i=1}^{n} u_i \partial h_i(x) : (u_1,\ldots,u_n)\in\partial g(h(x))\right\}. \tag{2.24}$$

Moreover, equality holds in (2.24) if one of the following is satisfied:
(i) g is subdifferentially regular at h(x), each h_i is subdifferentially regular at x and $\partial g(h(x)) \subset R^n_+$ ($R^n_+ = \{z\in R^n: z_i\geq 0$ for all i$\}$);
(ii) g is subdifferentially regular at h(x) and each h_i is continuously differentiable at x;
(iii) Each h_i is continuously differentiable at x and either g (or - g) is subdifferentially regular at h(x) or the Jacobian matrix of h at x is surjective;
(iv) n=1, g is continuously differentiable at h(x) or g (or - g) is subdifferentially regular at h(x) and h is continuously differentiable at x. In cases (ii) - (iv) the symbol "conv" is superfluous in (2.24). If (ii) holds then goh is subdifferentially regular at x.

Corollary 2.8. Suppose that f_1 and f_2 are locally Lipschitzian on R^N. For each $x\in R^N$ let $(f_1+f_2)(x)=f_1(x)+f_2(x)$, $(f_1f_2)(x)=f_1(x)f_2(x)$ and $(f_1/f_2)(x)=f_1(x)/f_2(x)$ if $f_2(x)\neq 0$. Then

$$\partial(f_1+f_2)(x) \subset \partial f_1(x)+\partial f_2(x), \tag{2.25a}$$

$$\partial(f_1f_2)(x) \subset f_2(x)\,\partial f_1(x)+f_1(x)\partial f_2(x), \tag{2.25b}$$

$$\partial(f_1/f_2)(x) \subset \frac{1}{(f_2(x))^2}\left[f_2(x)\partial f_1(x)-f_1(x)\partial f_2(x)\right]. \tag{2.25c}$$

Equality holds in (2.25a) if each f_i is subdifferentially regular at x, and in (2.25b) if in addition $f_i(x)\geq 0$.

Clarke (1975) established the following crucial relations between the subdifferential and the generalized directional derivatives of a locally Lipschitzian function f defined on R^N

$$f^o(x;d) = \max\{<g_f,d>: g_f \in \partial f(x)\} \text{ for all } x,d, \tag{2.26}$$

$$\partial f(x) = \{g_f \in R^N: <g_f,d> \le f^o(x;d) \text{ for all } d\} \text{ for all } x. \tag{2.27}$$

We shall now interpret these relations in geometric terms. In what follows let x be a fixed point in R^N.

First, suppose that f is continuously differentiable at x. From Lemma 2.3, (2.26) and (2.8) we have

$$\partial f(x) = \{\nabla f(x)\}, \tag{2.28a}$$

$$f^o(x;d) = f'(x;d) = <\nabla f(x),d> \quad \text{for all } d. \tag{2.28b}$$

Suppose that $\nabla f(x) \neq 0$. Then $\nabla f(x)$ corresponds to the hyperplane

$$H_{\nabla f} = \{(z,\beta) \in R^{N+1}: \beta = f(x) + <\nabla f(x),z-x>\}$$

being tangent to the <u>graph</u> of f

$$\text{graph } f = \{(z,\beta) \in R^{N+1}: \beta = f(z)\}$$

at the point $(x,f(x))$. Here β denotes the "vertical" coordinate of a point $(x,\beta) \in R^{N+1}$. Moreover, the hyperplane

$$H_C = \{z \in R^N : <\nabla f(x),z-x> = 0\}$$

is tangent at x to the <u>contour</u> of f at x

$$C = \{z \in R^N: f(z) = f(x)\}.$$

$\nabla f(x)$ is perpendicular to C at x and is the direction of steepest ascent for f at x. Define the following <u>linearization</u> of f at x

$$\bar{f}(z) = f(x) + <\nabla f(x),z-x> \quad \text{for all } z \text{ in } R^N \tag{2.29}$$

and observe that $\nabla\bar{f}(z) = \nabla f(x)$ for all z (x is fixed). Therefore this linearization has the same differentiability properties as f at x in the sense that

$$\partial\bar{f}(x) = \partial f(x), \tag{2.30a}$$

$$\bar{f}^o(x;d) = \bar{f}'(x;d) = f^o(x;d) \text{ for all } d, \tag{2.30b}$$

cf. (2.28). In particular, by (2.28a), (2.9) and (2.30b), for any $d \in R^N$ we have

$$f(x+td) = f(x) + t\overline{f}'(x;d) + o(t), \tag{2.31}$$

where $o(t)/t \to 0$ as $t \downarrow 0$. Moreover, the graph of $\overline{f}$ equals $H_{\nabla f}$, while the contour of $\overline{f}$ at x is equal to H_C. We conclude that linearizations based on $\partial f(\cdot) = \{\nabla f(\cdot)\}$ provide convenient differential approximations to f when f is mooth.

Next suppose that f is convex. Then f is locally Lipschitzian and ∂f is the subdifferential in the sense of convex analysis:

$$\partial f(x) = \{g_f \in R^N: f(z) \geq f(x) + \langle g_f, z-x\rangle \quad \text{for all } z\}. \tag{2.32}$$

The above relation says that each subgradient $g_f \in \partial f(x)$ defines a <u>linearization</u> of f at x

$$\overline{f}_{g_f}(z) = f(x) + \langle g_f, z-x\rangle \qquad \text{for all z in } R^N, \tag{2.33}$$

which is a <u>lower approximation</u> to f at x

$$f(x) = f_{g_f}(x), \tag{2.34a}$$

$$f(z) \geq f_{g_f}(z) \qquad \text{for all } z, \tag{2.34b}$$

and a hyperplane

$$H_{g_f} = \{(z,\beta) \in R^{N+1}: \beta = \overline{f}_{g_f}(z)\} \tag{2.35}$$

supporting the epigraph of f at (x,f(x)). Observe that

$$H_{g_f} = \text{graph } \overline{f}_{g_f}. \tag{2.36}$$

Also if $g_f \neq 0$ then the hyperplane

$$H_1 = \{z \in R^N: \langle g_f, z-x\rangle = 0\}$$

supports at x the <u>level set</u> $\{z \in R^N: f(z) \leq f(x)\}$. Such hyperplanes are nonunique when $\partial f(x)$ is not a singleton. However, one easily checks that a convex combination of supporting hyperplanes to epi f at (x,f(x)) is still a supporting hyperplane. This gives the reason for the symbol "conv" in the definition of $\partial f(x)$, see (2.13).

When f is convex but not continuously differentiable at x, relations of the form (2.30) and (2.31) do not, in general, hold if one replaces $\overline{f}$ with $\overline{f}_{g_f}$. Yet such relations may be extended as follows. Define the following approximation to f at x

$$\hat{f}(z) = \max\{\overline{f}_{g_f}(z) : g_f \in \partial f(x)\} \quad \text{for all } z \text{ in } R^N ,$$
$$\overline{f}_{g_f}(z) = f(x) + \langle g_f, z-x \rangle \quad \text{for each } g_f \in \partial f(x) \text{ and all } z \text{ in } R^N. \tag{2.37}$$

Observe that the "max" above is attained, because $\partial f(x)$ is a compact set by Lemma 2.2. By (2.34), $\hat{f}$ is a <u>lower approximation</u> to f at x

$$f(x) = \hat{f}(x), \tag{2.38a}$$

$$f(z) \geq \hat{f}(z) \quad \text{for all } z . \tag{2.38b}$$

The epigraph of $\hat{f}$ can be expressed in the form

$$\text{epi } \hat{f} = (x, f(x)) + K_f , \tag{2.39}$$

where

$$K_f = \{(d,\beta) \in R^{N+1} : \beta \geq \langle g_f, d \rangle \quad \text{for all } g_f \in \partial f(x)\} \tag{2.40}$$

is a closed convex <u>cone</u> (it contains all nonnegative multiples of its elements). Moreover, we deduce from (2.32) and (2.37) that the epigraph of $\hat{f}$, being an intersection of all the epigraphs of $\overline{f}_{g_f}$ containing epi f, is a <u>convex outer approximation</u> to the epigraph of f:

$$\text{epi } f \subset \text{epi } \hat{f}. \tag{2.41}$$

Observe that the convexity of $\hat{f}$ follows directly from (2.37) even when f is nonconvex, since

$$\begin{aligned} \hat{f}(\lambda z^1 + (1-\lambda) z^2) &= \max\{\lambda \overline{f}_{g_f}(z^1) + (1-\lambda)\overline{f}_{g_f}(z^2) : g_f \in \partial f(x)\} \\ &\leq \lambda \max\{\overline{f}_{g_f}(z^1) : g_f \in \partial f(x)\} + (1-\lambda)\max\{\overline{f}_{g_f}(z^2) : g_f \in \partial f(x)\} \\ &= \lambda \hat{f}(z^1) + (1-\lambda)\hat{f}(z^2) \end{aligned}$$

for all $\lambda \in [0,1]$ and $z^1, z^2 \in R^N$. If convexity fails, relations (2.34b), (2.38b) and (2.41) are no longer valid. However, $\hat{f}$ is still a useful approximation to f, as will be shown below.

<u>Lemma 2.9</u>. Suppose that $f: R^N \to R$ is locally Lipschitzian, $x \in R^N$ and $\hat{f}: R^N \to R$ is defined by (2.37). Then $\hat{f}$ is convex and subdifferentially regular on R^N, and

$$\partial \hat{f}(x) = \partial f(x), \tag{2.42a}$$

$$\hat{f}^{\circ}(x;d) = f^{\circ}(x;d) \quad \text{for all } d. \tag{2.42b}$$

Moreover, for each d in R^N one has

$$\hat{f}^o(x;d) = \hat{f}'(x;d) = \max\{\langle g_f,d\rangle : g_f \in \partial f(x)\}, \tag{2.43}$$

$$f(x+td) \le f(x) + t\hat{f}'(x,d) + o(t) \qquad \text{for all } t\ge 0, \tag{2.44}$$

where $o(t)/t\to 0$ as $t\to 0$.

Proof. The convexity of $\hat{f}$ was shown above. For each $g_f \in \partial f(x)$, $\bar{f}_{g_f}$ is continuously differentiable with $\partial\bar{f}_{g_f}(z) = \{g_f\}$ for all z. Therefore the compactness of $\partial f(x)$, (2.37), (2.34a) and Lemma 2.5 imply that $\hat{f}$ is subdifferentially regular and satisfies (2.43), and $\partial f(x) = \text{conv}\{g_f: g_f \in \partial f(x)\}$. The last relation and the convexity of $\partial f(x)$ yield (2.42a). Then (2.42b) follows from (2.43) and (2.26). In view of (2.5), (2.7), (2.42a) and (2.43), for each $d \in R^N$ we have

$$f(x+td) \le f(x)+tf^D(x;d)+o(t) \le f(x)+tf^o(x;d)+o(t) \le$$
$$\le f(x)+t\hat{f}'(x;d)+o(t)$$

for all $t\ge 0$, which proves (2.44). □

A basic question in nondifferentiable optimization is how to find a descent direction d for f at x that satisfies

$$f(x+td) < f(x) \qquad \text{for all small } t>0. \tag{2.45}$$

This problem is tackled in the following lemma.

Lemma 2.10. (i) Suppose that $f: R^N\to R$ is locally Lipschitzian, $x \in R^N$ and $d \in R^N$ satisfies

$$\max\{\langle g_f,d\rangle : g_f \in \partial f(x)\} < 0. \tag{2.46}$$

Then d is a descent direction for f at x.

(ii) Suppose that d is a descent direction for $\hat{f}$ at x, i.e.

$$\hat{f}(x+td) < \hat{f}(x) \qquad \text{for all small } t>0, \tag{2.47}$$

where $\hat{f}$ defined by (2.37) is an approximation to a locally Lipschitzian function $f: R^N\to R$ at x. Then d is also a descent direction for f at x. Moreover, d satisfies (2.46).

Proof. (i) From (2.46), (2.43) and (2.44), we have $\hat{f}'(x;d)<0$ and

$$f(x+td) \le f(x)+t[\hat{f}'(x;d)+o(t)/t] < f(x) \text{ for all small } t>0,$$

because $o(t)/t \to 0$ as $t \downarrow 0$.

(ii) Using (2.47), we choose $\hat{t}>0$ satisfying $\hat{f}(x+\hat{t}d)<\hat{f}(x)$. By Lemma 2.9, $\hat{f}$ is convex, hence

$$\hat{f}(x+td)=\hat{f}((1-\frac{t}{\hat{t}}x)+\frac{t}{\hat{t}}(x+\hat{t}d)) \le (1-\frac{t}{\hat{t}})\hat{f}(x)+\frac{t}{\hat{t}}\hat{f}(x+\hat{t}d)$$

and hence

$$[\hat{f}(x+td)-\hat{f}(x)]/t \le [\hat{f}(x+\hat{t}d)-\hat{f}(x)]/\hat{t}$$

for all $t \in [0,\hat{t}]$. Therefore $\hat{f}'(x;d) \le [\hat{f}(x+\hat{t}d)-\hat{f}(x)]/\hat{t}<0$, and (2.46) follows from (2.43). □

The above lemma will be used below in two schemes for finding descent directions. Relation (2.46) means that the set $\partial f(x)$ can be separated from the origin by a hyperplane. Since $\partial f(x)$ is a convex compact set, this is possible if and only if $0 \notin \partial f(x)$. Therefore we shall first state two auxiliary results.

Lemma 2.11. Suppose that $f: R^N \to R$ is convex. Then a point $\bar{x} \in R^N$ minimizes f, i.e. $f(\bar{x}) \le f(y)$ for all y, if and only if $0 \in \partial f(\bar{x})$.

Proof. This follows immediately from (2.32). □

Lemma 2.12. Suppose that $G \subset R^N$ is a convex compact set and let

$$\text{Nr } G = \arg\min\{|g| : g \in G\}$$

denote the point in G that is nearest to the origin (the projection of the origin onto G). Then $p \in G$ is $\text{Nr } G$ if and only if $\langle g,p\rangle \ge |p|^2$ for all $g \in G$.

Proof. We note that Nr G is well-defined, because the convex function $|\cdot|$ attains its unique minimum on the convex compact set G. Let $g \in G$, $0 \le t \le 1$; then $p+t(g-p) \in G$ and $|p+t(g-p)|^2=|p|^2+2t[\langle g,p\rangle-|p|^2]+t^2|g-p|^2$, which is less than $|p|^2$ for small t unless $\langle g,p\rangle \ge |p|^2$. □

The following lemma shows how one may find descent directions for nonsmooth functions.

Lemma 2.13. Consider a locally Lipschitzian function $f\colon R^N \to R$, a point $x \in R^N$ and an approximation $\hat{f}$ to f at x defined by (2.37). Let

$$\hat{p} = \mathrm{Nr}\, \partial f(x) = \arg\min\{\tfrac{1}{2}|g_f|^2 : g_f \in \partial f(x)\} \tag{2.48}$$

and let $\hat{d}$ denote a solution to the problem

$$\text{minimize } \hat{f}(x+d) + \tfrac{1}{2}|d|^2 \quad \text{over all } d \in R^N. \tag{2.49}$$

Then

(i) $\hat{d}$ exists, is uniquely determined and satisfies

$$-\hat{d} = \hat{p} \in \partial f(x), \tag{2.50}$$

$$\max\{\langle g_f, \hat{d}\rangle : g_f \in \partial f(x)\} = -|\hat{p}|^2, \tag{2.51}$$

$$\hat{f}(x+\hat{d}) = \hat{f}(x) - |\hat{p}|^2, \tag{2.52a}$$

$$\hat{f}(x+td) \le \hat{f}(x) - t|\hat{p}|^2 \quad \text{for all } t \in [0,1]; \tag{2.52b}$$

(ii) $d \neq 0$ if and only if $0 \notin \partial f(x)$;

(iii) $0 \in \partial f(x)$ if and only if x is a global minimum point for $\hat{f}$.

Proof (i) The objective function of (2.49) can be written as

$$\rho(d) = \hat{f}(x) + v(d) + \tfrac{1}{2}|d|^2,$$

$$v(d) = \max\{\langle g_f, d\rangle : g_f \in \partial f(x)\}.$$

Let $g_f \in \partial f(x)$ be fixed. Then, by the Cauchy-Schwarc inequality, $\rho(d) \ge f(x) + \langle g_f, d\rangle + \tfrac{1}{2}|d|^2 \ge \hat{f}(x) - |g_f||d| + \tfrac{1}{2}|d|^2 \to +\infty$, as $|d| \to \infty$, hence $\hat{d}$ exists. Since ρ is strictly convex, $\hat{d}$ is unique. In view of Lemma 2.11, Corollary 2.8 and Lemma 2.5, we have $0 \in \partial\rho(\hat{d}) = \hat{d} + \partial v(\hat{d})$ and

$$\partial v(\hat{d}) = \{g_f \in \partial f(x) : \langle g_f, \hat{d}\rangle = v(\hat{d})\},$$

hence there exists $\bar{p} \in \partial f(\bar{x})$ satisfying $\bar{p} = -\hat{d}$ and $v(\hat{d}) = \langle \bar{p}, \hat{d}\rangle = -|\bar{p}|^2$. Thus $\bar{p} \in \partial f(\bar{x})$ and $\langle g_f, \bar{p}\rangle \ge |\bar{p}|^2$ for all $g_f \in \partial f(x)$, therefore $\bar{p} = \hat{p}$ by Lemma 2.12. Combining the preceding relations, we establish (2.50), (2.51) and (2.52a). Then (2.52b) follows from (2.52a) and the convexity of $\hat{f}$.

(ii) This follows from (2.50) and (2.48).

(iii) By Lemma 2.9 and Lemma 2.11, $0 \in \partial f(x) = \partial\hat{f}(x)$ is equivalent to x minimizing the convex function $\hat{f}$. □

We conclude from Lemma 2.13 and Lemma 2.10 that if a point $x \in R^N$ satisfies $0 \notin \partial f(x)$, then one may find a descent direction for f at x, cf. (2.48), (2.52) and (2.47). In particular therefore, f cannot have local minima at such points. Thus we have derived the following necessary condition of optimality.

Lemma 2.14. If $\bar{x}$ is a local minimum point for a locally Lipschitzian function f, then $0 \in \partial f(\bar{x})$.

A point $\bar{x} \in R^N$ satisfying $0 \in \partial f(\bar{x})$ will be called stationary for f. Thus stationarity is necessary for optimality.

We may add that if f is strictly differentiable at x and $\nabla f(x) \neq 0$, then the direction $\hat{d} = -\nabla f(x)$ defined in Lemma 2.13 is the direction of steepest descent for f at x. In general, we have

$$f^{o}(x;\hat{d}/|\hat{d}|) = \min\{f^{o}(x;d):\ |d| \leq 1\},$$

see (Wolfe, 1975).

Consider the following constrained problem

$$\text{minimize } f(x), \quad \text{subject to } F_i(x) \leq 0 \quad \text{for } i=1,\ldots,m, \tag{2.53}$$

where the objective function f and the constraint functions F_i are real-valued functions defined on R^N, and $m \geq 1$ is finite. Define the total constraint function

$$F(x) = \max\{F_i(x):\ i=1,\ldots,m\} \tag{2.54}$$

and the feasible set for (2.53)

$$S = \{x \in R^N:\ F(x) \leq 0\}.$$

A point $\bar{x} \in S$ is a local solution to problem (2.53) if for all x in a neighborhood B of $\bar{x}$ one has $f(\bar{x}) \leq f(x)$ when $x \in S$.

We note that if $\bar{x}$ is a local solution of (2.53), then the function

$$H(x;\bar{x}) = \max\{f(x) - f(\bar{x}),\ F(x)\} \quad \text{for all } x \tag{2.55}$$

has a local (unconstrained) minimum at $\bar{x}$

$$H(\bar{x};\bar{x}) = 0\ , \tag{2.56}$$

for if $H(x;\bar{x})$ were smaller than 0 for some $x \in B$ then x would be feasible for (2.53) and strictly better than $\bar{x}$, which cannot be.

In what follows we assume that the problem functions of (2.53) are locally Lipschitzian.

We shall now derive necessary optimality conditions for problem (2.53). Define the point-to-set mappings

$$\tilde{\partial}F(x) = \text{conv}\{\partial F_i(x) : F_i(x) = F(x)\}, \tag{2.57}$$

$$\hat{M}(x) = \begin{cases} \partial f(x) & \text{if} \quad F(x)<0, \\ \text{conv}\{\partial f(x) \cup \partial F(x)\} & \text{if} \quad F(x)=0, \\ \partial F(x) & \text{if} \quad F(x)>0, \end{cases} \tag{2.58}$$

$$\tilde{M}(x) = \begin{cases} \partial f(x) & \text{if} \quad F(x)<0, \\ \text{conv}\{\partial f(x) \cup \tilde{\partial}F(x)\} & \text{if} \quad F(x)=0, \\ \tilde{\partial}F(x) & \text{if} \quad F(x)>0. \end{cases} \tag{2.59}$$

By Lemma 2.5, $F(\cdot)$ and $H(\cdot;x)$ are locally Lipschitzian and the above mappings satisfy

$$\partial F(x) \subset \tilde{\partial}F(x), \tag{2.60}$$

$$\partial H(x;x) \subset \hat{M}(x), \tag{2.61}$$

$$\hat{M}(x) \subset \tilde{M}(x), \tag{2.62}$$

where $\partial H(\cdot;x)$ denotes the subdifferential of $H(\cdot;x)$ for fixed x.

We have the following necessary condition of optimality.

Lemma 2.15. If $\bar{x}$ solves (2.53) locally, then

$$0 \in \partial H(\bar{x};\bar{x}), \tag{2.63}$$

$$0 \in \tilde{M}(\bar{x}). \tag{2.64}$$

In particular, there exist numbers μ_i, $i=0,\ldots,m$, satisfying

$$0 \in \mu_0 \partial f(\bar{x}) + \sum_{i=1}^{m} \mu_i \partial F_i(\bar{x}),$$

$$\mu_i \geq 0,\ i=0,\ldots,m, \quad \sum_{i=0}^{m} \mu_i = 1, \tag{2.65}$$

$$\mu_i F_i(\bar{x}) = 0 \ , \quad i=1,\ldots,m.$$

Proof. Since $\bar{x}$ must minimize $H(\cdot,\bar{x})$ locally, from Lemma 2.14 we obtain (2.63), which in turn implies (2.64) by (2.61) and (2.62). To see that (2.65) follows from (2.64), (2.59) and (2.57), note that $F(\bar{x}) \leq 0$ and that one may set $\mu_i=0$ if $F_i(\bar{x})<0$. □

A point $\bar{x} \in S$ is called stationary for f on S if it satisfies the necessary optimality condition (2.65), or equivalently (2.64).

The multipliers μ_i in (2.65) may be nonunique. In particular, one may have $\mu_o=0$. If $\mu_o=0$ then relation (2.65) reduces to

$$F(\bar{x})=0 \quad \text{and} \quad 0 \in \tilde{\partial}F(\bar{x}),$$

which describes only geometry of the feasible set at $\bar{x}$, without providing any information about the objective function. This degenerate case is eliminated if the Cottle constraint qualification holds at $\bar{x}$:

$$\text{either} \quad F(\bar{x}) < 0 \quad \text{or} \quad 0 \notin \tilde{\partial}F(\bar{x}). \tag{2.66}$$

If the constraint functions are convex and $\bar{x} \in S$, then (2.66) is equivalent to the Slater constraint qualification

$$F(\tilde{x}) < 0 \quad \text{for some} \quad \tilde{x} \in R^N. \tag{2.67}$$

This follows from the fact that in the convex case we have $\tilde{\partial}F=\partial F$, see Corollary 2.8, and the condition $0 \in \partial F(x)$ is equivalent to $F(x) \le F(y)$ for all y, see Lemma 2.11.

Relation (2.65) is known as the F. John necessary condition of optimality. It becomes the Kuhn-Tucker condition

$$0 \in \partial f(\bar{x}) + \sum_{i=1}^{m} \bar{\mu}_i \partial F_i(\bar{x}), \tag{2.68a}$$

$$\bar{\mu}_i \ge 0, \quad \bar{\mu}_i F_i(\bar{x})=0, \; i=1,\dots,m, \tag{2.68b}$$

when $\mu_o \ne 0$, since one may take $\bar{\mu}_i=\mu_i/\mu_o$. When problem (2.53) is convex, i.e. f and each F_i are convex functions, then the Kuhn-Tucker condition and the Slater constraint qualification yield the following sufficient condition for optimality.

Lemma 2.16. Suppose that problem (2.53) is convex and satisfies the Slater constraint qualification (2.67). Then the following are equivalent:

(i) $\bar{x}$ solves problem (2.53);

(ii) $\bar{x}$ satisfies

$$\min\{H(x;\bar{x}) : x \in R^N\} = H(\bar{x};\bar{x}) = 0\;; \tag{2.69}$$

(iii) $\bar{x}$ satisfies

$$0 \in \partial H(\bar{x};\bar{x}); \tag{2.70}$$

(iv) $\bar{x}$ is stationary for f on S;
(v) the Kuhn-Tucker condition (2.68) holds at $\bar{x} \in S$.

Proof. (a) As noted above, (i) implies (ii). Suppose that (2.69) holds, but $f(\hat{x})<f(\bar{x})$ for some $\hat{x}$ satisfying $F(\hat{x})\le 0$. Then

$$f(\hat{x}+t(\tilde{x}-\hat{x})) \le (1-t)f(\hat{x})+tf(\tilde{x}) < f(\bar{x}),$$

$$F(\hat{x}+t(\tilde{x}-\hat{x})) \le (1-t)F(\hat{x})+tF(\tilde{x}) < tF(\tilde{x})<0,$$

$$H(\hat{x}+t(\tilde{x}-\hat{x});\bar{x}) < 0 = H(\bar{x};\bar{x})$$

for sufficiently small $t>0$, which contradicts (2.69). Therefore (ii) implies (i).
(b) By convexity and Corollary 2.8, we have

$$\begin{aligned} \tilde{\partial}F(x) &= \partial F(x) \quad \text{for all } x,\\ \partial H(x;x) &= \hat{M}(x) = \tilde{M}(x) \quad \text{for all } x. \end{aligned} \tag{2.71}$$

By Lemma 2.11, $\bar{x}$ minimizes $H(\cdot;\bar{x})$ if and only if $0 \in \partial H(\bar{x};\bar{x})$. But $F(\bar{x})=H(\bar{x};\bar{x})>0$ and $0 \in \partial H(\bar{x};\bar{x})= \partial F(\bar{x})$ is impossible in view of (2.67). We conclude that (ii) and (iii) are equivalent.
(c) The equivalence of (iii) and (iv) follows from (2.71) and (b).
(d) As noted above, owing to (2.67) (iv) implies (v). As for the reverse implication, use (2.68) and let $\mu_i=1/(1+\sum_{j=1}^{m}\bar{\mu}_j)$ in (2.65). □

For unconstrained problems the necessary optimality conditon $0 \in \partial f(x)$ is equivalent to x being a minimum point for the convex first order approximation $\hat{f}$ to f at x, see Lemma 2.13 (iii). We shall now provide a similar interpretation of the stationarity condition (2.65) for the constrained problem (2.53) in terms of properties of the following convex approximation to problem (2.53) defined at each $x \in S$:

$$P(x): \text{minimize } \hat{f}(z), \text{ subject to } \hat{F}_i(z)\le 0, \quad i=1,\dots,m, \tag{2.72}$$

where for any fixed $x \in S$ and all $z \in R^N$

$$\hat{f}(x) = \max\{\bar{f}_g(z):\ g \in \partial f(x)\},$$

$$\bar{f}_g(z) = f(x) + \langle g,z-x\rangle \qquad \text{for each } g \in \partial f(x),$$

$$\hat{F}_i(z) = \max\{\bar{F}_{i,g}(z):\ g \in \partial F_i(x)\}, \qquad i=1,\dots,m, \tag{2.73}$$

$$\bar{F}_{i,g}(z) = F_i(x) + \langle g,z-x\rangle \qquad \text{for each } g \in \partial F_i(x), \quad i=1,\dots,m,$$

are convex first order approximations to the problem functions at x. Also let

$$\hat{F}(z) = \max\{\hat{F}_i(z):\ i=1,\dots,m\} \qquad \text{for all } z,$$

$$\hat{H}(z) = \max\{\hat{f}(x)-\hat{f}(x),\ \hat{F}(z)\} \qquad \text{for all } z, \tag{2.74}$$

denote convex first order approximations to $F(\cdot)$ and $H(\cdot;x)$, respectively, at x. Since $F(x)\leq 0$ by assumption, and $\hat{F}_i(x)=F_i(x)$ for all i, we have $\hat{F}(x)\leq 0$ and $\hat{H}(x)=0$, hence x is feasible for P(x). Also, as shown above, functions of the form (2.37) and (2.73) are convex. Thus P(x) is a convex problem.

Differential properties of functions of the form (2.73) were studied above. In particular, in view of Lemma 2.9, we have

$$\partial\hat{f}(x) = \partial f(x), \qquad \partial\hat{F}_i(x) = \partial F_i(x), \qquad i=1,\dots,m, \tag{2.75}$$

hence (2.73), Corollary 2.8, (2.57) and (2.59) yield

$$\partial\hat{F}(x) = \widetilde{\partial F}(x), \tag{2.76}$$

$$\partial\hat{H}(x) = \tilde{M}(x). \tag{2.77}$$

To study the relations between the original problem and its convex approximations we shall need the following concepts. We say that $d \in R^N$ is a <u>feasible direction</u> for S at x if

$$x+td \in S \qquad \text{for all small } t>0. \tag{2.78}$$

Note that S is closed, because F is continuous, hence (2.78) implies $x \in S$. Let

$$\bar{I}(x) = \{x \in R^N:\ F_i(x) = F(x) \geq 0\} \tag{2.79}$$

and observe that $\bar{I}(x)$ is empty when $F(x)<0$. Relation (2.78) is equivalent to the following

$$F_i(x+td) < 0 \qquad \text{for all small } t>0 \quad \text{and} \quad i \in \bar{I}(x), \tag{2.80}$$

since, by the continuity of F_i, we have $F_i(x+td)<0$ for small t if

$F_i(x) < 0$. One easily checks that if d is a descent direction for $H(\cdot;x)$ at $x \in S$, i.e.

$$H(x+td;x) < H(x;x)=0 \quad \text{for all small} \quad t > 0,$$

then d is also a feasible direction of descent for f at x relative to S, i.e. (2.45) and (2.80) are satisfied.

Using the above observations and arguing essentially as in the proof of Lemma 2.10, one may obtain the following sufficient condition for d to be a feasible descent direction for f at x relative to S.

<u>Lemma 2.17</u>. Under the above assumptions and conventions, if d is a descent direction for $\hat{H}$ at $x \in S$, i.e.

$$\hat{H}(x+td) < \hat{H}(x)=0 \quad \text{for all small} \quad t > 0,$$

then d is a feasible direction of descent for f at x relative to S. □

The following result demonstrates how one may find feasible descent directions.

<u>Lemma 2.18</u>. Consider a locally Lipschitzian problem (2.53) and its convex approximation $P(x)$ at $x \in S$ defined via (2.72) and (2.73). Let

$$\hat{p} = \operatorname{Nr} \tilde{M}(x), \tag{2.81}$$

and let $\hat{d}$ denote a solution to the problem

$$\text{minimize} \quad \hat{H}(x+d)+\tfrac{1}{2}|d|^2 \quad \text{over all} \quad d \quad \text{in} \quad R^N. \tag{2.82}$$

Then

(i) $\hat{d}$ exists, is unique and satisfies

$$-\hat{d} = \hat{p} \in \tilde{M}(x), \tag{2.83}$$

$$\hat{H}(x+d) = \hat{H}(x)-|\hat{p}|^2, \tag{2.84a}$$

$$\hat{H}(x+td) \le \hat{H}(x)-t|\hat{p}|^2 \quad \text{for all} \quad t \in [0,1]; \tag{2.84b}$$

(ii) $\hat{d} \neq 0$ if and only if $0 \in \tilde{M}(x)$;

(iii) $0 \in \tilde{M}(x)$ if and only if x is a minimum point for $\hat{H}$.

(iv) If additionally the Cottle constraint qualification is satisfied at x, i.e.

$$\text{either} \quad F(x)<0 \quad \text{or} \quad 0\in \tilde{\partial}F(x), \tag{2.85}$$

then x is stationary for f on S if and only x solves problem $P(x)$.

<u>Proof</u>. In view of the preceding results, the objective function of (2.82) can be written as

$$\phi(d)=\max\{<g,d>:\ g\in \tilde{M}(x)\}+\tfrac{1}{2}|d|^2 \quad \text{for all } d,$$

hence (i)-(iii) may be proved similarly to Lemma 2.13. Therefore, we shall only prove (iv). Since $\hat{F}(x)=F(x)$ and $\partial\hat{F}(x)=\tilde{\partial}F(x)$, see (2.76), we observe that (2.85) is also the Cottle constraint qualification for $\hat{F}$ at x, hence the Slater constraint qualification holds for the convex problem $P(x)$. Therefore we deduce from Lemma 2.16 and (2.77) that $x\in S$ solves problem $P(x)$ if and only if $0\in \hat{H}(x)=\tilde{M}(x)$, which proves (iv). □

We conclude from Lemma 2.17 and Lemma 2.18 that if a point $x\in S$ is nonstationary for f on S, then one may use convex first order approximations (2.73) and (2.74) for finding a feasible direction of descent for f at x. Moreover, if the Cottle constraint qualification is satisfied at all feasible points, then stationary points of the original problem are precisely the solutions of its convex first order approximations.

We end this section by recalling the notion of ε-subdifferential. If $f:R^N \to R$ is convex, $x\in R^N$ and $\varepsilon\geq 0$, then the <u>ε-subdifferential</u> of f at x is the convex set

$$\partial_\varepsilon f(x)=\{g_f\in R^N: f(z)\geq f(x)+<g_f,z-x>-\varepsilon \quad \text{for all } z\}. \tag{2.86}$$

Each element of $\partial_\varepsilon f(x)$ is called an <u>ε-subgradient</u> of f at x. Clearly, $\partial_0 f(x)=\partial f(x)$, see (2.32).

3. A Review of Existing Algorithms and Original Contributions of This Work

In this section we briefly review general properties of several existing algorithms for nonsmooth minimization. A fuller discussion of those algorithms is postponed to subsequent chapters. Our intention here is to motivate the need for the class of methods which is introduced in this work.

Throughout this work we assume that the functions of the problem

$$P: \text{minimize} \quad f(x), \text{ subject to } \quad F_i(x) \leq 0 \quad \text{for} \quad i=1,...,m$$

are locally Lipschitzian on R^N. Also, we place strict limitations on our ability to get information about the problem functions. Let us define the (total) constraint function

$$F(x)=\begin{cases} \max\{F_i(x): i=1,...,m\} & \text{if} \quad m \geq 1, \\ 0 \quad \text{if} \quad m=0, \end{cases}$$

and the feasible set of problem P

$$S = \{x \in R^N: F(x) \leq 0\}.$$

We assume that we have a subroutine that can evaluate $f(x)$ and a certain subgradient $g_f(x) \in \partial f(x)$ of f at each $x \in S$, and $F_i(x)$ and one subgradient $g_{F_i}(x) \in \partial F_i(x)$ for each $x \in S$ and $i=1,...,m$. We do not impose any further assumptions on the calculated subgradients. Such a limitation is realistic for problems of interest to us, where the determination of all elements of a subdifferential is either very expensive or just impossible, see (Wolfe, 1975). On the other hand, sometimes the objective function cannot be evaluated at infeasible points. Also at feasible points we require no knowledge of F_i, other than F_i being nonpositive. Such assumptions are common in the literature on nonsmooth optimization. However, for convenience we sometimes assume temporarily that g_f and g_{F_i} are defined at each $x \in R^N$.

Before discussing algorithms for nonsmooth minimization, let us recall basic ideas behind classical algorithms for solving smooth versions of problem P. Given a starting point $x^1 \in R^N$, an iterative method constructs a sequence of points $x^2, x^3, ...$ in R^N that is intended to converge to the required solution. An algorithm is a <u>feasible point method</u> if it generates a sequence $\{x^k\} \subset S$. If additionally $f(x^{k+1}) < f(x^k)$ for all k, then an algorithm is a <u>descent method</u>. A descent algorithm usually proceeds by searching from $x^k \in S$ along a direction d^k for a scalar stepsize $t^k > 0$ that gives a reduction in the objective function value $f(x^k+t^kd^k) < f(x^k)$ and the next feasible point $x^{k+1}=x^k+t^kd^k \in S$. Such a stepsize can be found if d^k is a descent direction

$$f(x^k+td^k) < f(x^k) \quad \text{for small} \quad t>0$$

which is also feasible

$$F_i(x^k+td^k) \le 0 \quad \text{for small} \quad t>0 \quad \text{and} \quad i=1,\dots,m.$$

A feasible descent direction d^k is usually found by solving an auxiliary optimization subproblem which approximates problem P in a neighborhood of x^k. The idea is that if d^k is a feasible descent direction for the subproblem then d^k should also be a feasible descent direction for problem P at x^k. To construct a suitable direction finding subproblem many algorithms use differentiation for linearizing the problem functions. When the problem functions are smooth, they can be approximated for x in some neighborhood of x^k by the following linearizations

$$\begin{aligned}&\bar f(x;x^k)=f(x^k)+\langle \nabla f(x^k),x-x^k\rangle,\\&\bar F_i(x;x^k)=F_i(x^k)+\langle \nabla F_i(x^k),x-x^k\rangle \quad \text{for} \quad i=1,\dots,m.\end{aligned} \tag{3.1}$$

Replacing the problem functions by their linearizations, we obtain the search direction finding subproblem

$$\text{minimize} \quad \bar f(x^k+d;x^k) \quad \text{subject to} \quad \bar F_i(x^k+d;x^k) \le 0 \quad \text{for} \quad i=1,\dots,m, \tag{3.2}$$

whose various modifications are used in many well-known algorithms, see (Bazaraa and Shetty, 1979; Pshenichny and Danilin, 1975). For instance, in the Pironneau and Polak (1973) method of feasible directions d^k is found from the solution (d^k,v^k) to the problem

$$\begin{aligned}&\text{minimize } \tfrac{1}{2}|d|^2+v \quad \text{over all} \quad (d,v)\in R^N \times R^1 \quad \text{satisfying}\\&\bar f(x^k+d;x^k)-f(x^k)\le v,\ \bar F_i(x^k+d;x^k)\le v \quad \text{for} \quad i=1,\dots,m.\end{aligned} \tag{3.3}$$

In particular, in such algorithms one usually has

$$\begin{aligned}&\bar f(x^k+d^k;x^k) < f(x^k),\\&\bar F_i(x^k+d^k;x^k) < 0 \quad \text{for} \quad i=1,\dots,m\end{aligned} \tag{3.4}$$

if x^k is nonstationary for problem P. Then

$$\begin{aligned}&f(x^k)+\langle \nabla f(x^k),d^k\rangle < f(x^k),\\&F_i(x^k)+\langle \nabla F_i(x^k),d^k\rangle < 0 \quad \text{for} \quad i=1,\dots,m\end{aligned} \tag{3.5}$$

and it follows from the differentiability of f and F_i that

$$f(x^k+td^k)=f(x^k)+t<\nabla f(x^k),d^k>+o(t)<f(x^k)$$

$$F_i(x^k+td^k)=(1-t)F_i(x^k)+t[F_i(x^k)+<\nabla F_i(x^k),d^k>]+o(t)<0$$

for small $t>0$ and $i=1,...,m$, because each $F_i(x^k)$ is nonpositive. We conclude that for smooth problems the linearizations (3.1) provide a sufficient condition (3.4) for d^k to be a feasible descent direction.

When the problem functions are nonsmooth, the linearizations

$$\begin{aligned}&\bar f(x;x^k)=f(x^k)+<g_f(x^k),x-x^k>,\\&\bar F_i(x;x^k)=F_i(x^k)+<g_{F_i}(x^k),x-x^k>\quad\text{for}\quad i=1,...,m\end{aligned}\tag{3.5}$$

may not suffice for assessing the behavior of the problem functions around x^k. For instance, consider an unconstrained problem and an analogue of the steepest descent direction

$$d^k=-g_f(x^k).$$

If $g_f(x^k)\neq 0$, e.g. x^k is nonstationary for f, then $\bar f(x^k+d^k;x^k)=f(x^k)-|g_f(x^k)|^2<f(x^k)$, so that (3.4) holds. But d^k need not be a descent direction for f at x^k, because we no longer have $f(x^k+td)=f(x^k)+t<g_f(x^k)d>+o(t)$ in the nondifferentiable case. This is shown by the example

$$f(x)=|x|\quad\text{for}\quad x\in R^1,\ x^k=0,\ g_f(x^k)=1\quad \partial f(x^k)=[-1,1],\tag{3.6}$$

in which no descent direction exists.

In general, for nonsmooth problems the information given by $f(x^k)$, $g_f(x^k)$, $F_i(x^k)$ and $g_{F_i}(x^k)$ for $i=1,...,m$ may not suffice for constructing a feasible descent direction at x^k. For this reason several descent algorithms that require the knowledge of full subdifferentials at x^k have been proposed; for an excellent survey see (Dixon and Gaviano, 1980). However, most existing descent algorithms for nonsmooth minimization must be regarded as theoretical or conceptual methods, since their optimization subproblems involved in computing a descent direction are constrained nondifferentiable problems, which generally cannot be carried out. Therefore, such methods will not be discussed here.

In view of the difficulties mentioned above, recently much research has been devoted to methods that do not maintain feasibility and descent at each iteration. Two examples of such methods are given below.

The <u>subgradient algorithms</u> developed mainly in the Soviet Union

(Gupal, 1979; Nurminski, 1979; Shor, 1979) are nondescent methods. An example of a typical iteration of the subgradient method for unconstrained minimization is given by

$$x^{k+1} = x^k - t^k g_f(x^k) \tag{3.7}$$

with scalar stepsizes $\{t^k\}$ satisfying

$$t^k \downarrow 0, \quad \sum_{k=1}^{\infty} t^k = +\infty, \quad t^{k+1}/t^k \longrightarrow 1.$$

Owing to their simplicity, the subgradient algorithms have no reliable stopping criteria for terminating the iteration when the current iterate is sufficiently close to the required solution. For instance, when f is smooth and $\{x^k\}$ converges to a minimum point $\bar{x}$ of f, then $g_f(x^k) \longrightarrow g_f(\bar{x}) = 0$. However, this need not occur when f is nondifferentiable. Therefore stopping criteria of the form $|g_f(x^k)| \le 10^{-6}$, which are customary in mathematical programming, are useless in nonsmooth optimization. More advanced implementations of the subgradient algorithms require interactive tuning of certain tolerances, which regulate stepsizes, during the calculations (Lemarechal, 1982; Nurmiński, 1979). The tuning requires much experimentation, but, when properly tuned to a given problem, the subgradient algorithms can be very efficient, see (Lemarechal, 1982; Shor, 1979).

To simplify notation below, we observe that an equivalent formulation of problem P is given by

$$P : \text{minimize} \quad f(x) \quad \text{subject to} \quad F(x) \le 0.$$

The formulation with F emphasizes the possibility of not keeping the functions F_i completely in view at all times, which is essential in classical algorithms for smooth versions of problem P. Also, as noted in the preceding section, the differential properties of problem P can be studied with the help of the mapping

$$\tilde{\partial}F(x) = \text{conv}\{g \in R^N : g \in \partial F_i(x) \quad \text{for some } i \text{ satisfying} \quad F_i(x) = F(x)\}$$

which satisfies $\partial F(x) \subset \tilde{\partial}F(x)$; $\tilde{\partial}F(x) = \partial F(x)$ when each F_i is convex. Therefore we assume below that we have a function g_F satisying $g_F(x) \in \tilde{\partial}F(x)$ for each $x \in R^N \setminus S$. For instance, $g_F(x) = g_{F_i}(x)$ if $i = 1, \dots, m$ is the smallest number satisfying $F_i(x) = F(x)$. For convenience, we also temporarily assume that $g_F(x) \in \tilde{\partial}F(x)$ at each $x \in R^N$. Thus $g_F(x) \in \partial F(x)$ when problem P is convex.

The Kelley (1960) cutting plane method is a nondescent method for solving convex problems, i.e. problems with convex functions f and F.

The method is based on the following crucial observation. For any fixed $y \in R^N$, define the <u>linearizations</u>

$$\begin{aligned} \bar{f}(x;y) &= f(y) + \langle g_f(y), x-y \rangle, \\ \bar{F}(x;y) &= F(y) + \langle g_F(y), x-y \rangle. \end{aligned} \tag{3.8}$$

Then

$$\begin{aligned} f(y) &= \bar{f}(y;y), \\ F(y) &= \bar{F}(y;y) \end{aligned} \tag{3.9}$$

and, since $g_f(y) \in \partial f(y)$ and $g_F(y) \in \partial F(y)$, for each $x \in R^N$ we have

$$\begin{aligned} f(x) &\geq \bar{f}(x;y), \\ F(x) &\geq \bar{F}(x;y). \end{aligned} \tag{3.10}$$

For points $y^1,...,y^k$ in R^N define the following piecewise linear (polyhedral) functions

$$\begin{aligned} \hat{f}^k(x) &= \max\{\bar{f}(x;y^j) : j=1,...,k\}. \\ \hat{F}^k(x) &= \max\{\bar{F}(x;y^j) : j=1,...,k\}. \end{aligned} \tag{3.11}$$

It follows from (3.9) and (3.10) that

$$\begin{aligned} f(x) &\geq \hat{f}^k(x), \\ F(x) &\geq \hat{F}^k(x), \end{aligned} \tag{3.12}$$

and

$$\begin{aligned} f(y^j) &= \hat{f}^k(y^j), \\ F(y^j) &= \hat{F}^k(y^j) \end{aligned} \tag{3.13}$$

for all x in R^N and $j=1,...,k$. Thus the polyhedral functions (3.11) are lower approximations to the problem functions. If the points y^j are close to any given point, say x^k, then such polyhedral functions can approximate the problem functions around x^k much more accurately than only the linearizations (3.5). This property will be frequently used in what follows.

At the k-th iteration of the cutting plane method, one sets $y^j = x^j$ for $j=1,...,k$ and uses (3.11) to calculate a direction d^k as a solution to the problem

$$\text{minimize} \quad \hat{f}^k(x^k+d) \quad \text{subject to} \quad \hat{F}^k(x^k+d) \leq 0. \tag{3.14}$$

By (3.8) and (3.11), this is equivalent to the following linear programming problem with respect to variables $(d,u) \in R^N \times R^1$

$$\begin{aligned} &\text{minimize} \quad u, \\ &\text{subject to} \quad f_j^k + \langle g_f^j, d \rangle \le u \ , \ j=1,\dots,k, \qquad (3.15) \\ &\qquad\qquad\quad F_j^k + \langle g_F^j, d \rangle \le 0 \ , \ j=1,\dots,k, \end{aligned}$$

where

$$\begin{aligned} f_j^k &= \bar{f}(x^k;y^j) \quad \text{and} \quad F_j^k = \bar{F}(x^k;y^j), \\ g_f^j &= g_f(y^j) \quad \text{and} \quad g_F^j = g_F(y^j) \end{aligned} \qquad (3.16)$$

for all $j=1,\dots,k$. Setting $x^{k+1}=y^{k+1}=x^k+d^k$ completes the k-th iteration.

An interesting feature of the cutting plane method is its use of linearizations provided by each newly generated point for improving the polyhedral approximations to the problem functions. In other words, the next search direction finding subproblem (3.15) is modified by appending the constraints generated by the latest linearizations. This idea is used in many algorithms for nonsmooth optimization.

Convergence of the cutting plane algorithm can be very slow (Wolfe, 1975). This is mainly due to the fact that $|d^k|$ may be so large that the point x^k+d^k is far from the points y^j, $j=1,\dots,k$. Then x^k+d^k is in the region where $\hat{f}^k$ and $\hat{F}^k$ poorly approximate f and F. Also subproblems (3.14) and (3.15) may have no solutions. These drawbacks can be eliminated by adding to the objective functions of (3.14) and (3.15) a penalizing term $1/2|d|^2$, which will prevent large values of $|d^k|$. Thus we obtain the following regularized modification of subproblem (3.14)

$$\text{minimize} \quad \hat{f}^k(x^k+d;x^k) + \tfrac{1}{2}|d|^2 \quad \text{subject to} \quad \hat{F}^k(x^k+d;x^k) \le 0, \qquad (3.17)$$

and its quadratic programming formulation

$$\text{minimize} \quad \tfrac{1}{2}|d|^2+u \qquad (3.18a)$$

$$\text{subject to} \quad f_j^k + \langle g_f^j, d \rangle \le u \ , \ j=1,\dots,k, \qquad (3.18b)$$

$$F_j^k + \langle g_F^j, d \rangle \le 0 \ , \ j=1,\dots,k. \qquad (3.18c)$$

Search direction finding subproblems like (3.18) were introduced by Lemarechal (1978) in an algorithm for unconstrained convex minimization. He also showed how to construct sequences of points $\{x^k\}$ and auxiliary points of the form

$$y^{k+1} = x^k + d^k \quad \text{for} \quad k=1,2,..., \tag{3.19}$$

with $y^1=x^1$ being an arbitrary starting point in R^N, so that his algorithm is a descent method in the sense that

$$f(x^{k+1}) < f(x^k) \quad \text{if} \quad x^{k+1} \neq x^k \text{ , for all k.} \tag{3.20}$$

This relaxed version of the usual requirement for a descent method, i.e. $f(x^{k+1}) < f(x^k)$ for all k, is much easier to attain in practice. The main idea consists in taking the trial point $y^{k+1}=x^k+d^k$ as x^{k+1} only if this leads to an improvement in the objective function value, i.e. $f(y^{k+1}) < f(x^k)$. This is called a serious step. Otherwise a null step is taken by setting $x^{k+1}=x^k$.

To analyze Lemarechal's (1978) line search rules in more detail, let (d^k,u^k) denote the solution of (3.18a)-(3.18b) and let

$$v^k = u^k - f(x^k). \tag{3.21}$$

Letting $v=u-f(x^k)$ we obtain the following subproblem equivalent to (3.18a)-(3.18b)

$$\begin{aligned} &\text{minimize} \quad \tfrac{1}{2}|d|^2+v \,, \\ &\text{subject to} \quad f_j^k - f(x^k) + \langle g_f^j, d \rangle \le v \;,\; j=1,...,k, \end{aligned} \tag{3.22}$$

whose solution (d^k,v^k) satisfies (3.21). By (3.12) and (3.16), we have

$$f(x^k) \geq \hat{f}^k(x^k) = \max\{f_j^k : j=1,...,k\}. \tag{3.23}$$

This shows that $(d,v)=(0,0)$ is feasible for (3.22). Hence the optimal value of (3.22) $\frac{1}{2}|d^k|^2+v^k \le \frac{1}{2}|0|^2+0=0$. Therefore

$$v^k < -\tfrac{1}{2}|d^k|^2 \le 0. \tag{3.24}$$

Thus v^k is nonpositive. If $v^k=0$ then x^k minimizes f and the algorithm can stop, as will be shown in Chapter 2. Therefore we may assume that $v^k < 0$. Since (d^k,u^k) solves (3.18a)-(3.18b), we have

$$u^k = \max\{f_j^k + \langle g_f^j, d^k \rangle : j=1,...,k\}, \tag{3.25}$$

hence (3.8), (3.16) and (3.21) give

$$u^k = \hat{f}^k(x^k+d^k), \tag{3.26a}$$

$$v^k = \hat{f}^k(x^k+d^k) - f(x^k). \tag{3.26b}$$

Thus $v^k < 0$ is an estimate of $f(x^k+d^k)-f(x^k)=f(y^{k+1})-f(x^k)$. If the actual reduction in the objective function value is within $m \cdot 100\%$ of the predicted value, i.e.

$$f(y^{k+1})-f(x^k) \leq mv^k, \tag{3.27}$$

where $m \in (0,1)$ is a fixed line search parameter, then the trial point y^{k+1} is accepted as the next iterate $x^{k+1}=y^{k+1}$. Otherwise the algorithm stays at $x^{k+1}=x^k$. In both cases $f(x^{k+1}) \leq f(x^k)$, since $m > 0$ and $v^k < 0$.

The following remarks on the above line search rules will be useful in what follows. The condition for a serious step of the form

$$f(x^{k+1}) \leq f(x^k)+mv^k,$$

instead of a simpler test $f(x^{k+1}) < f(x^k)$, prevents the algorithm from taking infinitely many serious steps without significantly reducing the objective value, which could impair convergence. On the other hand, at a null step we have $x^{k+1}=x^k$ and

$$f(y^{k+1}) > f(x^k)+mv^k > f(x^k)+v^k=u^k,$$

because $v^k < 0$, $m \in (0,1)$ and (3.21) holds. But $f(y^{k+1})=\bar{f}(y^{k+1};y^{k+1})$ from (3.9), hence the above inequality and the fact that $y^{k+1}=x^k+d^k=x^{k+1}+d^k$ yield $\bar{f}(x^{k+1}+d^k;y^{k+1}) > u^k$. Therefore we obtain from (3.11)

$$\hat{f}^{k+1}(x^{k+1}+d^k) > u^k. \tag{3.28a}$$

In view of (3.26a) and the fact that $x^{k+1}=x^k$, we have

$$\hat{f}^k(x^{k+1}+d^k)=u^k, \tag{3.28b}$$

$$\hat{f}^{k+1}(x^{k+1}+d^{k+1})=u^{k+1}. \tag{3.28c}$$

From (3.28) we conclude that after a null step the linearization obtained at the trial point y^{k+1} leads to significant modifications of both the next polyhedral approximation and the next search direction finding subproblem. Therefore eventually a serious step must be taken if the current point is not a solution.

The above algorithm of Lemarechal (1978) was extended by Mifflin (1982) to constrained convex problems as follows. The set

$$S^k = \{x \in R^N : \bar{F}(x;y^j) \leq 0 \quad \text{for} \quad j=1,\ldots,k\} \tag{3.29}$$

is an <u>outer polyhedral approximation</u> to the feasible set

$$S = \{x \in R^N : F(x) \le 0\},$$

that is

$$S \subset S^k. \qquad (3.30)$$

This follows easily from (3.10). If the auxiliary points $y^1,\ldots,y^k$ are near to x^k, then S^k is a close approximation to S in some neighborhood of x^k. However, the solution (d^k,u^k) of (3.18) would usually give x^k+d^k lying at some "corner" of S^k which is outside S. Therefore almost every trial point $y^{k+1}=x^k+d^k$ would be infeasible. For this reason, Mifflin (1982) obtains d^k from the solution (d^k,v^k) to the problem

$$\begin{aligned} &\text{minimize} \quad \tfrac{1}{2}|d|^2+v, \\ &\text{subject to} \quad f_j^k-f(x^k)+\langle g_f^j,d\rangle \le v, \quad j=1,\ldots,k, \\ &\qquad\qquad\quad F_j^k+\langle g_F^j,d\rangle \le v, \quad j=1,\ldots,k. \end{aligned} \qquad (3.31)$$

This subproblem is a blend of (3.18) and (3.22). Clearly,

$$\max\{F_j^k+\langle g_F^j,d^k\rangle : j=1,\ldots,k\} \le v^k. \qquad (3.32)$$

In Mifflin's algorithm one always has $x^k \in S$. Then (3.10) and (3.23) show that $(d,v)=(0,0)$ is feasible for (3.31), hence (3.24) also holds here. Aditionally, $v^k=0$ only if x^k solves problem P. Therefore, in general, one has

$$\max\{\bar{F}(x^k+d^k;y^j): \qquad j=1,\ldots,k\} \le v^k < 0 \qquad (3.33)$$

from (3.32), (3.16) and (3.8). Combining this with (3.29), we see that the trial point $y^{k+1}=x^k+d^k$ lies in the interior of S^k. Therefore we shall have $y^{k+1} \in S$ whenever S^k is sufficiently close to S around x^k.

The above described line search rules of Lemarechal (1978) need only a simple modification in the presence of the constraints. If the trial point y^{k+1} is feasible, then one may use the test (3.27) and proceed as above. If $y^{k+1} \notin S$ then a null step $x^{k+1}=x^k$ is declared. Thus $f(x^{k+1}) \le f(x^k)$ and $x^k \in S$ for all k.

Whenever a null step results from

$$F(y^{k+1}) > 0$$

then

$$\hat{F}^{k+1}(x^{k+1}+d^k)=\bar{F}(y^{k+1};y^{k+1})=F(y^k) > v^k$$

from (3.9)-(3.13), and

$$v^{k+1} > \hat{F}^{k+1}(x^{k+1}+d^{k+1})$$

from (3.33) and (3.11). The above inequalities imply that $(d^{k+1},v^{k+1})\neq(d^k,v^k)$. We conclude that a null step due to infeasibility provides a significant modification of the polyhedral approximation to the constraint function. This explains why a feasible trial point is generated after finitely many null steps.

The following remark on Lemarechal's (1978) search direction finding subproblem (3.22)(see also (3.17) and (3.18a)-(3.18b)) will be useful in what follows. Observe that at the k-th iteration the j-th linearization

$$\bar{f}(x;y^j)=f(y^j)+\langle g_f(y^j),x-y^j\rangle \quad \text{for all} \quad x$$

can be written as

$$\bar{f}(x;y^j)=f(x^k)-\alpha_f(x^k,y^j)+\langle g_f(y^j),x-x^k\rangle, \tag{3.34}$$

where

$$\alpha_f(x^k,y^j)=f(x^k)-\bar{f}(x^k;y^j)\geq 0 \tag{3.35}$$

is the <u>linearization error</u> (nonnegative by (3.10)). Therefore the k-th polyhedral approximation $\hat{f}^k$ is also given by

$$\hat{f}^k(x)=\max\{f(x^k)-\alpha_f(x^k,y^j)+\langle g_f(y^j),x-x^k\rangle : j=1,\dots,k\}. \tag{3.36}$$

Since d^k is found by minimizing $\hat{f}^k(x^k+d)+\frac{1}{2}|d|^2$ over all d, we see that if the linearization error $\alpha_f(x^k,y^j)$ is large then it tends to make the subgradient $g_f(y^j)$ less active in the determination of the current search direction d^k. Indeed, it will be shown in Chapter 2 that

$$d^k = -\sum_{j=1}^{k}\lambda_j^k g_f(y^j), \tag{3.37}$$

where the numbers λ_j^k, j=1,...,k, are the Lagrange multipliers of (3.22) which solve the following dual of (3.22):

$$\underset{\lambda}{\text{minimize}} \quad \frac{1}{2}\Big|\sum_{j=1}^{k}\lambda_j g_f(y^j)\Big|^2+\sum_{j=1}^{k}\lambda_j\alpha_f(x^k,y^j), \tag{3.38}$$

$$\text{subject to} \quad \lambda_j\geq 0,\ j=1,\dots,k,\quad \sum_{j=1}^{k}\lambda_j=1.$$

Moreover, by (3.10) and (3.34)

$$f(x)\geq f(x^k)+\langle g_f(y^j),x-x^k\rangle-\alpha_f(x^k,y^j) \quad \text{for all} \quad x,$$

hence

$$g_f(y^j)\in\partial_\varepsilon f(x^k) \quad \text{for} \quad \varepsilon=\alpha_f(x^k,y^j)\geq 0, \tag{3.39}$$

which means that the value of $\alpha_f(x^k,y^j)$ indicates how far $g_f(y^j)$ is from $\partial f(x^k)$ ($g_f(y^j)\in\partial f(x^k)$ if $\alpha_f(x^k,y^j)=0$). Thus the algorithm of Lemarechal (1978) uses automatic weighing of the past subgradients on the basis of the corresponding linearization errors, which can be interpreted as <u>subgradient locality measures</u>.

One may distinguish three classes of descent methods for nonsmooth minimization. The classes differ by the form of their search direction finding subproblems and the associated line search rules. The first class originated with the above described algorithms of Lemarechal (1978) and Mifflin (1982). We shall now describe the remaining two classes, confining ourselves - for simplicity - to the unconstrained case.

The second class of methods stems from the algorithms for unconstrained convex minimization due to Lemarechal (1975) and Wolfe (1975). They use polyhedral approximations of the form

$$\hat{f}^k_{LW}(x)=\max\{f(x^k)+\langle g_f(y^j),x-x^k\rangle : j\in J^k_f\} \tag{3.40}$$

for some set $J^k_f\subset\{1,\ldots,k\}$, and choose the k-th direction d^k to

$$\text{minimize } \hat{f}_{LW}(x^k+d)+\tfrac{1}{2}|d|^2 \quad \text{over all } d\in R^N. \tag{3.41}$$

Comparing (3.36) and (3.40) we see that $\hat{f}^k_{LW}=\hat{f}^k$ if $J^k_f=\{1,\ldots,k\}$ and $\alpha_f(x^k,y^j)=0$ for all $j\in J^k_f$. In general, however, $\hat{f}^k_{LW}$ does not satisfy global relations of the form (3.12)-(3.13). Thus $\hat{f}^k_{LW}$ is a useful approximation to f around x^k only if the set J^k_f is chosen in a way that ensures for each $j\in J^k_f$ that y^j is close enough to x^k, so that the linearization error $\alpha_f(x^k,y^j)$ may be neglected. In view of (3.39), this means that in the algorithms of Lemarechal (1975) and Wolfe (1975) each past subgradient $g_f(y^j)$, $j\in J^k_f$, is treated as if it were a subgradient of f at the current point x^k. Hence, in order to provide another interpretation of (3.40) and (3.41), assume temporarily that $g_f(y^j)\in\partial f(x^k)$ for all $j\in J^k_f$. Define

$$\hat{f}(x)=\max\{f(x^k)+\langle \tilde{g}_f,x-x^k\rangle : \tilde{g}_f\in\partial f(x^k)\} \tag{3.42}$$

and let $\hat{d}$ denote a solution to the problem

$$\text{minimize } \hat{f}(x^k+d)+\tfrac{1}{2}|d|^2 \quad \text{over all } d\in R^N. \tag{3.43}$$

Then $\hat{f}^k_{LW}$ and subproblem (3.41) may be regarded as approximate versions of the "theoretical" constructions (3.42) and (3.43). Moreover, from Lemma 2.13 we deduce that

$$-\hat{d} = \hat{p}_f = \mathrm{Nr}\ \partial f(x^k), \tag{3.44a}$$

$$\max\{\langle \tilde{g}_f, \hat{d}\rangle : \tilde{g}_f \in \partial f(x^k)\} \le -|\hat{p}_f|^2, \tag{3.44b}$$

$$-d^k = p_f^k = \mathrm{Nr}\ \mathrm{conv}\{g_f(y^j) : j \in J_f^k\}, \tag{3.45a}$$

$$\max\{\langle g_f(y^j), d^k\rangle : j \in J_f^k\} \le -|p_f^k|^2. \tag{3.45b}$$

Thus $p_f^k = -d^k$ is found by projecting the origin onto the set $\mathrm{conv}\{g_f(y^j) : j \in J_f^k\}$, which approximates $\partial f(x^k)$. Moreover, as in (2.52), we have

$$\hat{f}_{LW}^k(x^k + td^k) \le f(x^k) - t|p_f^k|^2 \quad \text{for all} \quad t \in [0,1], \tag{3.46}$$

hence the value of $-|p_f^k|^2 = -|d^k|^2$ may be thought of as an approximate derivative of f at x^k in the direction d^k.

We may add that search direction finding subproblems of the form (3.41) are also used in the algorithms of Mifflin (1977b) and Polak, Mayne and Wardi (1983). A quadratic programming formulation of (3.41) is to find (d^k, v^k) to

$$\begin{aligned} &\text{minimize} \quad \tfrac{1}{2}|d|^2 + v, \\ &\text{subject to} \quad \langle g_f(y^j), d\rangle \le v, \ j \in J_f^k, \end{aligned} \tag{3.47}$$

with its dual

$$\begin{aligned} &\underset{\lambda}{\text{minimize}} \quad \tfrac{1}{2}\Big|\sum_{j \in J_f^k} \lambda_j g_f(y^j)\Big|^2, \\ &\text{subject to} \quad \lambda_j \ge 0, \ j \in J_f^k, \ \sum_{j \in J_f^k} \lambda_j = 1, \end{aligned} \tag{3.48}$$

corresponding to (3.45a)(cf.(3.38)). Moreover, $v^k = -|d^k|^2 = -|p_f^k|^2$.

Several strategies have been proposed for selecting the sets J_f^k so that f_{LW}^k is a local approximation to f at x^k. Mifflin (1977b) sets

$$J_f^k = \{j : \alpha_f(x^k, y^j) \le \delta^k\} \tag{3.49}$$

for a suitably chosen sequence $\delta^k \downarrow 0$. The algorithm of Lemarechal (1975) uses $y^j = x^j$ for $j \in J_f^k = \{1, \ldots, k\}$ until for some k

$$\alpha_f(x^1, x^k) \ge f(x^1) - f(x^k) + \varepsilon,$$

where $\varepsilon > 0$ is a parameter. Then the algorithm is <u>reset</u> by starting from the point $x^1 = x^k$ (with $g_f(x^1) = g_f(x^k)$ and $J_f^1 = \{1\}$). After sufficiently many resets one has $f(x^1) \approx f(x^k)$ and $\alpha_f(x^1, x^k) < 2\varepsilon$ between the resets, so that $g_f(y^j) \in \partial_{2\varepsilon} f(x^1)$ for all $j \in J_f^k$. The algorithm of Wolfe

(1975) uses $J_f^k=\{1,...,k\}$ until $|d^k|\le\varepsilon$, where $\varepsilon>0$ is a parameter. If $|d^k|\le\varepsilon$ then the algorithm stops provided that $\max\{|y^j-x^k|: j\in J_f^k\}\le\varepsilon$; otherwise x^k is taken as the new starting point x^1. Wolfe (1975) shows that his strategy makes the value of $\max\{|y^j-x^k|:j\in J_f^k\}$ arbitrarily small after sufficiently many resets. Another strategy (Mifflin, 1977b; Polak, Mayne and Wardi, 1983) is to set

$$J_f^k = \{j : |y^j-x^k|\le\delta^k\}, \tag{3.50}$$

where $\delta^k>0$ converges to zero. Such strategies will be discussed in detail in subsequent chapters.

To sum up, algorithms based on the polyhedral approximations (3.40) that neglect the linearization errors need suitable rules for reducing the past subgradient information, i.e. deleting the obsolete subgradients. Such rules should be implemented carefully, since any premature reduction slows down convergence until sufficiently many new subgradients are accumulated.

Lemarechal (1975) and Wolfe (1975) describe important modifications of their algorithms that require storing only finitely many subgradients. The modification consists in setting (cf.(3.45a))

$$-d^k = p_f^k = \mathrm{Nr}\ \mathrm{conv}\left[\{p_f^{k-1}\}\cup\{g_f(y^j): j\in J_f^k\}\right] \tag{3.51}$$

between each two consecutive resets (with $p_f^0=g_f(y^1)$ for $k=1$). The vector p_f^{k-1}, satisfying

$$p_f^{k-1}\in\mathrm{conv}\{g_f(y^j) : j=1,...,k-1\},$$

carries over from the previous iteration the relevant past subgradient information. In this case J_f^k may be selected subject only to the requirement

$$k\in J_f^k,$$

e.g. one may set $J_f^k=\{k\}$. The use of (3.51) corresponds to setting

$$\hat f_{LW}^k(x)=\max\{f(x^k)+\langle p_f^{k-1},x-x^k\rangle, f(x^k)+\langle g_f(y^j),x-x^k\rangle : j\in J_f^k\}$$

in subproblem (3.41), and appending an additional constraint

$$\langle p^{k-1},d\rangle\le v$$

in subproblem (3.47). Thus for direction finding p^{k-1} is treated as any past subgradient. Therefore we may call it the (k-1)-st aggregate subgradient.

We now pass to the line search rules used in Wolfe (1975) and (Mifflin, 1977b). To this end recall that the Lemarechal (1978) algorithm described above generates sequences related by

$$x^{k+1} = x^k + t_L^k d^k, \qquad y^{k+1} = x^k + t_R^k d^k, \tag{3.52}$$

with $t_L^k=1$ at serious steps, $t_L^k=0$ at null steps, and $t_R^k=1$ for all k. Moreover, at each step we have

$$f(x^{k+1}) \le f(x^k) + mt_L^k v^k, \tag{3.53}$$

and if a null step occurs at the k-th iteration then

$$f(y^{k+1}) - f(x^{k+1}) > mv^k. \tag{3.54}$$

The above relations follow from the criterion (3.27) and the fact that $t_L^k=1$ at a serious step, while a null step occurs with $t_L^k=0$ and $x^{k+1}=x^k$. At a null step we also have $y^{k+1}=x^k+d^k=x^{k+1}+d^k$, hence $y^{k+1}-x^{k+1}=d^k$ and

$$\begin{aligned} f(y^{k+1})-f(x^{k+1}) &= f(y^{k+1}) + \langle g_f(y^{k+1}), x^{k+1}-y^{k+1} \rangle - f(x^{k+1}) + \\ &\quad + \langle g_f(y^{k+1}), y^{k+1}-x^{k+1} \rangle = \\ &= -\alpha_f(x^{k+1},y^{k+1}) + \langle g_f(y^{k+1}), d^k \rangle \end{aligned}$$

from (3.35), therefore (3.54) can be written as

$$-\alpha_f(x^{k+1},y^{k+1}) + \langle g_f(y^{k+1}), d^k \rangle > mv^k. \tag{3.55}$$

We have shown above that the direction finding subproblems in the Wolfe (1975) algorithm can essentially be obtained from those in (Lemarechal, 1978) by neglecting the linearization errors. Now, if we assume that $\alpha_f(x^{k+1},y^{k+1})=0$ in (3.55) then we obtain

$$\langle g_f(y^{k+1}), d^k \rangle > mv^k, \tag{3.56}$$

which is essentially the criterion used in (Wolfe, 1975). To ensure that the value of the linearization error $\alpha_f(x^{k+1},y^{k+1})$ is sufficiently small, Wolfe (1975) imposed an additional condition

$$|y^{k+1}-x^{k+1}| \le \delta^k \tag{3.57a}$$

for some sequence $\delta^k \downarrow 0$. In fact, he used the following modification of

(3.53) and (3.56)

$$f(x^{k+1}) \le f(x^k)+m_L t_L^k v^k, \tag{3.57b}$$

$$\langle g_f(y^{k+1}),d^k \rangle \ge m_R v^k, \tag{3.57c}$$

where m_L and m_R are fixed line search parameters satisfying $0 < m_L < m_R < 1$. Line search procedures for finding stepsizes t_L^k and t_R^k satisfying $0 \le t_L^k \le t_R^k$, (3.52) and (3.57) can be found in (Mifflin, 1977b; Wolfe, 1975).

We may add that the criterion (3.57c) ensures that the new subgradient $g_f(y^{k+1})$ will significantly modify the next polyhedral approximation $\hat{f}_{LW}^{k+1}$ and the corresponding direction finding subproblem. This follows from the fact that if $k+1 \in J_f^{k+1}$ then (d^{k+1},v^{k+1}), being the solution of the $(k+1)$-st subproblem (3.47), satisfies

$$\langle g_f(y^{k+1}),d^{k+1} \rangle \le v^{k+1}.$$

Combining this with (3.56c) and the fact that $m_R v^k > v^k$ since $v^k = -|d^k|^2 < 0$ and $m_R \in (0,1)$ at line searches, we obtain $(d^{k+1},v^{k+1}) \ne (d^k,v^k)$.

In algorithms based on (3.45)(or (3.51)) and (3.57c) the value of $|d^k|=|p_f^k|$ decreases at each iteration, provided that no reduction of the past subgradient information occurs. To see this, note that (3.45) says that $p_f^k \ne 0$ defines a hyperplane separating $\mathrm{conv}\{g_f(y^j): j \in J_f^k\}$ from the null vector

$$\langle g_f(y^j),p_f^k \rangle \ge |p_f^k|^2 \quad \text{for all} \quad j \in J_f^k,$$

whereas (3.57c), written as

$$\langle g_f(y^{k+1}),p_f^k \rangle \le m_R|p_f^k|^2,$$

means that $g_f(y^{k+1})$ lies in the open halfspace containing the origin. It follows that the next separating haperplane, corresponding to $J_f^{k+1} = J_f^k \cup \{k+1\}$, must be closer to the null vector, i.e. $|p_f^{k+1}| < |p_f^k|$. Thus eventually the direction degenerates (one can have $d^k=0$), which provides another motivation for resetting strategies.

To sum up, the second class of algorithms discussed above (Lemarechal, 1975; Mifflin, 1977b; Polak, Mayne and Wardi, 1983; Wolfe,1975), which neglect the linearization errors at search direction finding, need rules for discarding obsolete subgradients. This is in contrast with the first class (Lemarechal, 1978; Mifflin, 1982), in which the linearization errors automatically weigh the past subgradients, cf.(3.38) and (3.48).

We shall now review the third class of methods, which is intermediate between the two classes discussed above.It contains so-called boundle methods (Lemarechal, 1976; Lemarechal, Strodiot and Bihain, 1981; Strodiot, Nguyen and Heukemes, 1983). At the k-th iteration of the algorithms based on relation (3.45) the set

$$G^k = \operatorname{conv}\{g_f(y^j) : j=1,\ldots,k\}$$

was supposed to approximate $\partial f(x^k)$. In bundle methods the convex polyhedron

$$G^k(\varepsilon)=\{\sum_{j=1}^{k} \lambda_j g_f(y^j) : \lambda_j \geq 0,\ j=1,\ldots,k,\ \sum_{j=1}^{k} \lambda_j \alpha_f(x^k,y^j) \leq \varepsilon\} \tag{3.58}$$

is used for approximating $\partial_\varepsilon f(x^k)$, where $\varepsilon > 0$. Indeed, using (3.39) and taking convex combinations we obtain

$$f(x)=\sum_{j=1}^{k} \lambda_j f(x) \geq f(x^k)+ \langle \sum_{j=1}^{k} \lambda_j g_f(y^j), x-x^k \rangle - \sum_{j=1}^{k} \lambda_j \alpha_f(x^k,y^j)$$

for all x, which shows that

$$G^k(\varepsilon) \subset \partial_\varepsilon f(x^k). \tag{3.59}$$

If we now choose $\varepsilon^k > 0$ and substitute G^k by $G^k(\varepsilon^k)$ in (3.45), we obtain the k-th direction d^k of the bundle methods as follows:

$$-d^k = p_f^k = \operatorname{Nr} G^k(\varepsilon^k), \tag{3.60}$$

with

$$\langle \tilde{g}, p_f^k \rangle \geq |p_f^k|^2 \quad \text{for all} \quad \tilde{g} \in G^k(\varepsilon^k) \tag{3.61}$$

Thus d^k can be computed by finding multipliers λ_j^k, j=1,...,k, to

$$\begin{aligned} &\text{minimize} \quad \tfrac{1}{2}\Big|\sum_{j=1}^{k} \lambda_j g_f(y^j)\Big|^2, \\ &\text{subject to} \quad \lambda_j \geq 0,\ j=1,\ldots,k,\ \sum_{j=1}^{k} \lambda_j = 1, \\ &\qquad\qquad \sum_{j=1}^{k} \lambda_j \alpha_f(x^k,y^j) \leq \varepsilon^k, \end{aligned} \tag{3.62}$$

and setting

$$d^k = -p_f^k = -\sum_{j=1}^{k} \lambda_j^k g_f(y^j). \tag{3.63}$$

Search direction finding via (3.60) can be motivated as follows. Suppose that we want to find a direction d such that $f(x^k+d)<f(x^k)-\varepsilon^k$. Then, since

$$f(x^k+d) \geq f(x^k) + \langle\tilde{g},d\rangle - \varepsilon^k \quad \text{for all} \quad \tilde{g} \in \partial_{\varepsilon^k} f(x^k)$$

and $G^k(\varepsilon^k) \subset \partial_{\varepsilon^k} f(x^k)$, d must satisfy

$$\langle\tilde{g},d\rangle < 0 \quad \text{for all } \tilde{g} \in G^k(\varepsilon^k),$$

i.e. we must find a hyperplane separating $G^k(\varepsilon^k)$ from the origin. The best such hyperplane is defined by p_f^k, cf. (3.60) and (3.61). Note also that if $p_f^k=0$ then $f(x) \geq f(x^k)-\varepsilon^k$ for all x, which means that the value of ε^k should be decreased.

Observe that if $\alpha_f(x^k,y^j) \leq \varepsilon^k$ for all j, then subproblem (3.62) is equivalent to (3.48). For smaller values of ε^k the last constraint of (3.62) tends to make the subgradients with larger linearization errors contribute less to d^k, since the corresponding multipliers must be smaller, cf. (3.63). Thus the weighing of the past subgradients depends on the value of ε^k. Since it is difficult to design convergent rules for automatic choice of the value of ε^k (Lemarechal, 1980), this is the main drawback of bundle methods in comparison with the first class of methods based on polyhedral approximations.

Lemarechal, Strodiot and Bihain (1981) have proposed a bundle method that requires storing only a finite number, say $M_g \geq 1$, of the past subgradients. Suppose that at the k-th iteration we have the (k-1)-st aggregate subgradient $(p_f^{k-1}, f_p^k) \in R^N \times R$, satisfying

$$(p_f^{k-1}, f_p^k) \in \text{conv}\{(g_f(y^j), \bar{f}(x^k;y^j) : j=1,\ldots,k-1\}.$$

Then, since

$$f(x) \geq \bar{f}(x^k;y^j) + \langle g_f(y^j), \; x-x^k\rangle \quad \text{for all } x,$$

we have

$$f(x) \geq f_p^k + \langle p_f^{k-1}, \; x-x^k\rangle \quad \text{for all } x,$$

hence

$$p_f^{k-1} \in \partial_\varepsilon f(x^k) \quad \text{for} \quad \varepsilon=\alpha_p^k,$$

where

$$\alpha_p^k = f(x^k) - f_p^k.$$

Subproblem (3.62) is replaced by the following one: find values of multipliers λ_j^k, $j \in J_f^k$, and λ_p^k to

$$\text{minimize } \frac{1}{2}\Big|\sum_{j\in J_f^k} \lambda_j g_f(y^j) + \lambda_p p_f^{k-1}\Big|^2,$$

$$\text{subject to } \lambda_j \geq 0,\ j\in J_f^k,\ \lambda_p^k \geq 0,\ \sum_{j\in J_f^k} \lambda_j + \lambda_p = 1, \tag{3.64}$$

$$\sum_{j\in J_f^k} \lambda_j \alpha_f(x^k,y^j) + \lambda_p \alpha_p^k \leq \varepsilon^k ,$$

and (3.63) by

$$d^k = -p_f^k = -\Big(\sum_{j\in J_f^k} \lambda_j^k g_f(y^j) + \lambda_p p_f^{k-1}\Big).$$

Thus for search direction finding (p_f^{k-1}, f_p^k) is treated as any "ordinary" vector $(g_f(y^j), \bar{f}(x^k;y^j))$. The algorithm uses resets for selecting the sets J_f^k. Between any two resets, one sets $J_f^{k+1} = J_f^k \cup \{k+1\}$. When J_f^k has M_g elements, the algorithm is reset by setting $J_f^{k+1}=\{k+1\}$. Of course, such strategy is not very efficient when M_g is small, since then too frequent reduction of the subgradient information hinders convergence.

Line search criteria of bundle methods are essentially of the form (3.57) with the additional requirement

$$\alpha_f(x^{k+1},y^{k+1}) < \varepsilon^{k+1}$$

ensuring that subproblem (3.62) (or (3.63)) is always feasible.

Up till now we have dealt mainly with the three classes of algorithms for convex problems. We shall now review extensions of these algorithms to the nonconvex case.

As shown above, in the convex case the algorithms of the first class (Lemarechal, 1978; Mifflin, 1982) have much clearer interpretation than the remaining methods. This is mainly due to the global properties (3.12) and (3.13) of their polyhedral approximations, which make it possible to weigh the past subgradients by the corresponding linearization errors. Of course, such global properties no longer hold in the nonconvex case. For this reason, Mifflin (1982) proposed the following subgradient locality measure

$$\alpha_f(x,y) = \max\{f(x) - \bar{f}(x;y), \gamma|x-y|^2\}, \tag{3.65}$$

where $\gamma>0$ is a parameter (γ can be set to zero when f is convex). The value of $\alpha_f(x,y)$ indicates how far $g_f(y)$ is from $\partial f(x)$. Note that if f is convex and $\gamma=0$ then $\alpha_f(x^k,y^j)$ defined by (3.65) reduces to the linearization error $f(x^k)-\bar{f}(x^k,y^j)$, as in (3.35). Therefore in (Mifflin, 1982) the k-th polyhedral approximation $\hat{f}^k$ is defined via (3.36) and (3.65), and (3.55) is still used at line searches. As before, d^k minimizes $\hat{f}^k(x^k+d)+\frac{1}{2}|d|^2$ over all d. The line search is more complicated in the nonconvex case, since (3.54) need not imply (3.55).

As far as the second class is concerned, observe that we have interpreted the direction finding subproblems and line search rules of these algorithms only in terms of local properties of the corresponding polyhedral approximations $\hat{f}^k_{LW}$, with no reference to convexity. This explains why these approximations were used by Mifflin (1977b) and Polak, Mayne and Wardi (1983) also for nonconvex problems, with subgradient deletion rules based on (3.50) localizing the approximations.

The third class (the bundle methods) has been extended by Lemarechal, Strodiot and Bihain (1981) by using the subgradient locality measures $\alpha_f(x^k,y^j)$ defined by (3.65) in subproblem (3.62). In connection with subproblem (3.64), they have also considered using the "path lengths"

$$s_j^k = |y^j-x^j| + \sum_{i=j}^{k-1} |x^{i+1}- x^i|$$

instead of $|x^k-y^j|$ in the definition of $\alpha_f(x^k,y^j)$. Then the points y^j need not be stored, since $s_j^{k+1} = s_j^k + |x^{k+1}- x^k|$.

So far we have concentrated on describing the algorithms. We shall now comment on the known results on their convergence and computational efficiency.

The algorithms of the first class discussed above have a potential for fast convergence (Lemarechal and Mifflin, 1982). However, at the k-th iteration they have k linear inequalities in their quadratic programming subproblems. This would present serious problems with storage and computation after a large number of iterations. Under additional boundedness assumptions on the generated sequences of points and the corresponding subgradients, these algorithms have at least one stationary accumulation point.

The second class of methods require bounded storage and use simple quadratic programming subproblems, but seem to converge slowly in practice (Lemarechal, 1982). As for convergence, Polak, Mayne and Wardi (1983) have modified the line search rules of the earlier versions so as to obtain global convergence in the sense that each of the algorithm's accumulation points is stationary.

The bundle method of Lemarechal, Strodiot and Bihain (1981), which is representative of the third class, requires bounded storage. Numerical experiments (Lemarechal, 1982) indicate that the method usually converges much more rapidly than the algorithms of the second class. However, no global convergence of that method seems to have been established in the nonconvex case.

Of course, much more work remains to be done before practical efficiency of each class of algorithms is fully assessed.

In view of the advantages and drawbacks of existing methods, our aim has been to construct methods for nonsmooth minimization which are characterized by

(a) applicability - the algorithms should use only general properties of problem P so as to be applicable to a broad class of practical problems;

(b) implementability - the algorithms should not require unbounded storage or an infinite number of arithmetic operations per iteration;

(c) reliability - a guarantee should exist, at least in the form of a proof of convergence, that the algorithms can find (approximate) solutions to a broad class of problems;

(d) efficiency - ability to provide satisfactory approximate solutions with minimal computational effort.

As far as efficiency is concerned, we note that function evaluations in the problems of interest to us are very time-consuming. Therefore, even relatively complex algorithms are admissible, provided that the computational overhead incurred in their auxiliary operations is smaller than the gain from a decrease in the number of function evaluations. For this reason the algorithms that are the subject of this work are rather complex and will be described in detail in subsequent chapters. Here we want to comment on their relations with the methods discussed so far.

In this work we shall present new versions, modifications and extensions of algorithms belonging to all the three existing classes of methods for nonsmooth optimization. We shall concentrate mainly on the first class of algorithms, since it seems to be particularly promising.

In Chapter 2 we extend the first class by describing aggregate subgradient methods for unconstrained convex minimization. In order to provide upper bounds on the amount of the past subgradient information which is stored and processed during the calculations, we give basic rules for selecting and aggregating the past subgradient information.

In Chapter 3 and Chapter 4 we show that the methods of Chapter 2 can be extended to the nonconvex case in two fundamentally different ways. The first strategy consists in modifying the polyhedral approximations by using subgradient locality measures of the form (3.65). The second, alternative strategy is to use subgradient deletion rules for localizing the polyhedral approximations. This approach is adopted in Chapter 4, where we also show that it leads to new algorithms belonging to the second class of methods that neglect the linearization errors.It will be seen that the two approaches to the nonconvex case result in significantly different algorithms.

In Chapter 5 and Chapter 6 we extend the preceding methods to inequality constrained problems. We present feasible point methods in which the past subgradient information about the problem functions is separately accumulated in two aggregate subgradients, one corresponding to the objective function and the other to the constraints. The methods differ in their line search rules and treatment of nonconvexity.

In Chapter 7 we apply our techniques of subgradient selection and aggregation to the third class of algorithms,obtaining new versions of bundle methods that require bounded storate. We also give bundle methods for problems with nonlinear inequality constraints, while up till now bundle methods for only linearly constrained problems have been considered.

We shall present apparently novel techniques for analyzing convergence of algorithms for nonsmooth optimization. In the absence of convexity, we will content ourselves with finding stationary points for problem P, i.e. points which satisfy the F.John necessary optimality condition for problem P, see Section 2. For each algorithm introduced in this work we prove that it is globally convergent in the sense that all its accumulation points are stationary. In the convex case, each of our algorithms generates a minimizing sequence of points, which in addition converges to a solution of problem P whenever problem P has any solution. Moreover, the convergence is finite in the piecewise linear case.

We may add that the algorithms discussed in this monograph are first-order methods. Some research is currently being done to obtain faster convergence ; see Auslender (1982), Demyanov, Lemarechal and Zowe (1985), Hiriart-Urruty (1983), Lemarechal and Mifflin (1982), Lemarechal and

Strodiot (1985), Lemarechal and Zowe (1983), Mifflin (1983 and 1984). This research is not discussed here, for our purpose is to establish some general convergence theory in the higher dimensional and constrained case.

CHAPTER 2

Aggregate Subgradient Methods for Unconstrained Convex Minimization

1. Introduction

In this chapter we consider the problem of minimizing a convex, not necessarily differentiable, function $f: R^N \to R$. We introduce a class of readily implementable algorithms, differing in complexity and efficiency, and analyze their convergence under no additional assumption on f. Each of the algorithms generates a minimizing sequence of points; if f attains its minimum then this sequence converges to a minimum point of f. Particular members of this algorithm class terminate when f happens to be piecewise linear (Kiwiel,1983).

The algorithms presented are descent methods which combine a generalized cutting plane idea with quadratic approximation. They can be interpreted as an extension of Pshenichny's method of linearizations (Pshenichny and Danilin,1975) to the nonsmooth case. Stemming from the pioneering algorithm of Lemarechal (1978), they differ from it in the updating of search direction finding subproblems. More specifically, instead of using all previously computed subgradients in quadratic programming subproblems, the methods use an aggregate subgradient, which is a convex combination of the past subgradients. It is recursively updated in a way that preserves that part of the past subgradient information which is essential for convergence.

In Section 2 we derive basic versions of the methods, comparing them with the algorithms of Pshenichny and Lemarechal. A formal description of an algorithmic procedure is given in Section 3. Its global convergence is demonstrated in Section 4, where we also introduce certain concepts that will be useful for analyzing subsequent extensions. In Section 5 we study convergence of methods with subgradient selection. Section 6 is devoted to the piecewise linear case. Further modifications of the methods are described in Section 7.

2. Derivation of the Algorithm Class

In this section we derive a class of methods for minimizing a convex function $f: R^N \to R$. To deal with nondifferentiability of f, the methods construct polyhedral approximations to f with the help of previously evaluated subgradients of f. To this end we introduce two general strategies for selecting and aggregating the past subgradient information. Such strategies enable one to impose a uniform upper bound on the amount of storage and work per iteration without impairing convergence. Our detailed description should help the reader to devise his or her own strategies that are tailored to particular optimization problems.

Since the algorithms to be described have structural relationship with Pshenichny's method of linearizations (Pshenichny and Danilin, 1975), we shall now review this method. To this end, suppose momentarily that

$$f(x) = \max\{f_j(x) : j \in J\} \quad \text{for all } x, \tag{2.1}$$

where each f_j is a convex function with continuous gradient ∇f_j on R^N, and J is finite. Given the k-th approximation to a solution $x^k \in R^N$, the method of linearizations finds a search direction d_p^k from the solution $(d_p^k, u_p^k) \in R^{N+1}$ to the following problem

$$\begin{aligned} &\text{minimize } \tfrac{1}{2}|d|^2 + u, \\ &\text{subject to } f_j(x^k) + \langle \nabla f_j(x^k), d\rangle \le u, \; j \in J . \end{aligned} \tag{2.2}$$

The above subproblem may be interpreted as a local first order approximation to the problem of minimizing $f(x^k+d)$ over all $d \in R^N$. Indeed, let us introduce the following polyhedral approximation to f at x^k

$$\hat{f}_p^k(z) = \max\{f_j(x^k) + \langle \nabla f_j(x^k), z-x^k\rangle : j \in J\} \quad \text{for all } z. \tag{2.3}$$

Then subproblem (2.2) is equivalent to the following

$$\text{minimize } \hat{f}_p^k(x^k+d) + \tfrac{1}{2}|d|^2 \quad \text{over all } d \in R^N, \tag{2.4}$$

and we have

$$u_p^k = \hat{f}_p^k(x^k+d_p^k). \tag{2.5}$$

At first sight it may appear that a more natural way of finding a search direction by minimizing $\hat{f}_p^k(x^k+d)$ over all d could be better. However, the latter problem may have no solution; moreover, $\hat{f}_p(x^k+d)$ is a doubtful approximation to $f(x^k+d)$ if $|d|$ is large. This gives the reason for the regularizing penalty term $\frac{1}{2}|d|^2$ in (2.4).

The next point $x^{k+1}=x^k+t^k d_p^k$ is found by searching for a stepsize $t^k>0$ satisfying

$$f(x^k+t^k d_p^k) < f(x^k) + m t^k v_p^k , \tag{2.6}$$

where $m \in (0,1)$ is a fixed line search parameter and

$$v_p^k = u_p^k - f(x^k). \tag{2.7}$$

More specifically, t^k is the largest number from the sequence $\{1, \frac{1}{2}, \frac{1}{4}, \ldots\}$ that satisfies (2.7). Such a positive number exists if $v_p^k < 0$. This follows from the fact that

$$\begin{aligned} f'(x^k; d_p^k) &= \max\{<\nabla f_k(x^k), d_p^k> : f_j(x^k) = f(x^k)\} \\ &\le \max\{f_j(x^k) - f(x^k) + <\nabla f_j(x^k), d^k> : j \in J\} \\ &\le \max\{f_j(x^k) + <\nabla f_j(x^k), d^k> : j \in J\} - f(x^k) \\ &\le u_p^k - f(x^k) = v_p^k , \end{aligned}$$

see Corollary 1.2.6, (2.3), (2.5) and (2.7). Therefore

$$\lim_{i \to \infty} [f(x^k + (\tfrac{1}{2})^i d_p^k) - f(x^k)| / (\tfrac{1}{2})^i = f'(x^k; d_p^k) \le v_p^k < m v_p^k$$

if $m \in (0,1)$ and $v_p^k < 0$, which shows that (2.6) must hold if t^k is sufficiently small. On the other hand, if $v_p^k \ge 0$ then the method of linearizations stops, because x^k is stationary for f.

In fact, Pshenichny defined v_p^k in (2.6) as $-|d_p^k|^2$, which is slightly larger than v_p^k given by (2.7), and assumed that the gradients of f_j are Lipschitz continuous. However, it is easy to prove that the above version of the method of linearizations is globally convergent when each f_i is continuously differentiable, and that the rate of convergence is at least linear under standard second order sufficiency conditions even when f is nonconvex, see (Kiwiel,1981a). Moreover, if all the functions f_j are affine, then the method finds a solution in a finite number of iterations. Therefore it seems worthwile to extend this method to more general nondifferentiable problems.

Although our methods will not require the special form (2.1) of the objective function, they are in fact based on a similar, but implicit, representation

$$f(x) = \max\{f(y) + <g_y, x-y> : g_y \in \partial f(y),\ y \in R^N\}, \tag{2.8}$$

which is due to convexity. Since we do not assume the availibility of the whole subdifferential $\partial f(y)$ at each y in R^N, the methods will use approximate versions of (2.8) constructed as follows.

We suppose that we have a subroutine that can evaluate a subgradient $g_f(x) \in \partial f(x)$ at each $x \in R^N$. Suppose that at the k-th iteration of the algorithm we have the current point $x^k \in R^N$ together with some auxiliary points y^j and subgradients $g^j = g_f(y^j)$ for $j \in J^k$, where J^k is a nonempty subset of $\{1,...,k\}$. Define the linearizations

$$f_j(x) = f(y^j) + \langle g_j, x-y^j \rangle , \quad j \in J^k , \tag{2.9}$$

and the current polyhedral approximation to f

$$\hat{f}_s^k(x) = \max\{f_j(x): j \in J^k\} . \tag{2.10}$$

Comparing (2.1) with (2.9) and (2.10), we see that an application of one step of the method of linearizations to $\hat{f}_s^k$ at x^k leads to the following search direction finding subproblem

$$\begin{aligned} &\text{minimize } \tfrac{1}{2} |d|^2 + u , \\ &\text{subject to } f_j^k + \langle g^j, d \rangle \le u, \quad j \in J^k , \end{aligned} \tag{2.11}$$

where

$$f_j^k = f_j(x^k) \quad \text{for } j=1,...,k. \tag{2.12}$$

The above subproblem may be interpreted as a local approximation to the problem of minimizing $\hat{f}_s^k(x^k+d)$ over all d, and hence to the problem of minimizing $f(x^k+d)$ over d. Let (d^k, u^k) denote the solution of (2.11), and let

$$v^k = u^k - f(x^k). \tag{2.13}$$

Then, as in (2.5) and (2.7), we have

$$u^k = \hat{f}_s^k(x^k+d^k), \tag{2.14}$$

$$v^k = \hat{f}_s^k(x^k+d^k) - f(x^k). \tag{2.15}$$

Moreover, (2.8) implies that $f(x^k) \ge \hat{f}_s^k(x^k)$, hence (2.15) and the convexity of $\hat{f}_s^k$ yield

$$\begin{aligned} \hat{f}_s^k(x^k+td^k) &\le (1-t)\hat{f}_s^k(x^k) + t\, \hat{f}_s^k(x^k+d^k) \\ &\le f(x^k) + tv^k \qquad \text{for all } t \in [0,1]. \end{aligned} \tag{2.17}$$

Therefore v^k may be interpreted as an <u>approximate directional derivative</u> of f at x^k in the direction d^k. We shall show later that $v^k<0$ if x^k is nonstationary for f. Thus we may assume that $v^k<0$. In general, we may have $f'(x^k;d^k)>v^k$, because $\hat{f}_s^k$ may poorly approximate f. For this reason, the line search rule of the method of linearizations needs the following modifications.

We assume that $m \in (0,1)$ and $\bar{t} \in (0,1]$ are fixed line search parameters. First we shall search for the largest number t_L^k in $\{1, \frac{1}{2}, \frac{1}{4}, \ldots\}$ that satisfies

$$f(x^k+t_L^k d^k) \le f(x^k) + mt_L^k v^k \tag{2.18}$$

and $t_L^k \ge \bar{t}$. This involves a finite number of function evalutions, because $\bar{t}>0$ (only one if $\bar{t}=1$). If such a number $t_L^k>0$ exists, we shall set $x^{k+1}=x^k+t_L^k d^k$ and $y^{k+1}=x^{k+1}$ (a <u>serious step</u>). Otherwise we have to accept a <u>null step</u> by setting $x^{k+1}=x^k$. In this case we also know a number $t_R^k \in [\bar{t},1]$ satisfying

$$f(x^k+t_R^k d^k) > f(x^k) + mt_R^k v^k.$$

Therefore at a null step we shall set $y^{k+1}=x^k+t_R^k d^k$, because this new trial point will define a linearization f_{k+1} by (2.9) that satisfies

$$f_{k+1}(x^k+d^k) > f(x^k) + mv^k, \tag{2.19}$$

see Section 7. Comparing (2.10), (2.14), (2.15) and (2.19), and using the fact that $x^{k+1}=x^k$, $v^k<0$ and $m \in (0,1)$, we deduce that after a null step we have

$$\hat{f}_s^{k+1}(x^{k+1}+d^k) > u^k,$$

$$\hat{f}_s^{k+1}(x^{k+1}+d^{k+1}) = u^{k+1}$$

provided that

$$k+1 \in J^{k+1}.$$

Thus after a null step the linearization from the trial point y^{k+1} will modify both the next polyhedral approximation and the next search direction finding subproblem.

We shall now show how to choose the next subgradient index set J^{k+1}. As noted above, we should have $k+1 \in J^{k+1}$, which is satisfied if

$$J^{k+1} = \hat{J}^k \cup \{k+1\} \tag{2.20}$$

for some set $\hat{J}^k \subset J^k$. The obvious choice $\hat{J}^k=J^k$, suggested by the cutting plane methods (Cheney and Goldstein,1959;Kelley,1960), would present serious problems with storage and computation after a large number of iterations. We may add that such a choice, i.e.

$$J^k = \{1,\dots,k\} \qquad \text{for all } k,$$

is used by Lemarechal (1978) and Mifflin (1982). It is therefore important to be able to construct $\hat{J}^k$ in a way that permits dropping some of the linear inequalities in subproblem (2.11). Part (iv) of the following lemma suggests a strategy analogous to constraint dropping strategies of the cutting plane methods due to Eaves and Zangwill (1971) and Topkis (1970a,1970b,1982). The other parts are taken from (Lemarechal,1978; Wierzbicki,1982).

Lemma 2.1 (i) The unique solution (d^k,u^k) of subproblem (2.11) always exists.

(ii) (d^k,u^k) solves (2.11) if and only if there exists Lagrange multipliers $\lambda_j^k \in R^N$, $j \in J^k$, and a vector $p^k \in R^N$ satisfying

$$\lambda_j^k \geq 0,\ j \in J^k, \quad \sum_{j \in J^k} \lambda_j^k = 1, \tag{2.21a}$$

$$[f_j^k + \langle g^j, d^k\rangle - u^k]\lambda_j^k = 0,\ j \in J^k, \tag{2.21b}$$

$$p^k = \sum_{j \in J^k} \lambda_j^k g^j, \tag{2.21c}$$

$$d^k = -p^k, \tag{2.21d}$$

$$u^k = -\{|p^k|^2 - \sum_{j \in J^k} \lambda_j^k f_j^k\}, \tag{2.21e}$$

$$f_j^k + \langle g^j, d^k\rangle \leq u^k,\ j \in J^k. \tag{2.21f}$$

(iii) The multipliers λ_j^k, $j \in J^k$, satisfy (2.21) if and only if they solve the following dual subproblem

$$\begin{aligned} &\underset{\lambda}{\text{minimize}}\ \tfrac{1}{2}\Big|\sum_{j \in J^k} \lambda_j g^j\Big|^2 - \sum_{j \in J^k} \lambda_j f_j^k\,, \\ &\text{subject to } \lambda_j \geq 0,\ j \in J^k, \quad \sum_{j \in J^k} \lambda_j = 1\,. \end{aligned} \tag{2.22}$$

(iv) There exists a solution λ_j^k, $j \in J^k$, of subproblem (2.22) such that the set

$$\hat{J}^k = \{j \in J^k : \lambda_j^k > 0\} \tag{2.23a}$$

satisfies

$$|\hat{J}^k| \le N+1. \tag{2.23b}$$

Such a solution can be obtained by solving the following linear programming problem by the simplex method:

$$\begin{aligned} &\underset{\lambda}{\text{minimize}} - \sum_{j \in J^k} \lambda_j f_j^k , \\ &\text{subject to } \sum_{j \in J^k} \lambda_j = 1 , \\ &\qquad \sum_{j \in J^k} \lambda_j g^j = p^k , \\ &\qquad \lambda_j \ge 0, \ j \in J^k , \end{aligned} \tag{2.24}$$

where $p^k = -d^k$. Moreover, (d^k, u^k) solves the following reduced subproblem

$$\begin{aligned} &\text{minimize } \tfrac{1}{2}|d|^2 + u , \\ &\text{subject to } f_j^k + \langle g^j, d\rangle \le u, \ j \in \hat{J}^k . \end{aligned} \tag{2.25}$$

<u>Proof</u>. (i) Subproblem (2.11) is equivalent to the following problem

$$\text{minimize}\left[\phi(d) = \tfrac{1}{2}|d|^2 + u_s(d)\right] \quad \text{over all } d,$$

where $u_s(d) = \max\{f_j^k + \langle g^j, d\rangle : j \in J^k\}$. The function ϕ is strictly convex and satisfies $\phi(d) \to +\infty$ if $|d| \to \infty$, because $u_s(d) \ge f_j^k + \langle g^j, d\rangle \ge f_j^k - |g^j||d|$ for any $j \in J^k$ and $d \in R^N$. Therefore d^k, the unique minimum point of ϕ, exists and satisfies $u^k = u_s(d^k)$.
(ii) Subproblem (2.11) is convex and satisfies the Slater constraint qualification (let $\tilde{d}=0$ and $\tilde{v} = \max\{f_j^k : j \in J^k\}+1$). Therefore we deduce from Lemma 1.2.16 that (2.21) is the Kuhn-Tucker condition for (d^k, u^k) to solve (2.11).
(iii) One may check that (2.21) is the Kuhn-Tucker condition for λ_j^k, $j \in J^k$, to be a solution of (2.22). Although subproblem (2.22) does not satisfy the Slater constraint qualification, the Kuhn-Tucker condition (2.21) is both necessary and sufficient here, because the constraints of (2.22) are linear; see, e.g. (Bazaraa and Shetty, 1979).
(iv) Since any solution λ_j^k, $j \in J^k$, of (2.22) must satisfy (2.21c) and $p^k = -d^k$ is unique, we deduce that (2.22) and (2.24) have a common, nonempty set of solutions. The simplex method will find an optimal basic solution of (2.24) with no more than N+1 strictly positive components (Dantzig, 1963). Thus we get the desired multipliers satisfy-

ing (2.23). Since these multipliers also solve (2.22), parts (ii) - (iii)of the lemma imply (2.21). Therefore one may use (2.23a) and part (ii) of the lemma to complete the proof. □

We shall use the above lemma to design two different constraint dropping strategies, which will yield different algorithms. Both choices of constraints for the next direction finding subproblem are based on the following <u>generalized cutting plane idea</u>: having solved the current search direction finding subproblem, construct an auxiliary reduced subproblem which yields the same solution. Then obtain the next search direction finding subproblem by appending to the auxiliary subproblem the constraint generated by the new subgradient.

The first application of this principle makes use of Lemma 2.1(iv). Subproblem (2.25) is the desired auxiliary problem, which is equivalent to the original subproblem (2.11). Therefore the choice of J^{k+1} specified by (2.20) and (2.23) conforms with the above cutting plane concept. Observe that this amounts to discarding those subgradients g^j which have null multipliers λ_j^k. Such subgradients do not contribute to the current direction d^k, see (2.21d) and (2.21c).

The above strategy, based on (2.23), may be termed the <u>subgradient selection strategy</u>. It leads to implementable algorithms that require storage of at most N+1 past subgradients. However, this may pose serious difficulties if N is large. Therefore we describe below another strategy that overcomes this drawback.

The second strategy may be termed the <u>subgradient aggregation strategy</u>, because it aggregates the constraints generated by the past subgradients. The auxiliary subproblem is constructed by forming a surrogate constraint based on the Lagrange multipliers of the original subproblem. Let λ_j^k, $j\in J^k$, denote any Lagrange multipliers of (2.11), which do not necessarily satisfy (2.23), and let

$$\tilde{f}_p^k = \sum_{j\in J^k} \lambda_j^k f_j^k \ . \tag{2.26}$$

Combining this with (2.21c), we obtain

$$(p^k,\tilde{f}_p^k) = \sum_{j\in J^k} \lambda_j^k (g^j,f_j^k) \ . \tag{2.27}$$

The following lemma describes the auxiliary subproblem of the subgradient aggregation strategy.

Lemma 2.2. Subproblem (2.11) is equivalent to the following reduced subproblem:

$$\text{minimize } \tfrac{1}{2}|d|^2 + u ,$$

$$\text{subject to } \tilde{f}_p^k + \langle p^k, d\rangle \le u , \tag{2.28}$$

$$f_j^k + \langle g^j, d\rangle \le u, \; j \in \hat{J}^k ,$$

where $\hat{J}^k$ is any subset of J^k (possibly empty).

Proof. Let $\tilde{\lambda}_p^k=1$, $\tilde{\lambda}_j^k=0$, $j \in \hat{J}^k$. From (2.21) and (2.27),

$$\tilde{\lambda}_p^k \ge 0, \; \tilde{\lambda}_j^k \ge 0, \; j \in \hat{J}^k, \quad \tilde{\lambda}_p^k + \sum_{j \in J^k} \tilde{\lambda}_j^k = 1,$$

$$[\tilde{f}_p^k + \langle p^k, d^k\rangle - u^k]\, \tilde{\lambda}_p^k = 0 ,$$

$$[f_j^k + \langle g^j, d^k\rangle - u^k]\, \tilde{\lambda}_j^k = 0, \quad j \in \hat{J}^k ,$$

$$p^k = \tilde{\lambda}_p^k p^k + \sum_{j \in J^k} \tilde{\lambda}_j^k g^j ,$$

$$d^k = -p^k,$$

$$u^k = -\{|p^k|^2 - \tilde{\lambda}_p^k \tilde{f}_p^k - \sum_{j \in J^k} \tilde{\lambda}_j^k f_j^k\},$$

$$\tilde{f}_p^k + \langle p^k, d^k\rangle \le u^k, \quad f_j^k + \langle g^j, d^k\rangle \le u^k, \quad j \in \hat{J}^k .$$

Subproblem (2.28) is of the form (2.11). Therefore the above relations and Lemma 2.1(ii) imply that (d^k,u^k) solves (2.28). Hence Lemma 2.1(i) yields that (2.11) and (2.28) have the same unique solution (d^k,u^k).□

We shall now interpret the above results in terms of linearizations and polyhedral approximations to f. Each subgradient $g^j \in \partial f(y^j)$ provides global information about the objective function, since the corresponding linearization f_j, given by (2.9), satisfies

$$f(x) \ge f_j(x) \qquad \text{for all } x . \tag{2.29}$$

Since $f_j^k = f_j(x^k)$, we have

$$f_j(x) = f_j^k + \langle g^j, x-x^k\rangle \qquad \text{for all } x \tag{2.30}$$

and (2.29) becomes

$$f(x) \ge f_j^k + \langle g^j, x-x^k\rangle \qquad \text{for all } x , \tag{2.31}$$

and j=1,...,k. Observe that the linearizations can be updated recursively:

$$f_j^{k+1} = f_j(x^{k+1}) = f_j^k + \langle g^j, x^{k+1} - x^k \rangle, \tag{2.32}$$

so the points $\{y^j\}$ need not be stored. Summing up, we see that at iteration k the subgradient information collected at the j-th iteration consists of the (N+1)-vector (g^j, f_j^k), which generates the corresponding linearization (2.30) and the constraint in subproblem (2.11), and yields the bound (2.31). Therefore we shall refer to (g^j, f_j^k) as the j-th subgradient of f at the k-th iteration, for any j=1,...,k. We also note that in terms of the <u>selective polyhedral approximation</u> to f at x^k

$$\hat{f}_s^k(x) = \max\{f_j^k + \langle g^j, x-x^k \rangle : j \in J^k\} \quad \text{for all } x, \tag{2.33}$$

the search direction finding subproblem (2.11) can be written as

$$\text{minimize } \hat{f}_s^k(x^k+d) + \tfrac{1}{2}|d|^2 \quad \text{over all } d. \tag{2.34}$$

Proceeding in the same spirit, we may associate with the first constraint of the reduced subproblem (2.28) the following <u>aggregate linearization</u>

$$\tilde{f}^k(x) = \tilde{f}_p^k + \langle p^k, x-x^k \rangle \quad \text{for all } x \tag{2.35}$$

and call the associated (N+1)-vector $(p^k, \tilde{f}_p^k)$ the <u>aggregate subgradient</u> of f at the k-th iteration. In view of Lemma 2.2, the aggregate subgradient $(p^k, \tilde{f}_p^k)$ embodies all the past subgradient information that is essential for the k-th search direction finding, since an equivalent formulation of subproblems (2.28) and (2.11) is

$$\text{minimize } \tilde{f}^k(x^k+d) + \tfrac{1}{2}|d|^2 \quad \text{over all } d. \tag{2.36}$$

Therefore one may use the aggregate linearization (2.35) for search direction finding at the next point x^{k+1}, where

$$\tilde{f}^k(x) = f_p^{k+1} + \langle p^k, x-x^{k+1} \rangle \quad \text{for all } x \tag{2.37}$$

with f_p^{k+1} defined similarly to (2.32):

$$f_p^{k+1} = \tilde{f}^k(x^{k+1}) = \tilde{f}_p^k + \langle p^k, x^{k+1} - x^k \rangle. \tag{2.38}$$

Thus at the (k+1)-st iteration the linearization (2.35) is generated by the updated aggregate subgradient (p^k, f_p^{k+1}).

Our use for aggregation of multipliers that form convex combinations, cf. (2.21a), yields the following useful property.

Lemma 2.3. The aggregate linearization (2.35) defined by (2.27) is a convex combination of the linearizations (2.29). Moreover

$$f(x) \geq \tilde{f}^k(x) \quad \text{for all } x. \tag{2.39}$$

Proof. From (2.35), (2.27) and (2.30)

$$\tilde{f}^k(x) = \sum_{j\in J^k} \lambda_j^k f_j^k + \left< \sum_{j\in J^k} \lambda_j^k g^j,\ x-x^k \right> =$$

$$= \sum_{j\in J^k} \lambda_j^k [f_j^k + \langle g^j, x-x^k\rangle] = \sum_{j\in J^k} \lambda_j^k f_j(x)$$

for each x. The above relations, (2.21a) and (2.29) yield (2.39). □

Following the generalized cutting plane concept introduced above, we obtain the next search direction finding subproblem of the method with aggregation in two steps. First, we use aggregation for deriving the auxiliary subproblem (2.28) and the aggregate linearization (2.35). Next, we update the linearizations according to (2.32) and (2.38) and append the new constraint generated by the latest subgradient $g^{k+1}=g(y^{k+1})$. Thus the next subproblem becomes: find (d^{k+1},u^{k+1}) to

$$\begin{aligned} &\text{minimize } \tfrac{1}{2}|d|^2 + u, \\ &\text{subject to } f_j^{k+1} + \langle g^j, d\rangle \leq u, \quad j\in J^{k+1} = \hat{J}^k \cup \{k+1\}, \\ &\qquad\qquad f_p^{k+1} + \langle p^k, d\rangle \leq u. \end{aligned} \tag{2.40}$$

Of course, the above subproblem need not be equivalent to the(k+1)-st subproblem (2.11), e.g. we may have $\hat{J}^k=\emptyset$ in (2.40), hence the resulting algorithms will differ. However, in order to stress their similarities, we denote the corresponding variables by the same symbols.

Since the second step in the above derivation of (2.40) required no reference to subproblem (2.11), if we now show how to aggregate subproblem (2.40) we shall in fact define recursively the aggregate subgradient method that does not need the points y^j, $j=1,...,k-1$, at the k-th iteration. This is quite easy if one observes that (2.40) is similar to (2.11). Consequently, one can aggregate subproblem (2.40) in essentially the same manner as shown above for subproblem (2.11).

In this way we arrive at the following description of consecutive aggregate subproblems. Let $(d^k,u^k)\in R^{N+1}$ denote the solution to the following k-th aggregate search direction finding subproblem (cf. (2.40)):

$$\text{minimize } \tfrac{1}{2}|d|^2 + u ,$$
$$\text{subject to } f_j^k + \langle g^j,d\rangle \le u,\ j\in J^k, \qquad (2.41)$$
$$f_p^k + \langle p^{k-1},d\rangle \le u.$$

In order to be able to use (2.41) for k=1, we shall initialize the method by choosing $x^1\in R^N$ and setting $y^1=x^1$ and

$$p^0 = g^1 = g(y^1), \quad f_p^1 = f_1^1 = f(y^1), \quad J^1 = \{1\}. \qquad (2.42)$$

Let λ_j^k, $j\in J^k$, and λ_p^k denote any Lagrange multipliers of (2.41). Since subproblem (2.41) is of the form (2.11), Lemma 2.1 implies that these multipliers satisfy

$$\lambda_j^k\ge 0,\ j\in J^k,\ \lambda_p^k\ge 0,\quad \sum_{j\in J^k}\lambda_j^k + \lambda_p^k = 1, \qquad (2.43a)$$

$$[f_j^k + \langle g^j,d^k\rangle - u^k]\lambda_j^k = 0,\quad j\in J^k, \qquad (2.43b)$$

$$[f_p^k + \langle p^{k-1},d^k\rangle - u^k]\ \lambda_p^k = 0, \qquad (2.43c)$$

$$p^k = \sum_{j\in J^k}\lambda_j^k g^j + \lambda_p^k p^{k-1}, \qquad (2.43d)$$

$$d^k = -p^k\ , \qquad (2.43e)$$

$$u^k = -\{|p^k|^2 - \sum_{j\in J^k}\lambda_j^k f_j^k - \lambda_p^k f_p^k\}. \qquad (2.43f)$$

Similarly to (2.26), we define the value of the current aggregate linearization

$$\tilde f_p^k = \sum_{j\in J^k}\lambda_j^k f_j^k + \lambda_p^k f_p^k, \qquad (2.44)$$

and obtain analogously to (2.27)

$$(p^k,\tilde f_p^k) = \sum_{j\in J^k}\lambda_j^k(g^j,f_j^k) + \lambda_p^k(p^{k-1},f_p^k). \qquad (2.45)$$

As above, we shall use (2.38) to update the aggregate linearization (2.35) when $x^{k+1}\neq x^k$. This completes the derivation of the method with subgradient aggregation.

We may add that for the method with aggregation Lemma 2.2 can be rephrased as follows: subproblem (2.28) is equivalent to subproblem

(2.41), which in turn is equivalent to the problem

$$\text{minimize } \hat{f}_a^k(x^k+d) + \tfrac{1}{2}|d|^2 \quad \text{over all } d, \tag{2.46}$$

where $\hat{f}_a^k$ is the k-th <u>aggregate polyhedral approximation</u> to f:

$$\hat{f}_a^k(x) = \max\{\tilde{f}^{k-1}(x), f_j(x):\ j\in J^k\}=$$

$$= \max\{f_p^k + \langle p^{k-1}, x-x^k\rangle,\ f_j^k + \langle g^j, x-x^k\rangle:\ j\in J^k\}. \tag{2.47}$$

In Section 4 we shall show that Lemma 2.3 holds also for the method with aggregation.

<u>Remark 2.4</u>. Convergence of the method which uses the aggregate subproblems (2.41) with $J^k=\{k\}$ can be slow, since only two linearizations may provide insufficient approximation to the nondifferentiable objective function. Using more subgradients for search direction finding enhances faster convergence, but at the cost of increased storage and work per iteration. To strike a balance, one may use the following strategy. Let $M_g\geq 2$ denote a user-supplied bound on the number of subgradients (including the aggregate subgradient) that the algorithm may use for each search direction finding. Then one may choose the set J^{k+1} on the basis of the k-th Lagrange multipliers subject to the following requirements:

$$J^{k+1} = \{k+1\}\cup \hat{J}^k, \tag{2.48a}$$

$$\hat{J}^k\subset\{j\in J^k:\ \lambda_j^k>0\}, \tag{2.48b}$$

$$|\hat{J}^k|\leq M_g - 2, \tag{2.48c}$$

with $\hat{J}^k$ containing the largest indices corresponding to $\lambda_j^k>0$. This ensures that the most "active" subgradients will not be prematurely discarded.

<u>Remark 2.5</u>. If the objective function is of the form

$$f(x) = \max\{\tilde{f}_j(x):\ j\in J\}$$

and it is possible to calculate some subgradients $g^{k,j}\in\partial\tilde{f}_j(x^k)$, $j\in J$, then one may increase the efficiency of the above methods by appending the constraints

$$\tilde{f}_j(x^k) + \langle g^{k,j}, d\rangle\leq u,\ j\in J, \tag{2.49}$$

to the search direction finding subproblems (2.11) and (2.41), for each k. One may also replace the set J in (2.49) with the set of indices of ε-active functions $J(\varepsilon)=\{j\in J: \tilde{f}_j(x^k)\geq f(x^k)-\varepsilon\}$, for some $\varepsilon\geq 0$. Clearly, aggregation can be used for the resulting augmented subproblems under an appropriate change in notation. The subsequent convergence results remain valid for such modifications.

The above remarks suggest that there is much freedom within the framework of subgradient selection and aggregation strategies for constructing particular algorithms taylored to special classes of problems.

We end this section by commenting on the relations of the above described methods with other algorithms. As shown above, the methods generalize Pshenichny's method of linearizations for minimax problems, which in turn extends the classical method of steepest descent for minimizing smooth functions. On the other hand, subproblems (2.11) and (2.41) may be seen as reduced versions of the Lemarechal (1978) subproblems with $J^k=\{1,\ldots,k\}$, cf. (1.3.22).

Remark 2.6. We recall from Section 1.3 that an approximation to f at x^k

$$\hat{f}(x) = \max\{f(x^k) + \langle g,x-x^k\rangle : g\in\partial f(x^k)\} \tag{2.50a}$$

may yield a descent direction for f at x^k as a solution to the problem

$$\text{minimize } \hat{f}(x^k+d) + \tfrac{1}{2}|d|^2 \quad \text{over all } d, \tag{2.50b}$$

provided that x^k is nonstationary, see Lemma 1.2.13. Comparing (2.50) with (2.33), (2.34), (2.46) and (2.47), we arrive at the following interpretation of the methods described above. Instead of using the "complete" approximations (2.50a), which would require the knowledge of all the linearizations associated with the current point x^k via $\partial f(x^k)$, the methods use approximations (2.33) and (2.47) which result from linearizations calculated at many trial points around x^k.

3. The Basic Algorithm

We now have all the necessary ingredients to state the simplest version of the aggregate subgradient method for solving the problem in question. Its more efficient modifications are discussed in subsequent sections.

Algorithm 3.1.

Step 0 (Initialization). Select the starting point $x^1 \in R^N$ and set $y^1 = x^1$. Choose a final accuracy tolerance $\varepsilon_s \geq 0$ and a line search parameter $m \in (0,1)$. Set $p^0 = g^1 = g_f(y^1)$, $f_p^1 = f_1^1 = f(y^1)$ and $J^1 = \{1\}$. Set the counters $k=1$, $l=0$ and $k(0)=1$.

Step 1 (Direction finding). Find multipliers λ_j^k, $j \in J^k$, and λ_p^k that solve the following k-th dual subproblem

$$\text{minimize } \frac{1}{2} \Big| \sum_{j \in J^k} \lambda_j g^j + \lambda_p p^{k-1} \Big|^2 - \sum_{j \in J^k} \lambda_j f_j^k - \lambda_p f_p^k ,$$
$$\text{subject to } \lambda_j \geq 0,\ j \in J^k,\ \lambda_p \geq 0, \quad \sum_{j \in J^k} \lambda_j + \lambda_p = 1. \tag{3.1}$$

Calculate the aggregate subgradient $(p^k, \tilde{f}_p^k)$ by (2.45). Set $d^k = -p^k$ and

$$v^k = -\{|p^k|^2 + f(x^k) - \tilde{f}_p^k\}. \tag{3.2}$$

Step 2 (Stopping criterion). Set

$$w^k = \tfrac{1}{2}|p^k|^2 + f(x^k) - \tilde{f}_p^k . \tag{3.3}$$

If $w^k \leq \varepsilon_s$ terminate; otherwise, go to Step 3.

Step 3 (Line search). Set $y^{k+1} = x^k + d$. If

$$f(x^k + d^k) \leq f(x^k) + m v^k \tag{3.4}$$

then set $t_L^k = 1$ (a serious step), set $k(l+1) = k+1$ and increase l by 1; otherwise, i.e. if (3.4) is violated, set $t_L^k = 0$ (a null step).

Step 4 (Linearization updating). Set $x^{k+1} = x^k + t_L^k d^k$. Choose a set $\hat{J}^k \subset \{1,\dots,k\}$ and calculate the linearization values f_j^{k+1}, $j \in \hat{J}^k$, and f_p^{k+1} by (2.32) and (2.38). Evaluate $g^{k+1} = g_f(y^{k+1})$ and

$$f_{k+1}^{k+1} = f(y^{k+1}) + \langle g^{k+1}, x^{k+1} - y^{k+1} \rangle . \tag{3.5}$$

Set $J^{k+1} = \hat{J}^k \cup \{k+1\}$. Increase k by 1 and go to Step 1.

Remark 3.2. It follows from Lemma 2.1 that in Algorithm 3.1 (d^k, u^k) solves the primal subproblem (2.41), where u^k is given by (2.43f), and that λ_j^k, $j \in J^k$, and λ_p^k are the associated Lagrange multipliers. Thus one may equivalently solve subproblem (2.41) in Step 1 of the above algorithm.

Remark 3.3. In Step 2 of Algorithm 3.1 we always have

$$p^k \in \partial_\varepsilon f(x^k) \quad \text{and} \quad |p^k| \le (2\varepsilon)^{1/2} \quad \text{for } \varepsilon = w^k, \tag{3.6a}$$

see the next section. Therefore

$$f(x) \ge f(x^k) + \langle p^k, x - x^k \rangle - w^k \ge f(x^k) - |p^k||x - x^k| - w^k \ge$$

$$\ge f(x^k) - (2w^k)^{1/2}\, |x - x^k| - w^k \tag{3.6b}$$

for each x in R^N. It follows that if f has a minimum point $\bar{x}$, then

$$\min\{f(x): x \in R^N\} = f(\bar{x}) \ge f(x^k) - (2w^k)^{1/2}\, |\bar{x} - x^k| - w^k.$$

The above estimate justifies the stopping criterion in Step 2 of the algorithm.

4. Convergence of the Basic Algorithm

In this section we shall show that each sequence $\{x^k\}$ generated by Algorithm 3.1 is minimizing, i.e. $f(x^k) \downarrow \inf\{f(x): x \in R^N\}$, and that $\{x^k\}$ converges to a minimum point of f, whenever f attains its minimum. Naturally, convergence results assume that the final accuracy tolerance ε_s is set to zero. For convenience, we precede the main results by several lemmas that analyze the properties of Algorithm 3.1.

We start by showing that the aggregate subgradients are convex combinations of the past subgradients.

Lemma 4.1. Suppose that Algorithm 3.1 did not stop before the n-th iteration, $n \ge 1$. Then for each k=1,...,n there exist multipliers $\tilde{\lambda}_j^k$, j=1,...,k, satisfying

$$(p^k, \tilde{f}_p^k) = \sum_{j=1}^{k} \tilde{\lambda}_j^k (g^j, f_j^k), \tag{4.1a}$$

$$\tilde{\lambda}_j^k \ge 0, \; j=1,\dots,k, \quad \sum_{j=1}^{k} \tilde{\lambda}_j^k = 1. \tag{4.1b}$$

Moreover, for each k satisfying $1 < k \le n$ one has

$$(p^{k-1}, f_p^k) = \sum_{j=1}^{k-1} \tilde{\lambda}_j^{k-1} (g^j, f_j^k), \tag{4.2a}$$

$$\tilde{\lambda}_j^{k-1} \ge 0, \; j=1,\dots,k-1, \quad \sum_{j=1}^{k-1} \tilde{\lambda}_j^{k-1} = 1. \tag{4.2b}$$

Proof. The proof will proceed by induction. Let

$$\tilde{\lambda}_1^1 = 1, \tag{4.3a}$$

$$\lambda_j^k = 0 \quad \text{for each} \quad j\in\{1,\ldots,k\} \setminus J^k \quad \text{and} \quad k\geq 1, \tag{4.3b}$$

$$\tilde{\lambda}_j^k = \lambda_j^k + \lambda_p^k \tilde{\lambda}_j^{k-1} \quad \text{for} \quad j=1,\ldots,k-1, \; \tilde{\lambda}_k^k = \lambda_k^k, \text{ for } k\geq 1, \tag{4.3c}$$

If k=1 then (4.1) follows from (4.3a) and (2.45), because $(p^1,\tilde{f}_p^1) = (g^1,f_1^1) = (p^0,f_p^1)$. Therefore (2.32) and (2.38) yield (4.2) for k=2. Suppose that (4.2) holds for some k=n>2. Then, since

$$(p^k,\tilde{f}_p^k) = \sum_{j=1}^{k} \lambda_j^k(g^j,f_j^k) + \lambda_p^k(p^{k-1},f_p^k),$$

$$\lambda_j^k\geq 0,\ j=1,\ldots,k,\ \lambda_p^k\geq 0,\quad \sum_{j=1}^{k} \lambda_j^k + \lambda_p^k = 1$$

by (2.43a), (2.45) and (4.3b), we obtain

$$(p^k,\tilde{f}_p^k) = \sum_{j=1}^{k} \lambda_j^k(g^j,f_j^k) + \lambda_p^k \sum_{j=1}^{k-1} \tilde{\lambda}_j^{k-1}(g^j,f_j^k) =$$

$$= \lambda_k^k(g^k,f_k^k) + \sum_{j=1}^{k-1} (\lambda_j^k + \lambda_p^k \tilde{\lambda}_j^{k-1})(g^j,f_j^k) =$$

$$= \sum_{j=1}^{k} \tilde{\lambda}_j^k (g^j,f_j^k),$$

$$\tilde{\lambda}_j^k\geq 0 \quad \text{for} \quad j=1,\ldots,k,$$

$$\sum_{j=1}^{k} \tilde{\lambda}_j^k = \tilde{\lambda}_k^k + \sum_{j=1}^{k-1} \tilde{\lambda}_j^k = \lambda_k^k + \sum_{j=1}^{k-1} (\lambda_j^k + \lambda_p^k \tilde{\lambda}_j^{k-1}) =$$

$$= \sum_{j=1}^{k} \lambda_j^k + \lambda_p^k \sum_{j=1}^{k-1} \tilde{\lambda}_j^{k-1} = \sum_{j=1}^{k} \lambda_j^k + \lambda_p^k = 1,$$

which yields (4.1) for k=n. Next,

$$f_p^{k+1} = \tilde{f}_p^k + \langle p^k,x^{k+1}-x^k\rangle = \sum_{j=1}^{k} \tilde{\lambda}_j^k f_j^k + \langle \sum_{j=1}^{k} \tilde{\lambda}_j^k g^j,x^{k+1}-x^k\rangle =$$

$$= \sum_{j=1}^{k} \tilde{\lambda}_j^k[f_j^k + \langle g^j,x^{k+1}- x^k\rangle] = \sum_{j=1}^{k} \tilde{\lambda}_j^k f_j^{k+1}$$

from (2.38), (4.1a) and (2.32). Therefore (4.2) holds for k=n+1, and the induction step is complete. □

Our convergence analysis hinges on the interpretation of the past subgradients and the aggregate subgradients in terms of ε-subgradients of the objective function. In the following, suppose that Algorithm 3.1 did not terminate before the k-th iteration, for some $k \geq 1$. Define the linearization errors

$$\alpha_j^k = f(x^k) - f_j^k , \quad j=1,\ldots,k, \tag{4.4a}$$

$$\alpha_p^k = f(x^k) - f_p^k , \tag{4.4b}$$

$$\tilde{\alpha}_p^k = f(x^k) - \tilde{f}_p^k , \tag{4.4c}$$

which may be associated with the subgradients g^j, p^{k-1} and p^k as follows.

Lemma. 4.2. At the k-th iteration of Algorithm 3.1, one has

$$g^j \in \partial_\varepsilon f(x^k) \quad \text{for } \varepsilon = \alpha_j^k , \quad j=1,\ldots,k, \tag{4.5a}$$

$$p^{k-1} \in \partial_\varepsilon f(x^k) \quad \text{for } \varepsilon = \alpha_p^k , \tag{4.6b}$$

$$p^k \in \partial_\varepsilon f(x^k) \quad \text{for } \varepsilon = \tilde{\alpha}_p^k , \tag{4.5c}$$

$$\alpha_j^k,\ \alpha_p^k,\ \tilde{\alpha}_p^k \geq 0, \quad j=1,\ldots,k. \tag{4.5d}$$

Proof. From (2.31) amd (4,4a), for each x in R^N we have

$$f(x) \geq f(x^k) + \langle g^j, x-x^k\rangle - [f(x^k) - f_p^k]$$

$$\geq f(x^k) + \langle g^j, x-x^k\rangle - \alpha_j^k , \tag{4.6}$$

hence (4.5a) follows from the definition of the ε-subdifferential, see (1.2.86). Setting $x=x^k$ in (4.6), we obtain $\alpha_j^k \geq 0$. By (4.1) and (4.6),

$$f(x) = \sum_{j=1}^{k} \tilde{\lambda}_j^k f(x) \geq \langle \sum_{j=1}^{k} \tilde{\lambda}_j^k g^j, x-x^k\rangle + \sum_{j=1}^{k} \tilde{\lambda}_j^k f_j^k =$$

$$= \langle p^k, x-x^k\rangle + \tilde{f}_p^k = f(x^k) + \langle p^k, x-x^k\rangle - \tilde{\alpha}_p^k ,$$

which proves (4.5c). Setting $x=x^k$, we get $\tilde{\alpha}_p^k \geq 0$. The rest follows similarly from (4.2) and (4.6). □

Remark 4.3. In view of (4.5), the values of α_j^k, α_p^k and $\tilde{\alpha}_p^k$ indicate the distance from g^j, p^{k-1} and p^k to the subdifferential of f at at x, respectively. For instance, the value of $\tilde{\alpha}_p^k \geq 0$ indicates how much p^k differs from being a member of $\partial f(x^k)$; if $\tilde{\alpha}_p^k = 0$ we have $p^k \in \partial f(x^k)$.

The following result will justify the stopping criterion of the algorithm.

Lemma 4.4. At the k-th iteration of Algorithm 3.1, one has

$$w^k = \frac{1}{2} |p^k|^2 + \tilde{\alpha}_p^k , \tag{4.7a}$$

$$v^k = -\{|p^k|^2 + \tilde{\alpha}_p^k\}, \tag{4.7b}$$

$$v^k \leq - w^k \leq 0. \tag{4.7c}$$

Proof. This follows immediately from (3.2), (3.3), (4.4c) and (4.5d).□

Remark 4.5. The variable w^k may be termed a stationarity measure of the current point x^k, for each k, because $\frac{1}{2}|p^k|^2$ indicates how much p^k differs from the null vector and $\tilde{\alpha}_{p_k}^k$ measures the distance from p^k to $\partial f(x^k)$ (x^k is stationary if $0 \in \partial f(x^k)$). The estimates (3.6), which follow from (4.5c) and (4.7a), show that x^k is approximately optimal when the value of w^k is small.

In what follows we assume that the final accuracy tolerance ε_s is set to zero. Since the algorithm stops if and only if $w^k \leq \varepsilon_s$, (4.7c) and (3.6) yield

Lemma 4.5. If Algorithm 3.1 terminates at the k-th iteration, then x^k is a minimum point of f.

From now on we suppose that the algorithm does not terminate, i.e. $w^k > 0$ for all k. Since the line search rules imply that we always have

$$f(x^{k+1}) \leq f(x^k) + m t_L^k v^k \tag{4.8}$$

with $m>0$ and $t_L^k \geq 0$, the fact that $v^k \leq -w^k < 0$ (see (4.7c)) yields that the sequence $\{f(x^k)\}$ is nonincreasing.

The next result states a fundamental property of the stationarity measures $\{w^k\}$.

Lemma 4.6. Suppose that there exist an infinite set $K \subset \{1,2,\ldots\}$ and a point $\bar{x} \in R^N$ satisfying $x^k \xrightarrow{K} \bar{x}$ and $w^k \xrightarrow{K} 0$. Then $\bar{x}$ minimizes f.

Proof. Passing to the limit in (3.6b) with $k \to \infty$, $k \in K$, we obtain $f(x) \geq f(\bar{x})$ for each x in R^N. □

We shall need the following auxiliary result.

Lemma 4.7. Suppose that the sequence $\{f(x^k)\}$ is bounded from below, i.e. $f(x^k) \geq c$ for some fixed c and all k. Then

$$\sum_{k=1}^{\infty} \{t_L^k |p^k|^2 + t_L^k \tilde{\alpha}_p^k\} \leq [f(x^1) - c]/m. \tag{4.9}$$

Proof. It follows from (4.8) that

$$f(x^1) - f(x^k) = f(x^1) - f(x^2) + \ldots + f(x^{k-1}) - f(x^k)$$

$$\geq m \sum_{i=1}^{k-1} t_L^i(-v^i).$$

Dividing the above inequality by $m>0$, letting k approach infinity and using (4.7b) and the assumption that $f(x^k) \geq c$, we obtain the desired relation (4.9). □

Note that the rules of Step 3 of Algorithm 3.1 imply

$$x^k = x^{k(l)} \quad \text{for } k=k(l),k(l)+1,\ldots,k(l+1)-1, \tag{4.10}$$

where we set $k(l+1)=\infty$ if the number l of serious steps stays bounded, i.e. if $x^k = x^{k(l)}$ for some fixed l and all $k \geq k(l)$.

The case of infinitely many serious steps is analyzed in the following lemma.

Lemma 4.8. Suppose that there exist an infinite set $L \subset \{1,2,...\}$ and a point $\bar{x} \in R^N$ such that $x^{k(l)} \longrightarrow \bar{x}$ as $l \to \infty$, $l \in L$. Then $\bar{x}$ is a minimum point of f.

Proof. Let $k = \{k(l+1)-1:\ l \in L\}$. Observe that the line search rules imply $t_L^k=1$ for all $k \in K$, while (4.10) yields

$$x^k \xrightarrow{K} \bar{x} \ . \tag{4.11a}$$

Since $\{f(x^k)\}$ is nonincreasing, (4.11a) and the continuity of f imply $f(x^k) \downarrow f(\bar{x})$. Then Lemma 4.7, (4.7) and the fact that $t_L^k=1$ for all $k \in K$ yield

$$w^k \xrightarrow{K} 0 . \tag{4.11b}$$

In view of Lemma 4.6, (4.11) yields the desired conclusion. □

In order to show that the stationarity measures $\{w^k\}$ tend to zero in the case of a finite number of serious steps, we have to analyze the dual search direction finding subproblems.

Lemma 4.9. At the k-th iteration of Algorithm 3.1, $k \geq 1$, w^k is the optimal value of the following problem

$$\begin{aligned} &\underset{\lambda}{\text{minimize}}\ \tfrac{1}{2}\ |\sum_{j \in J^k} \lambda_j g^j + \lambda_p p^{k-1}|^2 + \sum_{j \in J^k} \lambda_j \alpha_j^k + \lambda_p \alpha_p^k \ , \\ &\text{subject to } \lambda_j \geq 0,\ j \in J^k,\ \lambda_p \geq 0,\ \sum_{j \in J^k} \lambda_j + \lambda_p = 1 \ , \end{aligned} \tag{4.12}$$

which is equivalent to subproblem (3.1).

Proof. For each λ satisfying the constraints of (4.12),

$$\sum_{j \in J^k} \lambda_j \alpha_j^k + \lambda_p \alpha_p^k = f(x^k) - \sum_{j \in J^k} \lambda_j f_j^k - \lambda_p f_p^k$$

from (4.4a) and (4.4b), which proves the equivalence of (3.1) and (4.12). Since λ_j^k, $j \in J^k$, and λ_p^k solve (3.1), the optimal value of (4.12) is

$$\tfrac{1}{2}|\sum_{j \in J^k} \lambda_j^k g^j + \lambda_p^k p^k|^2 + f(x^k) - \sum_{j \in J^k} \lambda_j^k f_j^k - \lambda_p^k f_p^k =$$

$$= \tfrac{1}{2}|p^k|^2 + f(x^k) - \tilde{f}_p^k = w^k$$

from (2.45) and (3.3). □

The following result, which describes problems similar to (4.12), will be frequently used in subsequent chapters. It generalizes Lemma 4.10 in (Mifflin,1977b).

Lemma 4.10. Suppose that N-vectors p, g and d and numbers $m \in (0,1)$, C, v, w, $\tilde{\alpha}_p \ge 0$ and α satisfy

$$d = -p \,, \tag{4.13a}$$

$$w = \tfrac{1}{2}|p|^2 + \tilde{\alpha}_p \,, \tag{4.13b}$$

$$v = -\{|p|^2 + \tilde{\alpha}_p\}, \tag{4.13c}$$

$$-\alpha + \langle g,d\rangle \ge mv \,, \tag{4.13d}$$

$$C \ge \max\{|p|,|g|, \tilde{\alpha}_p, 1\}. \tag{4.13e}$$

Let

$$Q(\nu) = \tfrac{1}{2}|(1-\nu)p + \nu g|^2 + (1-\nu)\tilde{\alpha}_p + \nu\alpha \quad \text{for} \quad \nu \in R, \tag{4.14a}$$

$$\bar{w} = \min\{Q(\nu): \nu \in [0,1]\}. \tag{4.14b}$$

Then

$$\bar{w} \le \phi_C(w), \tag{4.15}$$

where

$$\phi_C(t) = t - (1-m)^2 t^2/(8C^2). \tag{4.16}$$

Proof. Simple calculations yield

$$Q(\nu) = \tfrac{1}{2}\nu^2 |p-g|^2 + \nu[\langle p,g\rangle - |p|^2] + \nu(\alpha - \tilde{\alpha}_p) + w. \tag{4.17}$$

From (4.13a,c,d)

$$\langle p,g\rangle \le m\{|p|^2 + \tilde{\alpha}_p\} - \alpha,$$

hence (4.17) yields

$$Q(\nu) \le \tfrac{1}{2}\nu^2|p-g|^2 - \nu(1-m)[|p|^2 + \tilde{\alpha}_p] + w$$

for all $\nu \ge 0$. Since $m \in (0,1)$ and $|p|^2 \ge 0$, we obtain

$$Q(\nu) \le \tfrac{1}{2}\nu^2|p-g|^2 - \nu(1-m)w + w \quad \text{for all } \nu \in [0,1].$$

By (4.13e), $|p-g|^2 \le (|p|+|g|)^2 \le 4C^2$, hence

$$Q(\nu) \le 2C^2\nu^2 - (1-m)w\,\nu + w \quad \text{for all } \nu \in [0,1]. \tag{4.18}$$

Denoting the right side of (4.18) by $q(\nu)$, we check that q is minimized by $\bar{\nu} = (1-m)w/(4C^2) \le (1-m)[C^2/2+C]/(4C^2) \le 1$, yielding

$$q(\bar{\nu}) = w - (1-m)^2 w^2/(8C^2). \tag{4.19}$$

Since $\bar{\nu} \in [0,1]$, (4.14b), (4.18) and (4.19) complete the proof. □

Applying the above lemma to Algorithm 3.1, we obtain

<u>Lemma 4.11</u>. Suppose that $t_L^{k-1}=0$ for some $k>1$. Then

$$-\alpha_k^k + \langle g^k, d^{k-1}\rangle \ge m v^{k-1}, \tag{4.20}$$

$$w^k \le \phi_{C^k}(w^{k-1}), \tag{4.21}$$

where ϕ_C is given by (4.16) and C^k is any number satisfying

$$C^k \ge \max\{|p^{k-1}|,\ |g^k|,\ \tilde{\alpha}_p^{k-1}, 1\}. \tag{4.22}$$

<u>Proof</u>.(i) $t_L^{k-1}=0$ if

$$f(y^k) - f(x^{k-1}) > m v^{k-1}.$$

Then $x^k = x^{k-1} + t_L^{k-1}d^{k-1} = x^{k-1}$ and $y^k = x^{k-1} + d^{k-1}$, so

$$-\alpha_k^k + \langle g^k, d^{k-1}\rangle = -\left[f(x^k) - f(y^k) - \langle g^k, x^k - y^k\rangle\right] + \langle g^k, d^{k-1}\rangle =$$

$$= f(y^k) - f(x^k).$$

Combining the above relations, we obtain (4.20).

(ii) Define the multipliers

$$\lambda_k(\nu) = \nu,\ \lambda_j(\nu)=0,\ j \in J^k \setminus \{k\},\ \lambda_p(\nu) = 1-\nu \tag{4.23}$$

for each $\nu \in [0,1]$. Note that $k \in J^k$ by the rules of Step 4. Moreover, for each ν

$$\sum_{j\in J^k} \lambda_j(\nu)g^j + \lambda_p(\nu)p^{k-1} = (1-\nu)p^{k-1} + \nu g^k,$$

$$\sum_{j\in J^k} \lambda_j(\nu)\alpha_j^k + \lambda_p(\nu)\alpha_p^k = (1-\nu)\tilde{\alpha}_p^{k-1} + \nu\alpha_k^k. \tag{4.24}$$

This follows from the fact that $\alpha_p^k=f(x^k)-f_p^k=f(x^{k-1})-\tilde{f}_p^{k-1}$ if $x^k=x^{k-1}$, see (4.4) and (2.38). Since for each $\nu\in[0,1]$ the multipliers (4.23) are feasible for (4.12), we deduce from (4.24) that w^k, the optimal value of (4.12), cannot exceed the optimal value of the following problem

$$\text{minimize } \tfrac{1}{2}|(1-\nu)p^{k-1} + \nu g^k|^2 + (1-\nu)\tilde{\alpha}_p^{k-1} + \nu\alpha_k^k,$$
$$\text{subject to } \nu\in[0,1]. \tag{4.25}$$

Therefore we obtain the desired conclusion from Lemma 4.10, (2.43e), (4.7a), (4.7b), (4.20) and (4.22). □

The following result deals with the case of a finite number of serious steps of the algorithm.

Lemma 4.12. Suppose that the number l of serious steps of Algorithm 3.1 stays bounded, i.e. $x^k=x^{k(l)}$ for all $k\ge k(l)$. Then the point $\bar{x}=x^{k(l)}$ minimizes f.

Proof. Suppose that $t_L^k=0$ for all $k\ge\bar{k}$ and some fixed $\bar{k}$. From Lemma 4.11, we have

$$0 \le w^k \le w^{k-1} \quad \text{for all } k\ge\bar{k}. \tag{4.26a}$$

In particular $\frac{1}{2}|p^k|^2 + \tilde{\alpha}_p^k = w^k \le w^{\bar{k}}$ for all $k\ge\bar{k}$, hence there exists a constant $C_1<\infty$ satisfying

$$\max\{|p^{k-1}|,\tilde{\alpha}_p^{k-1},1\} \le C_1 \quad \text{for all } k\ge\bar{k}. \tag{4.26b}$$

Since $y^k=x^{k-1}+d^{k-1}=\bar{x}+d^{k-1}=\bar{x}-p^{k-1}$ for all $k\ge\bar{k}$, it follows from (4.26b) that the sequence $\{y^k\}$ is bounded. Therefore from the local boundedness of ∂f (see Lemma 1.2.2) follows the existence of a constant $C_2<\infty$ satisfying $|g^k|=|g_f(y^k)|\le C_2$ for all $k\ge\bar{k}$. By (4.26), this gives

$$\max\{|p^{k-1}|,|g^k|,\tilde{\alpha}_p^{k-1},1\} \le C \quad \text{for all } k\ge\bar{k},$$

where $C=\max\{C_1,C_2\}$. Thus (4.22) holds for constant $C^k=C$ and all $k\ge\bar{k}$.

Consequently, (4.26a), (4.21), (4.16) and the fact that $m \in (0,1)$ is fixed imply $w^k \downarrow 0$. Combining this with Lemma 4.6 and the fact that $x^k = \bar{x}$ for all $k \geq \bar{k}$, we complete the proof. □

Combining Lemma 4.8 with Lemma 4.12 and using (4.10), we obtain

Theorem 4.13. Every accumulation point of the sequence $\{x^k\}$ generated by Algorithm 3.1 is a minimum point for f.

The following lemma provides a sufficient condition for $\{x^k\}$ to have accumulation points.

Lemma 4.14. Suppose that a point $\hat{x} \in R^N$ satisfies $f(\hat{x}) \leq f(x^k)$ for all k. Then the sequence $\{x^k\}$ is bounded and

$$|x-x^k|^2 \leq |\hat{x}-x^n|^2 + \sum_{i=n}^{k} \{|x^{i+1}-x^i|^2 + 2t_L^i \tilde{\alpha}_p^i\} \quad \text{for } k>n\geq 1, \tag{4.27a}$$

$$\sum_{i=n}^{\infty} \{|x^{i+1}-x^i|^2 + 2t_L^i \tilde{\alpha}_p^i\} \to 0 \quad \text{as } n\to\infty. \tag{4.27b}$$

Proof. From (4.5c), $0 \geq f(\hat{x})-f(x^k) \geq \langle p^k, \hat{x}-x^k \rangle - \tilde{\alpha}_p^k$, hence

$$\langle p^k, x-x^k \rangle \leq \tilde{\alpha}_p^k \quad \text{for all } k. \tag{4.28}$$

Since we always have $x^{k+1}-x^k = t_L^k d^k = -t_L^k p^k$ and $t_L^k \geq 0$, (4.28) implies $-\langle \hat{x}-x^k,\ x^{k+1}-x^k \rangle \leq t_L^k \tilde{\alpha}_p^k$. Therefore

$$|\hat{x}-x^{k+1}|^2 = |\hat{x}-x^k|^2 - 2\langle \hat{x}-x^k, x^{k+1}-x^k \rangle + |x^{k+1}-x^k|^2$$

$$\leq |\hat{x}-x^k|^2 + |x^{k+1}-x^k|^2 + 2t_L^k \tilde{\alpha}_p^k,$$

which yields (4.27a). Next, we always have $|x^{k+1}-x^k|^2 = (t_L^k)^2 |p^k|^2 \leq t_L^k |p^k|^2$, because $t_L^k \in [0,1]$, hence Lemma 4.7 implies

$$\sum_{k=1}^{\infty} \{|x^{k+1}-x^k|^2 + 2t_L^k \tilde{\alpha}_p^k\} < \infty.$$

This proves both the boundedness of $\{x^k\}$ (by (4.27a)) and (4.27b). □

We now state the principal result. Let

$$\bar{X} = \{\bar{x} \in R^N: f(\bar{x}) \leq f(x) \quad \text{for all } x\}.$$

Theorem 4.15. If the solution set $\bar{X}$ is nonempty, then each sequence $\{x^k\}$ calculated by Algorithm 3.1 converges to some point $\bar{x} \in \bar{X}$.

Proof. If $\hat{x} \in \bar{X}$ then $f(\hat{x}) \leq f(x^k)$ for all k, hence Lemma 4.14 implies the boundedness of $\{x^k\}$. By Theorem 4.13, $\{x^k\}$ has an accumulation point $\bar{x} \in \bar{X}$. It remains to show that $x^k \to \bar{x}$. Take any $\delta > 0$. Since $f(\bar{x}) \leq f(x^k)$ for all k, Lemma 4.14 implies that there exists a number n_1 such that

$$|\bar{x}-x^k|^2 \leq |\bar{x}-x^n|^2 + \delta/2 \quad \text{for all } k>n>n_1. \tag{4.29}$$

Since $\bar{x}$ is an accumulation point of $\{x^k\}$, there exists a number $n>n_1$ such that $|\bar{x}-x^n|^2 \leq \delta/2$. Then (4.29) yields $|\bar{x}-x^k|^2 \leq \delta$ for all $k>n$. Since δ was arbitrary, this proves $x^k \to \bar{x}$ as $k \to \infty$. □

Even when f has no minimum points, we still have the following result.

Theorem 4.16. Each sequence $\{x^k\}$ constructed by Algorithm 3.1 is minimizing:

$$f(x^k) \downarrow \inf \{f(x): x \in R^N\}.$$

Proof. In view of Theorem 4.15, it suffices to consider the case of an empty $\bar{X}$. Let $\{z^i\}$ be a minimizing sequence, i.e. $f(z^i) \downarrow \inf\{f(x): x \in R^N\}$ and $f(z^i) > f(z^{i+1})$ for all i. To obtain a contradiction, suppose that for some fixed index i $f(z^i) \leq f(x^k)$ for all k. Then Lemma 4.14 and Theorem 4.13 imply the existence of some $\bar{x} \in \bar{X}$, which contradicts the emptiness of $\bar{X}$. Therefore $f(x^k) < f(z^i)$ for every fixed i and large k ($\{f(x^k)\}$ is nonincreasing), hence $\{x^k\}$ minimizes f. □

The next result provides further substantiation of the stopping criterion.

Lemma 4.17. Suppose that $\{x^k\}$ is a sequence generated by Algorithm 3.1 satisfying $f(x^k) \geq c$ for a fixed number c and all k. Then

$$w^k \to 0 \quad \text{as} \quad k \to \infty. \tag{4.30}$$

Proof. If Algorithm 3.1 executes a finite number of serious steps, then (4.30) follows from the proof of Lemma 4.12. Suppose that $l \to \infty$. By the monotonicity of $\{f(x^k)\}$, we have $f(x^k) \downarrow \bar{c}$ as $k \to \infty$, where $\bar{c} \geq c$. Therefore

$$f(x^{k(l)}) - f(x^{k(l-1)}) \to 0 \quad \text{as } l \to \infty. \tag{4.31}$$

Observe that $-\langle p^k, x^{k+1} - x^k \rangle = -\langle p^k, -t_L^k p^k \rangle = t_L^k |p^k|^2$ for all k, hence (4.9), (4.10) and (4.31), yield

$$f(x^{k(l)}) - f(x^{k(l-1)}) - \langle p^{k(l)-1}, x^{k(l)} - x^{k(l)-1} \rangle \to 0 \text{ as } l \to \infty. \tag{4.32a}$$

From the proof of Lemma 4.8 we deduce that

$$w^{k(l)-1} \to 0 \quad \text{as} \quad l \to \infty. \tag{4.32b}$$

Arguing as in the proof of Lemma 4.11, we deduce that

$$\sum_{j \in J^k} \lambda_j(0) g^j + \lambda_p(0) p^{k-1} = p^{k-1},$$

$$\sum_{j \in J^k} \lambda_j(0) \alpha_j^k + \lambda_p(0) \alpha_p^k = \alpha_p^k ,$$

see (4.23) and (4.24), and that

$$w^k \leq \tfrac{1}{2}|p^{k-1}|^2 + \alpha_p^k = w^{k-1} + \alpha_p^k - \tilde{\alpha}_p^{k-1} \quad \text{for all k.} \tag{4.32c}$$

By (2.38) and (4.4),

$$\alpha_p^k - \tilde{\alpha}_p^{k-1} = f(x^k) - f(x^{k-1}) - \langle p^{k-1}, x^k - x^{k-1} \rangle \text{ for all k.} \tag{4.32d}$$

From (4.10) and (4.32c,d),

$$w^k \leq w^{k-1} \quad \text{for all } k(l) < k \leq k(l+1)-1 \quad \text{and} \quad l \geq 1. \tag{4.32e}$$

Using (4.32), we obtain (4.30). □

Corollary 4.18. Suppose that $\inf\{f(x): x \in R^N\} > -\infty$. Then Algorithm 3.1 terminates if its final accuracy tolerance ε_s is positive.

5. The Method with Subgradient Selection

In this section we analyze the method that uses subgradient selection, as specified in Section 2 by (2.20) and (2.23).

To save space, we shall use the notation of Algorithm 3.1 with certain modifications. Algorithm 5.1 is obtained from Algorithm 3.1 by replacing Step 1 with the following.

Step 1′ (Direction finding). Find multipliers λ_j^k, $j \in J^k$, that solve the k-th dual search direction finding subproblem (2.22), and a set $\hat{J}^k = \{j \in J^k : \lambda_j^k > 0\}$ satisfying $|\hat{J}^k| \le N+1$. Calculate the agregate subgradient $(p^k, \tilde{f}_p^k)$ by (2.27). Set $d^k = -p^k$ and $v^k = -\{|p^k|^2 + f(x^k) - f_p^k\}$.

Thus in Algorithm 5.1 the index set $\hat{J}^k$ of the retained past subgradients is chosen by direction finding, whereas in Algorithm 3.1 $\hat{J}^k$ may be determined arbitrarily. We also note that in Algorithm 5.1 there is no need for recursive updating of the aggregate subgradients, since they are calculated directly from the past subgradients retained at each iteration.

Remark 5.2. In view of the results of Section 2, in Algorithm 5.1 (d^k, u^k) solves the k-th primal subproblem (2.11), where u^k is given by (2.21e), for any k. λ_j^k, $j \in J^k$, are the associated Lagrange multipliers. Therefore, one may equivalently solve the primal subproblem (2.11) and then obtain the Lagrange multipliers as described is Lemma 2.1(iv). We may add that most quadratic programming subroutines applied to (2.11) or (2.22) will automatically calculate Lagrange multipliers satisfying (2.23), i.e. at most N+1 nonzero multipliers. This follows from the fact that the primal problem (2.11) has N+1 variables, whereas in the dual subproblem (2.22) at most N+1 vectors of the form $(g^j, 1)$ can be linearly independent, i.e. at most N+1 vectors g^j can be affinely independent (see also (Kiwiel,1983)).

Remark 5.3. The requirement of Algorithm 5.1 that the set $\hat{J}^k$ should contain at most N+1 indices can be modified as follows. Let $M_g \ge N+2$ denote the maximum number of the past subgradients that Algorithm 5.1 may use for each search direction finding. Then one may choose the Lagrange multipliers λ_j^k and the set J^{k+1} subject only to the following requirements

$$J^{k+1} \supset \{k+1\} \cup \{j \in J^k : \lambda_j^k > 0\},$$
$$|J^{k+1}| \le M_g, \tag{5.1}$$

for all k. In view of Lemma 2.1, this is always possible if $M_g \ge N+2$. Also the extensions discussed in Remark 2.5 are covered by the subsequent analysis. In particular, setting $M_g = +\infty$ and $J^{k+1} = \{1,\dots,k+1\}$ for all k, we see that (5.1) is satisfied. Therefore the analysis below applies also to the method of Lemarechal (1978), which uses $J^k = \{1,\dots,k\}$ for all k.

Global convergence of Algorithm 5.1 can be demonstrated by an appropriate modification of the results of Section 4. To save space, we provide here only an outline of required results.

Define the aggregate linearization by (2.35) and the linearization errors by (4.4). Also let

$$\tilde{\lambda}_j^k = \lambda_j^k,\ j \in J^k, \quad \tilde{\lambda}_j^k = 0, \quad j \in \{1,\dots,k\} \setminus J^k.$$

Then Lemma 4.1 follows directly from the definition of $(p^k, \tilde{f}_p^k)$. Moreover, it is straightforward to check that all the results from Lemma 4.2 to Lemma 4.8 also hold for Algorithm 5.1. Lemma 4.9 is substituted by

Lemma 5.4. At the k-th iteration of Algorithm 5.1, $k \ge 1$, w^k is the optimal value of the following problem

$$\text{minimize } \tfrac{1}{2} \Big| \sum_{j \in J^k} \lambda_j g^j \Big|^2 + \sum_{j \in J^k} \lambda_j \alpha_j^k,$$
$$\text{subject to } \lambda_j \ge 0,\ j \in J^k, \quad \sum_{j \in J^k} \lambda_j = 1. \tag{5.2}$$

In the proof of Lemma 4.11, replace (4.23) by

$$\lambda_k(\nu) = \nu, \quad \lambda_j(\nu) = (1-\nu)\lambda_j^{k-1}, \quad j \in \hat{J}^{k-1}, \tag{5.3}$$

for each $\nu \in [0,1]$, and note that (2.21a), (2.23a), (2.27), (2.32) and (2.38) imply

$$\lambda_j^{k-1} \geq 0, \quad j \in \hat{J}^{k-1}, \qquad \sum_{j \in J^{k-1}} \lambda_j^{k-1} = 1,$$

$$(p^{k-1}, \tilde{f}_p^{k-1}) = \sum_{j \in \hat{J}^{k-1}} \lambda_j^{k-1} (g^j, f_j^{k-1}),$$

$$f_p^k = \tilde{f}_p^{k-1},$$

$$\sum_{j \in J^k} \lambda_j(\nu) g^j = (1-\nu) p^{k-1} + \nu g^k, \tag{5.4a}$$

$$\sum_{j \in J^k} \lambda_j(\nu) \alpha_j^k = (1-\nu) \tilde{\alpha}_p^{k-1} + \nu \alpha_k^k, \tag{5.4b}$$

$$\lambda_j(\nu) \geq 0, \; j \in J^k, \qquad \sum_{j \in J^k} \lambda_j(\nu) = 1, \tag{5.4c}$$

for each $\nu \in [0,1]$, if $t_L^{k-1}=0$ $(x^k=x^{k-1})$. Since (5.4c) means that the multipliers (5.3) are feasible in (5.2) for each $\nu \in [0,1]$, we compare (4.25), (5.2) and (5.4) to deduce that the optimal value of (5.2) is not greater than the optimal value of (4.25). This observation suffices for the proofs of all the remaining results of Section 4 also for Algorithm 5.1.

Remark 5.5. It should be clear by now that the above approach to convergence analysis can be applied to methods that use more subgradients for search direction finding, cf. Remark 2.5 and Remark 5.3. For instance, if the sets J^k are chosen subject to the requirement (5.1), then one may replace (5.3) with the following definition

$$\begin{aligned} &\lambda_k(\nu)=\nu, \; \lambda_j(\nu)=(1-\nu)\lambda_j^{k-1}, \quad j \in \{i \in J^{k-1}: \lambda_i^{k-1}>0\}, \\ &\lambda_j(\nu)=0, \quad j \in \{1,\ldots,k-1\} \setminus \{i \in J^{k-1}: \lambda_i^{k-1}>0\}. \end{aligned} \tag{5.5}$$

We shall now show that methods with subgradient selection may be interpreted as regularized versions of the cutting plane method, see Section 1.3. Let us recall that

$$f_j(x) = f(y^j) + \langle g^j, x-y^j \rangle, \quad j=1,2,\ldots, \tag{5.6}$$

$$\hat{f}_s^k(x) = \max\{f_j(x): j \in J^k\}, \quad k=1,2,\ldots, \tag{5.7a}$$

and define the reduced polyhedral approximation to f at x^k

$$\hat{f}_r(x) = \max\{f_j(x) : j \in \hat{J}^k\} \tag{5.7b}$$

for all k. Let

$$X_s^k = \text{Argmin } \hat{f}_s^k = \{x \in R^N : \hat{f}_s^k(x) \le \hat{f}_s^k(y) \text{ for all } y\}, \tag{5.8a}$$

$$X_r^k = \text{Argmin } \hat{f}_r^k, \tag{5.8b}$$

denote the optimal sets of $\hat{f}_s^k$ and $\hat{f}_r^k$, respectively, for any k. Let

$$v_s^k(d) = \hat{f}_s^k(x^k+d) - f(x^k) \quad \text{for all } d, \tag{5.9a}$$

$$D^k = \text{Argmin } v_s^k, \tag{5.9b}$$

for all k. Clearly,

$$D^k = \underset{d}{\text{Argmin}} \; \hat{f}_s^k(x^k+d). \tag{5.9c}$$

We also recall from Section 2 that at each iteration of Algorithm 5.1 one has

$$d^k = \underset{d}{\text{argmin}} \; \{\hat{f}_s^k(x^k+d) + \tfrac{1}{2}|d|^2\}, \tag{5.10a}$$

$$y^{k+1} = x^k+d^k = \underset{y}{\text{argmin}} \; \{\hat{f}_s^k(y) + \tfrac{1}{2}|y-x^k|^2\}, \tag{5.10b}$$

see (2.36), where "arg" denotes the unique element of "Arg", if any.

We may compare (5.10b) with the cutting plane algorithm as follows. The k-th iteration of the cutting plane method that uses a polyhedral approximation of the form (5.7a) would calculate the next trial point y_C^{k+1} as a solution to the problem

$$\text{minimize } \hat{f}_s^k(y) \quad \text{over all } y, \tag{5.11a}$$

i.e. y_C^{k+1} is any point satisfying

$$y_C^{k+1} \in X_s^k, \tag{5.11b}$$

see (5.8a). We shall now show that under certain conditions also

$$y^{k+1} \in X_s^k. \tag{5.12}$$

Lemma 5.6. (i) At the k-th iteration of Algorithm 5.1 one has

$$f_j^k + \langle g^j, d^k \rangle \le u^k, \quad j \in J^k, \tag{5.13a}$$

$$f_j^k + \langle g^j, d^k \rangle = u^k, \quad j \in \hat{J}^k, \tag{5.13b}$$

$$\hat{f}_s^k(x^k+d^k) = \hat{f}_r^k(x^k+d^k) = u^k, \tag{5.13c}$$

$$v_s^k(d^k) = v^k = u^k - f(x^k). \tag{5.13d}$$

(ii) If additionally the subgradients g^j, $j \in \hat{J}^k$, are positively linearly dependent, i.e.

$$0 \in \mathrm{conv}\{g^j: j \in \hat{J}^k\}, \tag{5.14}$$

then

$$y^{k+1} = \Pr_{X_s^k} x^k = \Pr_{X_r^k} x^k, \tag{5.15a}$$

$$d^k = \Pr_{D^k} 0, \tag{5.15b}$$

where $\Pr_X x$ denotes the projection of x on X.

Proof. (i) (5.13a,b,c) follow from (2.21b), (2.23a) and (5.7). By (5.13c) and (5.9a), we have $v_s^k(d^k)=u^k-f(x^k)$. Combining this with (2.21e), (2.26) and (3.2), we obtain (5.13d).

(ii) Since

$$\partial\hat{f}_s^k(x^k+d^k) = \mathrm{conv}\{g^j: f_j^k + \langle g^j,d^k\rangle = \hat{f}_s^k(x^k+d^k)\}$$

(see Corollary 1.2.6), we deduce from (5.13b) and (5.14) that

$$0 \in \partial\hat{f}_s^k(x^k+d^k) \quad \text{and} \quad \hat{f}_s^k(x^k+d) = u^k.$$

It follows that $y^{k+1}=x^k+d^k$ minimizes the convex function $\hat{f}_s^k$, i.e. $y^{k+1} \in X_s^k$, and $\hat{f}_s^k(y)=u^k$ for all $y \in X_s^k$. Therefore (5.10b) can be formulated as

$$\begin{aligned} y^{k+1} &= \operatorname*{argmin}_y \{\hat{f}_s^k(y) + \tfrac{1}{2}|y-x^k|^2\} = \\ &= \operatorname*{argmin}_{y\in X_s^k} \{u^k + \tfrac{1}{2}|y-x^k|^2\} \\ &= \operatorname*{argmin}_{y\in X_s^k} |y-x^k| = \Pr_{X_s^k} x^k. \end{aligned}$$

Since

$$\partial\hat{f}_r^k(y^{k+1}) = \mathrm{conv}\{g^j: f_j^k + \langle g^j,y^{k+1}-x^k\rangle = \hat{f}_r^k(y^{k+1}),\quad j \in \hat{J}^k\},$$

we similarly deduce that $y^{k+1} \in X_r^k$ and $y^{k+1} = \Pr_{X_r^k} x^k$. Then (5.15b) follows from (5.15a), (5.8) and (5.9c). $\square$

To interpret the above result, consider the following condition

$$0 \in \operatorname{conv}\{g^j : j \in J^k\}, \tag{5.16}$$

which is slightly weaker than (5.14). One may show that (5.16) is equivalent to

$$\inf\{\hat{f}_s^k(x) : x \in R^N\} > -\infty,$$

which in turn is equivalent to X_s^k being nonempty. Similarly, (5.14) is equivalent to nonemptiness of X_r^k. The cutting plane method chooses any point in X_s^k as the next trial point, so (5.16) must hold. In Algorithm 5.1 the augmentation of the subproblem objective function with the regularizing term $\frac{1}{2}|d|^2$ uniquely determines y^{k+1} as the point in X_r^k nearest to x^k (if X_r^k is nonempty).

We conclude this section by remarking that results similar to Lemma 5.6 may be obtained also for methods that use more subgradients for search direction finding than Algorithm 5.1, cf. Remark 2.5 and Remark 5.3. Such results are crucial for showing that Algorithm 5.1 terminates when f happens to be piecewise linear (Kiwiel, 1983).

6. Finite Convergence for Piecewise Linear Functions

In this section we show that many versions of the method with subgradient selection are finite, terminating methods for minimizing piecewise linear functions.

The problem of finite convergence of nonsmooth optimization algorithms in the polyhedral case is interesting for at least two reasons. First, piecewise linear functions frequently arise in applications (Lasdon,1970;Shor,1979;Wolfe,1975), e.g. in decomposition methods for large scale linear programming problems (Nurminski,1981). Secondly, many objective functions can often be well approximated in the vicinity of minimum points by piecewise linear functions of the form (2.33) (Madsen and Schjaer-Jacbsen,1979). Then the finite termination property of an algorithm in the polyhedral case ensures fast convergence.

The above problem was analyzed for descent methods by Wolfe (1975) and Mifflin (1977b). They tried to modify their algorithms to ensure finite termination for polyhedral functions. To this end they changed line search rules, demanding exact minimization of $f(x^k+td^k)$ over all $t \geq 0$, and required storing all the past subgradients. This led to finite termination of the algorithms in (Mifflin,1977b;Wolfe,1975) only for Powell's function

$$f(x) = \max\{\langle a^i,x\rangle: i=1,\dots,5\}, \quad x\in R^2,$$
$$a^i = \left(\cos \frac{2\pi(i-1)}{5}, \sin \frac{2\pi(i-1)}{5}\right)^T, \quad i=1,\dots,5. \tag{6.1}$$

No guarantee exists that such modified algorithms are convergent for other functions.

Throughout this section we assume that the minimized function f is piecewise linear, i.e.

$$f(x) = \max\{\bar{f}_i(x): i\in I\}, \quad x\in R^N,$$
$$\bar{f}_i(x) = \langle a^i,x\rangle - b^i, \quad a^i\in R^N, \quad b^i\in R, \quad i\in I, \tag{6.2}$$

where the index set I is finite. Since the subdifferential of f at x is given by (see Corollary 1.2.6):

$$\partial f(x) = \mathrm{conv}\{a^i: i\in I(x)\}, \tag{6.3}$$

$$I(x) = \{i\in I: \bar{f}_i(x) = f(x)\}, \tag{6.4}$$

we assume that Algorithm 5.1 applied to f uses subgradients of the form

$$g^k = a^{i(k)} \in \{a^i: i\in I(y^k)\} \quad \text{for all } k, \tag{6.5a}$$

i.e. $g^k=a^{i(k)}\in \partial f(y^k)$ and

$$\bar{f}_{i(k)}(y^k) = f(y^k) \quad \text{for all } k. \tag{6.5b}$$

We also assume that Argmin $f\neq\emptyset$. By the results of Section 5, we know that Algorithm 5.1 either terminates, finding some $\bar{x}\in$ Argmin f in a finite number of iterations, or it generates an infinite sequence $\{x^k\}$ such that

$$x^k \to \bar{x} \quad \text{as } k\to\infty, \tag{6.6}$$

$$f(x^k)\downarrow f(\bar{x}) \quad \text{as } k\to\infty, \tag{6.7}$$

$$w^k\to 0 \quad \text{as } k\to\infty, \tag{6.8}$$

for some $\bar{x}\in$ Argminf. In the former case there is nothing to prove as far as finite convergence is concerned. Therefore we shall initially suppose that Algorithm 5.1 does not stop, and then show that in fact this is impossible if f satisfies a certain condition given below.

Thus suppose that $\{x^k\}$ is infinite, so that (6.6) - (6.8) are satisfied.

We start by collecting some useful results. By (6.8), (4.7) and (2.21d), we have

$$v^k \to 0, \quad d^k \to 0 \quad \text{and} \quad p^k \to 0 \quad \text{as } k \to \infty. \tag{6.9}$$

From (6.2), (6.5) and (2.9) we always have

$$f_j(x) = \overline{f}_{i(j)}(x) \quad \text{for all } x, \tag{6.10}$$

hence $\alpha_j^k = f(x^k) - f_j(x^k)$ can be expressed as

$$\alpha_j^k = f(x^k) - \overline{f}_{i(j)}(x^k), \tag{6.11}$$

and we have

$$-\alpha_j^k + \langle g^j, d^k \rangle = \overline{f}_{i(j)}(x^k + d^k) - f(x^k). \tag{6.12}$$

Define the sequence of sets

$$I^k = \{i(j) : j \in \hat{J}^k\} \quad \text{for all } k, \tag{6.13}$$

and note that each I^k has at most N+1 elements, since so has J^k. Asymptotic properties of $\{I^k\}$ are described in

<u>Lemma 6.1</u>. There exists a number n_I satisfying

$$I^k \subset I(\overline{x}) \quad \text{for all } k \geq n_I. \tag{6.14}$$

<u>Proof</u>. Let $i \in I$ be fixed and let $K^i = \{k : i \in I^k\}$. Then (6.2), (6.10), (5.13b) and (5.13d) imply

$$0 \leq f(x^k) - \overline{f}_i(x^k) = -v^k + \langle a^i, d^k \rangle$$

$$\leq -v^k + |a^i||d^k| \quad \text{for all } k \in K^i,$$

hence (6.9) implies that $i \in I(\overline{x})$ if K^i is infinite, because then $\overline{f}_i(x^k) \to \overline{f}_i(\overline{x}) = f(\overline{x})$. If the lemma were false, one could choose, since I is finite, an index $i \in I \setminus I(\overline{x})$ and a corresponding infinite set K^i. But then, by the above result, we would have $i \in I(\overline{x})$. This contradiction completes the proof. □

Let us introduce an auxiliary function ω defined on subsets $\tilde{I} \subset I$:

$$\omega(\tilde{I}) = \min\{|a| : a \in \operatorname{conv}\{a^i : i \in \tilde{I}\}\}. \tag{6.15}$$

Note that any minimum point $\hat{x}$ of f is characterized by

$$\omega(\hat{I}) = 0 \text{ for some } \hat{I} \subset I(\hat{x}), \tag{6.16}$$

since (6.16) means that $0 \in \partial f(\hat{x})$, cf. (6.3).

We shall now show that the selected past subgradients g^j, $j \in \hat{J}^k$, become positively linearly dependent for large k.

<u>Lemma 6.2</u>. There exists a number $n_\omega \geq n_I$ such that

$$\omega(I^k) = 0, \text{ i.e. } 0 \in \operatorname{conv}\{a^i : i \in I^k\}, \quad \text{for all } k \geq n_\omega.$$

<u>Proof</u>. Since ω has finitely many values (I is finite), the constant δ defined by $\delta = \min\{\delta(\tilde{I}) : \tilde{I} \in I \text{ and } \delta(\tilde{I}) > 0\}$ is positive, and $\omega(\tilde{I}) < \delta$ implies that $\omega(\tilde{I}) = 0$. By (6.5a), (6.13), (2.21a,c) and (2.23), we always have

$$\omega(I^k) \leq \Big| \sum_{j \in J^k} \lambda_j^k g^j \Big| = |p^k|.$$

Therefore (6.9) yields $\omega(I^k) \leq |p^k| < \delta$ for all sufficiently large k, and the assertion follows. □

Since the linearizations f_j used by the algorithm are in fact defined by the linear pieces $\bar{f}_{i(j)}$ (see (6.10)), the polyhedral approximations can be expressed as

$$\hat{f}_r^k(x) = \max\{\bar{f}_i(x) : i \in I^k\}, \tag{6.17a}$$

$$\hat{f}_s^{k+1}(x) = \max\{\bar{f}_i(x) : i \in I^k \cup \{i(k+1)\}\}, \tag{6.17b}$$

for any k. This follows from (5.7), (6.5), (6.13) and the fact that we always have $J^{k+1} = \hat{J}^k \cup \{k+1\}$. Combining Lemma 6.2 with Lemma 5.6, we get

Corollary 6.3. Relations (5.15) hold for all $k \geq n_\omega$, i.e. for sufficiently large k the point $y^{k+1}=x^k+d^k$ is the nearest point to x^k that minimizes the functions $\hat{f}_s^k$ and $\hat{f}_r^k$ given by (6.17). Moreover,

$$\hat{f}_s^k(y^{k+1}) = \hat{f}_r^k(y^{k+1}) = f(x^k) + v^k \quad \text{for all } k \geq n_\omega, \tag{6.18}$$

$$\hat{f}_r^k(y^{k+1}) \leq \hat{f}_s^k(x) \leq f(x) \quad \text{for all } x \text{ and } k \geq n_\omega. \tag{6.19}$$

Remark 6.4. By (6.18) and (6.19),

$$f(x^k) + v^k \leq \min\{f(x) : x \in R^N\} \leq f(x^k) \quad \text{for all } k \geq n_\omega, \tag{6.20}$$

which is similar to the global estimates employed by cutting plane algorithms.

We are now ready to prove

Theorem 6.5. If Algorithm 5.1 executes only a finite number of serious steps, then in fact it must terminate.

Proof. If $x^k=\bar{x}$ for all k large enough, then Lemma 6.1 implies that $I^k \subset I(\bar{x})$ for such k, hence (6.10),(6.11) and (6.13) yield

$$\alpha_j^{k+1}=0 \quad \text{for all } j \in \hat{J}^k \text{ and } k \text{ large enough.} \tag{6.21}$$

By Lemma 5.4, the multipliers λ_j^{k+1}, $j \in J^{k+1}=\hat{J}^k \cup \{k+1\}$, solve subproblem (5.2), hence (6.21) implies

$$w^{k+1} \leq \min\{\tfrac{1}{2}\}|\sum_{j \in \hat{J}^k} \lambda_j g^j|^2 : \lambda_j \geq 0,\ j \in \hat{J}^k,\ \sum_{j \in \hat{J}^k} \lambda_j = 1\}$$

for large k. In view of Lemma 6.2, the right side of the above inequality is equal to zero for large k, hence the algorithm must stop owing to $w^k=0$ for some k. □

In view of the above result, we assume below that Algorithm 5.1 executes an infinite number of serious steps. We shall need the following auxiliary results.

Lemma 6.6. If $l\to\infty$ then for all $k\geq n_\omega$ one has

$$f(y^{k+1}) > \hat{f}_r^k(y^{k+1}), \tag{6.22}$$

$$i(k+1)\notin I^k. \tag{6.23}$$

Proof. Suppose that $f(y^{k+1})\leq f(x^k)+v^k$ for some $k\geq n_\omega$. By (6.20), we have $f(y^{k+1}) = \min f$. On the other hand, the line search rules yield $x^{k+1}=y^{k+1}$, hence $f(x^{k+1})= \min f$. The next serious step must decrease the objective value, which contradicts $f(x^{k+1})= \min f$. Therefore we have $f(y^{k+1}) > f(x^k) + v^k$ for all $k\geq n_\omega$ and (6.18) yields

$$f(y^{k+1}) = \bar{f}_{i(k+1)}(y^{k+1}) > f(x^k) + v^k = \hat{f}_r^k(y^{k+1}) =$$

$$= \max \{\bar{f}_i(y^{k+1}): i\in I^k\},$$

which proves (6.23). □

Lemma 6.7. If $l\to\infty$ then for all $k\geq n_\omega$ one has

$$\bar{x} \in \operatorname{Argmin} \hat{f}_s^k \cap \operatorname{Argmin} \hat{f}_r^k, \tag{6.24}$$

$$\hat{f}_s^k(y^{k+1}) = \hat{f}_r^k(y^{k+1}) = \hat{f}_r^k(\bar{x}) = \min f, \tag{6.25}$$

$$y^{k+1} \notin \operatorname{Argmin} f. \tag{6.26}$$

Proof. By Lemma 6.1 and Lemma 6.2, we have $I^k\subset I(\bar{x})$ and $\omega(I^k)=0$. $I^k\subset I(\bar{x})$ and (6.17a) imply $\hat{f}_r^k(\bar{x})=\bar{f}_i(\bar{x})=f(\bar{x})$ for all $i\in I^k$. Thus we have $\hat{f}_r^k(\bar{x})=\bar{f}_i(\bar{x})$ for all $i\in I^k$ and $\omega(I^k)=0$. Therefore $\bar{x}\in \operatorname{Argmin} \hat{f}_r^k$, cf. (6.16) and (6.17a), and $\hat{f}_r^k(\bar{x})= \min\{\hat{f}_r^k(x): x\in R^N\} = f(\bar{x})= \min f$. Since $\bar{f}_i\leq f$ for all i and $\hat{f}_r^k\leq\hat{f}_s^k\leq f$, we obtain $f(\bar{x})=\hat{f}_r^k(\bar{x})\leq\hat{f}_s^k(\bar{x})\leq f(\bar{x})$ and $\hat{f}_s^k(y^{k+1})\geq\hat{f}_r^k(y^{k+1})\geq\hat{f}_r^k(\bar{x})$. Combining this with (6.19), we obtain (6.24) and (6.25). If we had $f(y^{k+1})= \min f$ for some $k\geq n_\omega$, this would contradict (6.22) and (6.25). □

Consider the following assumption on f and its minimum point $\bar{x}$.

Assumption 6.8. If $\tilde{I}\subset I(\bar{x})$ satisfies $\omega(\tilde{I})=0$ then at least one of the following conditions is satisfied:

(i) $\operatorname{Argmin}_x \max\{\bar{f}_i(x):\ i\in \tilde{I}\}\subset \operatorname{Argmin} f$;

(ii) rank $\{a^i:\ i\in \tilde{I}\}\ \geq N-1$ (the rank of the $N\times|\tilde{I}|$ matrix with columns a^i is greater than $N-2$) and $\operatorname{cone}\{a^i:\ i\in\tilde{I}\} = \operatorname{span}\{a^i:\ i\in\tilde{I}\}$ ("cone" and "span" denote the convex cone and the linear subspace generated by the vectors a^i, $i\in\tilde{I}$, respectively).

<u>Remark 6.9</u>. The well-known <u>Haar condition</u> at $\bar{x}$

$$\operatorname{rank}\{a^i:\ i\in\tilde{I}\} = \min\{N,|\tilde{I}|\} \quad \text{for all } \tilde{I}\in I(\bar{x}) \tag{6.27}$$

implies that Assumption 6.8 is satisfied. This follows from the fact that (6.27) and $\omega(\tilde{I})=0$ yield rank $\{a^i:\ i\in\tilde{I}\} = N$.

Under Assumption 6.8 one may describe geometric properties of the set of minimum points for $\hat{f}_r^k$ as follows.

<u>Lemma 6.9</u>. Suppose that Assumption 6.8(ii) is satisfied and $l\to\infty$. Then for all $k\geq n_\omega$ one has

$$\operatorname{Argmin} \hat{f}_r^k = \bar{x} + \operatorname{span}\{y^{k+1}-\bar{x}\} = \{\bar{x} + t(y^{k+1}-\bar{x}):\ t\in R\}, \tag{6.28a}$$

$$y^{k+1} \neq \bar{x}, \tag{6.28b}$$

i.e. Argmin $\hat{f}_r^k$ is the straight line passing through the different points y^{k+1} and $\bar{x}$.

<u>Proof</u>. Let $k\geq n_\omega$ be fixed, so that $I^k\subset I(\bar{x})$ and $\omega(I^k)=0$. Let $\tilde{b}^i=b^i+\min f$ for all $i\in I$. From Lemma 6.7 and Corollary 6.3 we deduce that

$$\operatorname{Argmin} \hat{f}_r^k = \{x:\ \langle a^i,x\rangle \leq \tilde{b}^i \quad \text{for all } i\in I^k\}, \tag{6.29a}$$

$$\langle a^i,y^{k+1}\rangle = \langle a^i,\bar{x}\rangle = \tilde{b}^i \quad \text{for all } i\in I^k, \tag{6.29b}$$

and that (6.28b) is satisfied. Using Assumption 6.8(ii) and a classical theorem on finite systems of linear inequalities (see, e.g. Theorem 4.1.10 in (Pshenichny,1980)), we deduce that any point $\check{x}$ satisfying $\langle a^i,\check{x}\rangle\leq 0$ for all $i\in I^k$ must satisfy $\langle a^i,\check{x}\rangle=0$ for all $i\in I^k$. Therefore it follows from (6.29) that

$$\operatorname{Argmin} \hat{f}_r^k = \bar{x} + \{y:\ \langle a^i,y\rangle = 0 \quad \text{for all } i\in I^k\}.$$

Let $\tilde{L}=\text{span}\{a^i : i\in I^k\}$ and let $\tilde{L}^\perp$ denote the orthogonal complement of the subspace $\tilde{L}$. Then Argmin $\tilde{f}_r^k = x+\tilde{L}^\perp$ and $0\neq \bar{x} - y^{k+1}\in \tilde{L}^\perp$ from (6.28b) and (6.26). Therefore to complete the proof it suffices to show that the dimension of $\tilde{L}^\perp$ dim $\tilde{L}^\perp=1$. We have $\tilde{L}+\tilde{L}^\perp=R^N$ and dim $\tilde{L}$+dim $\tilde{L}^\perp=N$. If rank$\{a^i : i\in I^k\}$ = dim $\tilde{L}$ = N then dim $\tilde{L}^\perp=0$, which contradicts $0\neq\bar{x}-y^{k+1}\in \tilde{L}^\perp$. Therefore rank$\{a^i : i\in I^k\}$= N-1 by Assumption 6.8(ii), and hence dim $\tilde{L}^\perp=1$. □

We may now prove the main result.

<u>Theorem 6.10</u>. If Assumption 6.8 is satisfied then Algorithm 5.1 terminates.

<u>Proof</u>. In view of Theorem 6.5 we need only consider the case of an infinite number of serious steps. Using Lemma 6.1 and Lemma 6.2, we choose a fixed index $k\geq n_\omega$ such that

$$I^k\cup I^{k+1}\subset I(\bar{x}) \quad \text{and } \omega(I^k)=0. \tag{6.30}$$

Thus we may use Assumption 6.8 with $\tilde{I}=I^k$. We have two cases:
(i) If Assumption 6.8(i) is satisfied then (6.19) yields $y^{k+1}\in$ Argmin f.
(ii) If Assumption 6.8(ii) holds then Argmin $\hat{f}_r^k$ is given by (6.29a).
On the other hand, (6.17), Corollary 6.3 and Lemma 6.7 imply

$$y^{k+2}\in \text{Argmin } \hat{f}_s^{k+1} = \text{Argmin } \hat{f}_r^k \cap \{x: \langle a^{i(k+1)}, x\rangle \leq \tilde{b}^{i(k+1)}\}. \tag{6.31}$$

By (6.22) and (6.25), $\langle a^{i(k+1)}, y^{k+1}\rangle > \tilde{b}^{i(k+1)}$, hence y^{k+1} lies outside the halfspace

$$H = \{x\in R^N : \langle a^{i(k+1)}, x\rangle \leq \tilde{b}^{i(k+1)}\}.$$

It is easy to check that if $i(k+1)\notin I^{k+1}$, then Corollary 6.3 implies $y^{k+2}=y^{k+1}$, and then (6.25) yields $\hat{f}_s^{k+1}(y^{k+1}) = \hat{f}_s^{k+1}(y^{k+2}) = \hat{f}_r^k(y^{k+2}) = \hat{f}_r^k(y^{k+1})$ = min f, so $\hat{f}_s^{k+1}(y^{k+1}) = \hat{f}_r^k(y^{k+1})$ = min f. But $\hat{f}_s^{k+1}(y^{k+1}) = f(y^{k+1}) \geq \hat{f}_r^k(y^{k+1})$, so we have $f(y^{k+1})$= min f, which contradicts (6.22). Therefore we must have $i(k+1)\in I^{k+1}\subset I(\bar{x})$. Then $\langle a^{i(k+1)},\bar{x}\rangle - b^{i(k+1)} = f(\bar{x})$ = min f, which means that $\bar{x}$ belongs to the boundary of H. Consequently, the straight line Argmin $\hat{f}_r^k$, given by (6.28a), intersects the boundary of H at $\bar{x}$. Similarly to (6.29b), we obtain $\langle a^i, y^{k+2}\rangle = \tilde{b}^i$ for all $i\in I^{k+1}$. Since $i(k+1)\in I^{k+1}$, we deduce that y^{k+2} belongs to the boundary of H, hence (6.31) yields $y^{k+2}=\bar{x}$.

In both cases considered above there exists $k \geq n_\omega$ such that $y^{k+1} \in \text{Argmin } f$. Since this contradicts (6.26), the proof is complete.

Remark 6.11. Powell's polyhedral example (6.1) satisfies the Haar condition and has a unique solution $\bar{x}=(0,0)^T$. Algorithm 5.1 finds $\bar{x}$ in a finite number of iterations and terminates from any starting point, using at most 3 subgradients for each search direction finding. Note that $|I(\bar{x})|=5$.

By allowing the algorithm to use more subgradients for search direction finding, we can ensure finite convergence even when Assumption 6.8 fails. Suppose that we know a number n such that

$$|I(\hat{x})| < n \quad \text{for all } \hat{x} \in \text{Argmin } f$$

and consider the following modification of Step 4 of Algorithm 5.1 for large k. If $\hat{J}^k$ has n^k elements, let J^{k+1} contain $\hat{J}^k \cup \{k+1\}$ together with $n-n^k-1$ largest elements from the set $\{1,\ldots,k\} \setminus \hat{J}^k$. One may use the preceding results, especially (6.23), to show that if $l\to\infty$ then for some k the set $\{i(j): j \in J^k\} \subset I(\bar{x})$ will have n elements. This contradiction proves finite termination of the modified algorithm. In particular, by choosing $n = |I|+1$ we demonstrate finite convergence of the algorithm of Lemarechal (1978) (with $J^k=\{1,\ldots,k\}$ for all k).

7. Line Search Modifications

In this section we give general line search rules that may be implemented in efficient procedures for stepsize selection.

The algorithms discussed so far used stepsizes $t_L^k \in \{0,1\}$ and $t_R^k=1$ for generating sequences

$$x^{k+1} = x^k+t_L^k d^k \quad \text{and} \quad y^{k+1}=x^k+t_R^k d^k \qquad \text{for } k=1,2,\ldots$$

from the starting point $x^1=y^1$. At the k-th iteration the objective function was evaluated only at $y^{k+1}=x^k+d^k$, and a serious step was taken if $f(x^k+d^k) \leq f(x^k)+ mv^k$. The requirement $t_L^k=1$ for a serious step may result in too many null steps. Therefore, following Lemarechal (1978), we may introduce a fixed threshold value $\bar{t}\in(0,1]$ for a serious step, say $\bar{t}=0.1$, and replace Step 3 in Algorithm 3.1 and Algorithm 5.1 by the following more general

<u>Step 3' (Line search)</u>. Select an auxiliary stepsize $t_R^k\in[\bar{t},1]$ and set $y^{k+1}=x^k+t_R^k d^k$. If

$$f(y^{k+1}) \le f(x^k) + mt_R^k v^k \tag{7.1}$$

then set $t_L^k=t_R^k$ (a serious step), set $k(l+1)=k+1$ and increase l by 1; otherwise, i.e. if (7.1) is violated, set $t_L^k=0$ (a null step).

Observe that if $\bar{t}=1$ then Step 3' reduces to Step 3. Also one may use $t_R^k=1$ as before. When $\bar{t}<1$, the search for a suitable value of $t_R^k\in[\bar{t},1]$ may use geometrical contraction, as described in Section 2, or interpolation based upon the value of $f(x^k+td^k)$ and $\langle g_f(x^k+td^k),d^k\rangle$ for trial values of $t>0$, and $f(x^k)$ and the approximate derivative v^k of f at x^k, corresponding to $t=0$. Many efficient procedures for executing Step 3' can be designed, see (Lemarechal, 1978 and 1981; Wierzbicki, 1978b; Wolfe, 1975 and 1978). The results of Section 6 indicate that an efficient line search procedure should try a unit stepsize in the neighborhood of a solution.

We shall now indicate the modifications necessary for the results of Section 4 and Section 5 to hold also for the algorithms with Step 3'. In the proof of Lemma 4.8 observe that $t_L^k\ge\bar{t}>0$ for all $k\in K$. Part(i) of the proof of Lemma 4.11 may be substituted by the following result, which is due to Lemarechal (1978).

<u>Lemma 7.1</u>. Suppose that a point $y=x^k+td^k$ satisfies $f(y)>f(x^k)+mtv^k$ for some $t\in(0,1]$. Let $g=g_f(y)\in\partial f(y)$ and $\alpha=f(x^k)-[f(y)+\langle g,x^k-y\rangle]$. Then

$$-\alpha + \langle g,d^k\rangle > mv^k.$$

<u>Proof</u>. By assumption, we have

$$-\alpha + \langle g,d^k\rangle = f(x^k) - f(y) - t\langle g,d^k\rangle + \langle g,d^k\rangle$$

$$> tmv^k + (1-t)\langle g,d^k\rangle.$$

By convexity $\alpha= f(x^k) - f(y) + t\langle g,d^k\rangle \ge 0$, hence

$$<g,d^k> \geq [f(y) - f(x^k)]/t > mv^k.$$

We conclude that $-\alpha + <g,d^k> tmv^k + (1-t)mv^k = mv^k$. $\square$

The rules of Step 3' and Lemma 6.1 imply

$$f(x^{k+1}) \leq f(x^k) + mt_L^k v^k, \tag{7.2a}$$

$$t_L^k = 0 \quad \text{if} \quad t_L^k < \bar{t}, \tag{7.2b}$$

$$-\alpha(x^{k+1},y^{k+1}) + <g_f(y^{k+1}),d^k> \geq mv^k \quad \text{if } t_L^k<\bar{t}, \tag{7.2c}$$

$$0 \leq t_L^k \leq t_R^k, \tag{7.2d}$$

$$t_R^k < \tilde{t}, \tag{7.2e}$$

where $\tilde{t}=1$ and

$$\alpha(x,y) = f(x) - [f(y) + <g_f(y), x-y>]$$

is the linearization error.

One may check that all the results of Section 4 and Section 5 hold also for algorithms that use stepsizes satisfying the criteria (7.2) on each iteration for some fixed positive values of the parameters $\bar{t}$ and $\tilde{t}$.

Mifflin (1982) used the criteria (7.2a,c,d,e) in his version of the Lemarechal (1978) method, but with $\bar{t}=\tilde{t}=+\infty$. In the next chapter we show that the convergence results of the present chapter remain valid for the line search criteria (7.2a), (7.2c) - (7.2e) with any fixed positive $\bar{t} \leq \tilde{t}$ (including $\bar{t}=\tilde{t}=+\infty$). We feel, however, that such generalizations, which allow for very short serious steps, are mainly of theoretical interest in the convex case. In practice, the use of (7.2b) with a small value of $\bar{t}$ means that instead of trying to make an insignificant step in the current direction, we prefer to find the next, hopefully better, search direction.

CHAPTER 3

Methods with Subgradient Locality Measures for Minimizing Nonconvex Functions

1. Introduction

In this chapter we consider the problem of minimizing a locally Lipschitzian function $f: R^N \to R$, which is not necessarily convex or differentiable. This problem abounds with applications and has been treated in many papers; see, e.g. (Goldstein,1977; Mifflin,1977b and 1982; Shor, 1979).

Several iterative algorithms for solving the problem in question have been proposed. Given a starting point $x^1 \in R^N$, they generate a sequence of points x^k, k=2,3,..., that is intended to converge to the required solution. Much attention has been devoted to descent methods (Dixon and Gaviano,1980; Goldstein,1977; Kiwiel,1981b; Lemarechal, Strodiot and Bihain,1981; Mifflin,1977b and 1982; Polak, Mayne and Wardi, 1983; Shor,1979). They obtain x^{k+1} by searching from x^k along a direction d^k for a scalar stepsize t^k that gives a reduction in the objective value: $f(x^k+t^kd^k)<f(x^k)$; then $x^{k+1}=x^k+t^kd^k$. Most existing descent algorithms for nonconvex minimization must be regarded as theoretical or conceptual methods (Dixon and Gaviano,1980), since their optimization subproblems involved in computing a descent direction are constrained nondifferentiable problems, which generally cannot be carried out. Only the algorithms in (Kiwiel,1981b; Lemarechal, Strodiot and Bihain,1981; Mifflin,1977b and 1982; Polak, Mayne and Wardi,1983) have quadratic programming subproblems. However, these algorithms are descent methods in the broader sense that

$$f(x^{k+1}) < f(x^k) \quad \text{if } x^{k+1} \neq x^k.$$

In this chapter we present readily implementable methods of descent, which extend the algorithms described in Chapter 2 to the nonconvex case. The methods are based on the Mifflin (1982) extension of an algorithm due to Lemarechal (1978). Our algorithms differ from the algorithm of Mifflin (1982) mainly in their rules for line searches and the updating of the search direction finding subproblems that are quadratic programming problems. At each iteration of the Lemarechal and Mifflin algorithm, every previously computed subgradient generates one linear inequality in the current search direction finding subproblem; there are k such inequalities at the k-th iteration. This would present serious problems with storage and computation after a large number of iterations. As in Chapter 2, we overcome this difficulty by introducing rules for selecting and aggregating the past subgradient information. This leads to the concept of aggregate subgradients in the nonconvex case.

A new aspect of the nonconvex case is the necessity to introduce so-called subgradient locality measures, which depend on distances from the current point to trial points at which past subgradients were computed. This is due to the local nature of the subdifferential in the nonconvex case and the resulting lack of appropriate generalizations of the notion of ε-subdifferential. The locality measures are used for weighing past subgradients at search direction finding, so that local subgradients contribute to the current direction more significantly than the obsolete ones.

In the absence of convexity, we will content ourselves with finding stationary points for f, i.e. points $\bar{x}$ that satisfy the necessary optimality condition $0 \in \partial f(\bar{x})$. We show that the methods are both readily implementable and globally convergent in the sense that all their accumulation points are stationary. This seems to be a novel result for descent methods that do not neglect linearization errors, cf. Section 1.3.

For convex f, the algorithms are extensions of the methods of Chapter 2, differing mainly by more general line search rules. In the convex case each algorithm generates a minimizing sequence of points, which converges to a solution whenever f attains its infimum.

In Section 2 we derive the methods. Section 3 contains a detailed description of the method with subgradient aggregation. Its convergence is analyzed in Section 4. Section 5 is devoted to the method with subgradient selection. In Section 6 the results of the preceding sections are used for establishing convergence of various modifications of the methods. In particular, we strengthen the existing results on the convergence of Mifflin's (1982) method.

2. Derivation of the Methods

In this section we derive two methods for minimizing a locally Lipschitzian function $f:R^N \to R$. We concentrate on the search direction finding subproblems, leaving other details to the next section.

In order to implement the methods, we suppose that we have a subroutine that can evaluate $f(x)$ and a function $g_f(x) \in \partial f(x)$ at each $x \in R^N$, i.e. an arbitrary subgradient $g_f(x)$ of f at x on which we cannot impose any further assumptions.

Given a starting point $x^1 \in R^N$, the algorithms will generate sequences of points $x^2, x^3, \ldots$ in R^N, search directions $\{d^k\} \subset R^N$ and nonnegative stepsizes $\{t_L^k\}$, related by

$$x^{k+1} = x^k + t_L^k d^k \quad \text{for } k=1,2,\ldots .$$

The algorithms are methods of descent in the sense that

$$f(x^{k+1}) < f(x^k) \quad \text{if } x^{k+1} \neq x^k, \quad \text{for all } k.$$

Due to nondifferentiability of f, only one subgradient $g_f(x^k)$ may not suffice for calculating a usable direction of descent for f at x^k; see Section 1.3. This would, in general, require the knowledge of the full subdifferential $\partial f(x^k)$, cf. Lemma 1.2.13. Therefore, following Lemarechal (1975), Wolfe (1975) and Mifflin (1977b), we shall use bundling of subgradients calculated at trial points

$$y^{k+1} = x^k + t_L^k d^k \quad \text{for } k=1,2,\ldots, \quad y^1 = x^1,$$

where the auxiliary stepsizes $t_R^k > 0$ satisfy $t_L^k \le t_R^k$, for all k. The two-point line search will detect discontinuities in the gradient of f. The algorithms evaluate subgradients

$$g^j = g_f(y^j) \quad \text{for } j=1,2,\ldots .$$

With each such subgradient we associate the corresponding linearization of f

$$f_j(x) = f(y^j) + \langle g^j, x-y^j \rangle \quad \text{for all } x. \tag{2.1}$$

In order to use a subgradient $g_f(y)$ and the corresponding linearization

$$\bar{f}(x;y) = f(y) + \langle g_f(y), x-y \rangle \quad \text{for all } x \tag{2.2}$$

for search direction finding at any $x \in R^N$, one needs a measure, say $\alpha(x,y) \ge 0$, that indicates how much the subgradient $g_f(y) \in \partial f(y)$ deviates from being a member of $\partial f(x)$; i.e. $\alpha(x,y)$ should measure the distance from $g_f(y)$ to $\partial f(x)$. In the convex case (see Remark 2.4.3), it suffices to take $\alpha(x,y)$ equal to the linearization error

$$f(x) - \bar{f}(x;y),$$

because we have

$$g_f(y) \in \partial_{\varepsilon} f(x) \quad \text{for } \varepsilon = f(x) - \bar{f}(x;y) \geq 0,$$

cf. Lemma 2.4.2. Thus one may have $g_f(y) \in \partial f(x)$ even when y is far from x, provided that the linearization error vanishes. This is no longer true when f is nonconvex; in particular, we may have $f(x)-\bar{f}(x;y)<0$. For this reason, Mifflin (1982) introduced measures of the form

$$\alpha_M(x,y) = \max\,\{f(x) - \bar{f}(x;y), \gamma|x-y|^2\}, \tag{2.3}$$

where γ is a positive parameter, which can be set to zero if f is convex. Clearly, $\alpha_M(x,y)\geq 0$ and $\alpha_M(x,y)=0$ implies $g_f(y)\in\partial f(x)$. Our methods will use the following subgradient locality measure

$$\alpha(x,y) = \max\{|f(x) - \bar{f}(x;y)|, \gamma|x-y|^2\} \tag{2.4}$$

with the convention that $\gamma=0$ if f is convex, and $\gamma>0$ in the nonconvex case. Of course, (2.3) and (2.4) are equivalent in the convex case, since then

$$\alpha(x,y) = \alpha_M(x,y) = f(x) - \bar{f}(x;y) \geq 0,$$

while our definition (2.4) puts more stress on the value of the linearization error for nonconvex f. This will allow for choosing a small value of the distance measure parameter γ.

The algorithm of Mifflin (1982) uses for search direction finding at the k-th iteration the following polyhedral approximation to f at x^k

$$\hat{f}_M^k(x) = \max\{f(x^k)-\alpha_M(x^k,y^j)+\langle g^j,x-x^k\rangle:\ j=1,\ldots,k\} \text{ for all } x. \tag{2.5}$$

The k-th search direction is obtained by solving the problem

$$\text{minimize } \hat{f}_M^k(x^k+d) + \tfrac{1}{2}|d|^2 \quad \text{over all } d \in R^N. \tag{2.6}$$

It is easy to observe that when f is convex and $\gamma=0$ then

$$\begin{aligned} f_j(x) &= \bar{f}(x;y^j) = \bar{f}(x^k;y^j) + \langle g^j,x-x^k\rangle = \\ &= f(x^k) - \alpha_M(x^k,y^j) + \langle g^j,x-x^k\rangle \end{aligned} \tag{2.7}$$

and

$$\hat{f}_M^k(x) = \max\{f_j(x):\ j=1,\ldots,k\} \quad \text{for all } x, \tag{2.8}$$

so that subproblem (2.6) reduces to a problem of the form (2.2.34) (if

$J^k = \{1,\ldots,k\}$ in (2.2.33)). In this case the Mifflin algorithm falls within the framework of the methods discussed in Chapter 2. In the nonconvex case, the term $\alpha_M(x^k,y^j)$ in (2.5) tends to make the subgradient $g_f(y^j)$ less active in the search direction finding if y^j is far from x^k. Of course, subproblem (2.6) can be formulated as a quadratic programming problem, cf. (2.2.34) and (2.2.11). Unfortunately, this problem will have k linear inequalities at the k-th iteration, which creates difficulties with storage and work per iteration.

For completeness, we recall below the line search criteria of Mifflin (1982):

$$f(x^{k+1}) \le f(x^k) + m_L t_L^k \hat{v}^k, \tag{2.9a}$$

$$-\alpha_M(x^{k+1},y^{k+1}) + \langle g_f(y^{k+1}),d^k\rangle \ge m_R \hat{v}^k, \tag{2.9b}$$

$$y^{k+1} = x^{k+1} = x^k \quad \text{if} \quad \langle g_f(x^k),\, d^k\rangle \ge m_R \hat{v}^k, \tag{2.9c}$$

where $x^{k+1} = x^k+t_L^k d^k$, $y^{k+1} = x^k+t_R^k d^k$, $0 \le t_L^k \le t_R^k$, and the variable

$$\hat{v}^k = \hat{f}_M^k(x^k+d^k) - f(x^k) < 0 \tag{2.10}$$

may be interpreted as an approximate directional derivative of f at x^k in the direction d^k. Here m_L and m_R are line search parameters satisfying

$$0 < m_L < m_R < 1.$$

One can relate the rules (2.9) to the the line search criteria (2.7.2) discussed in Section 2.7. We shall return to this subject in the next section and in Section 6.

The Mifflin algorithm requires the storage of points y^j for calculating the distances $|x^k-y^j|$ involved in $\alpha_M(x^k,y^j)$. This can be avoided by using the following upper estimates of $|x^k-y^j|$

$$s_j^k = |x^j-y^j| + \sum_{i=j}^{k-1} |x^{i+1}-x^i| \quad \text{for } j<k, \quad s_k^k = |x^k-y^k|, \tag{2.11}$$

which can be recursively updated according to the following formula

$$s_j^{k+1} = s_j^k + |x^{k+1}- x^k|. \tag{2.12}$$

We shall call s_j^k the j-th <u>distance measure</u> at the k-th iteration. Denoting

$$f_j^k = f_j(x^k) = \overline{f}(x^k;y^j) \tag{2.13}$$

and substituting $|x^k-y^j|$ with s_j^k in the definition of $\alpha(x^k,y^j)$, we obtain the following <u>subgradient locality measure</u>

$$\alpha_j^k = \max\{|f(x^k) - f_j^k|, \gamma(s_j^k)^2\}, \tag{2.14}$$

which indicates how far g^j is from $\partial f(x^k)$. For this reason, we shall call the triple

$$(g^j, f_j^k, s_j^k) \in R^N \times R \times R$$

the j-th subgradient of f at the k-th iteration, for all $j \le k$.

Since we want to extend the methods presented in Chapter 2, suppose momentarily that f is convex. Then the k-th search direction finding subproblem (2.2.11) of the method with subgradient selection can be written as

$$\begin{aligned} &\underset{(d,u)\in R^N\times R}{\text{minimize}} && \tfrac{1}{2}|d|^2 + u, \\ &\text{subject to} && f(x^k) - \alpha_j^k + \langle g^j, d\rangle \le u, \ j \in J^k, \end{aligned} \tag{2.15}$$

since

$$f(x^k) - \alpha_j^k = f(x^k) - [f(x^k) - f_j^k] = f_j^k \quad \text{for all } j,$$

because $\gamma=0$ and $\alpha_j^k=f(x^k)-f_j^k \ge 0$. The above problem can be formulated similarly to (2.6):

$$\text{minimize } \hat f_s^k(x^k+d) + \tfrac{1}{2}|d|^2 \quad \text{over all } d \in R^N \tag{2.16}$$

in terms of the k-th polyhedral approximation to f

$$\hat f_s^k(x) = \max\{f(x^k) - \alpha_j^k + \langle g^j, x-x^k\rangle : j \in J^k\}, \tag{2.17}$$

with the solution (d^k,u^k) of (2.15) satisfying

$$\hat f_s^k(x^k+d^k) = u^k. \tag{2.18}$$

Moreover, letting

$$\hat v^k = u^k - f(x^k), \tag{2.19}$$

we see that $(d^k, \hat v^k)$ is a solution to the following problem

$$\begin{aligned} &\underset{(d,\hat v)\in R^{N+1}}{\text{minimize}} && \tfrac{1}{2}|d|^2 + \hat v, \\ &\text{subject to} && -\alpha_j^k + \langle g^j, d\rangle \le \hat v, \ j \in J^k. \end{aligned} \tag{2.20}$$

Also the variable

$$\hat{v}^k = \hat{f}_s^k(x^k+d^k) - f(x^k) \tag{2.21}$$

approximates the derivative of f at x^k in the direction d^k.

To sum up, subproblem (2.20) reduces in the convex case to the search direction finding subproblem of the method with subgradient selection from Chapter 2. Moreover, we can use the subgradient selection rules developed in Chapter 2 for constructing sets $J^k \subset \{1,\ldots,k\}$ such that $|J^k| \le N+2$ for all k. These observations give ground for using subproblem (2.20) in the method with subgradient selection for nonconvex minimization, which will be described below. However, it turns out that the subgradient selection rules of Chapter 2 need to be modified in the nonconvex case. Since the same modifications will apply also to the subgradient aggregation rules, we shall first describe search direction finding subproblems based on subgradient aggregation.

To this end, let us suppose momentarily that f is convex and consider the k-th iteration of the method with subgradient aggregation (Algorithm 2.3.1) from Chapter 2. For search direction finding this method replaces the past subgradients (g^j,f_j^k), $j=1,\ldots,k-1$, by just one aggregate subgradient (p^{k-1},f_p^k), which is their convex combination

$$(p^{k-1},f_p^k) \in \operatorname{conv}\{(g^j,f_j^k):\ j=1,\ldots,k-1\} \tag{2.22}$$

calculated at the (k-1)-st iteration. The "ordinary" linearizations

$$f_j(x) = f_j^k + \langle g^j,x-x^k\rangle \quad \text{for all } x$$

and the (k-1)-st aggregate linearization

$$\tilde{f}^{k-1}(x) = f_p^k + \langle p^{k-1},x-x^k\rangle \quad \text{for all } x$$

satisfy

$$f(x) \ge f_j(x) \quad \text{for } j=1,\ldots,k,$$

$$f(x) \ge \tilde{f}^{k-1}(x)$$

for all $x \in R^N$ when f is convex. Therefore the k-th aggregate polyhedral approximation $\hat{f}_a^k$ to f, defined by choosing a set $J^k \subset \{1,\ldots,k\}$, $k \in J^k$, and setting

$$\hat{f}_a^k(x) = \max\{\hat{f}^{k-1}(x),\ f_j(x):\ j \in J^k\} \quad \text{for all } x \tag{2.23}$$

is a lower approximation to f,

$$f(x) \geq \hat{f}_a^k(x) \quad \text{for all } x.$$

Note that $\hat{f}_a^k$ can be expressed as

$$\hat{f}_a^k(x) = \max\{f(x^k) - [f(x^k) - f_p^k] + \langle p^{k-1}, x-x^k\rangle,$$
$$f(x^k) - [f(x^k) - f_j^k] + \langle g^j, x-x^k\rangle : j \in J^k\}, \qquad (2.24)$$

where the linearization errors satisfy

$$f(x^k) - f_j^k \geq 0 \quad \text{for } j \in J^k,$$
$$f(x^k) - f_p^k \geq 0. \qquad (2.25)$$

To extend the above construction to the nonconvex case, suppose that at the k-th iteration we have the (k-1)-st aggregate subgradient $(p^{k-1}, f_p^k, s_p^k) \in R^N \times R \times R$ that satisfies the following generalization of (2.22):

$$(p^{k-1}, f_p^k, s_p^k) = \sum_{j=1}^{k-1} \tilde{\lambda}_j^{k-1}(g^j, f_j^k, s_j^k), \qquad (2.26a)$$

$$\tilde{\lambda}_j^{k-1} \geq 0 \quad \text{for} \quad j=1,\dots,k-1, \quad \sum_{j=1}^{k-1} \tilde{\lambda}_j^{k-1} = 1. \qquad (2.26b)$$

Similarly to (2.14), define the following aggregate subgradient locality measure

$$\alpha_p^k = \max\{|f(x^k) - f_p^k|, \gamma(s_p^k)^2\}. \qquad (2.27)$$

The value of α_p^k indicates how far p^{k-1} is from $\partial f(x^k)$. Indeed, for convex f we have $(\gamma=0)$

$$\alpha_p^k = f(x^k) - f_p^k$$

by (2.25), while relation (2.26) implies

$$p^{k-1} \in \partial_\varepsilon f(x^k) \quad \text{for} \quad \varepsilon = f(x^k) - f_p^k,$$

as in Lemma 2.4.2. On the other hand, if the value of α_p^k is small and $\gamma>0$ then the value of

$$s_p^k = \sum_{j=1}^{k-1} \tilde{\lambda}_j^{k-1} s_j^k$$

is small, hence (2.26b) and the fact that $s_j^k \geq |x^k - y^j|$ imply that the value of $\tilde{\lambda}_j^{k-1}$ must be small if y^j is far from x^k, i.e. only local subgradients $g^j = g_f(y^j)$ with small values of s_j^k contribute significantly to

p^{k-1}. Therefore, in this case p^{k-1} is close to $\partial f(x^k)$ by the local upper semicontinuity of ∂f (see Lemma 1.2.2).

We may now define the k-th aggregate polyhedral approximation to f

$$\hat{f}_a^k(x) = \max\{f(x^k) - \alpha_p^k + \langle p^{k-1}, x-x^k\rangle,$$
$$f(x^k) - \alpha_j^k + \langle g^j, x-x^k\rangle : j \in J^k\} \quad \text{for all } x \tag{2.28}$$

and use it for finding the k-th search direction d^k that solves the problem

$$\text{minimize } \hat{f}_a^k(x^k+d) + \tfrac{1}{2}|d|^2 \quad \text{over all } d \in R^N. \tag{2.29}$$

Observe that for convex f the above definition of $\hat{f}_a^k$ reduces to (2.23) and (2.24). In effect, in the convex case we shall calculate d^k precisely as in the method with subgradient aggregation in Chapter 2, see (2.2.46) and (2.2.47).

Similarly to (2.20), one can find the solution of (2.29) by solving the following quadratic programming problem for $(d^k, \hat{v}^k) \in R^N \times R$:

$$\begin{aligned} &\underset{(d,\hat{v})\in R^{N+1}}{\text{minimize}} && \tfrac{1}{2}|d|^2 + \hat{v}, \\ &\text{subject to} && -\alpha_j^k + \langle g^j, d\rangle \le \hat{v};\ j \in J^k, \\ & && -\alpha_p^k + \langle p^{k-1}, d\rangle \le \hat{v}. \end{aligned} \tag{2.30}$$

Then the variable

$$\hat{v}^k = \hat{f}_a^k(x^k+d^k) - f(x^k), \tag{2.31}$$

which approximates the derivative of f at x^k in the direction d^k, can be used for line searching.

Next, we have to show how to update the aggregate subgradient recursively, i.e. so that if (2.26) is satisfied for some k then it also holds for k increased by 1. This is easy if one observes that subproblem (2.30) is of the form (2.2.41), hence the updating rules introduced in Section 2.2 can be used here. To this end, let λ_j^k, $j \in J^k$, and λ_p^k denote Lagrange multipliers of the k-th subproblem (2.30). By Lemma 2.2.1, these multipliers form a convex combination

$$\lambda_j^k \ge 0,\ j \in J^k,\ \lambda_p^k \ge 0, \quad \sum_{j \in J^k} \lambda_j^k + \lambda_p^k = 1.$$

Therefore, similarly to (2.2.45), we shall first calculate the current subgradient

$$(p^k,\tilde{f}_p^k,\tilde{s}_p^k) = \sum_{j\in J^k} \lambda_j^k(g^j,f_j^k,s_j^k) + \lambda_p^k(p^{k-1},f_p^k,s_p^k). \tag{2.31}$$

In view of (2.26), this leads to the following desirable property

$$(p^k,\tilde{f}_p^k,\tilde{s}_p^k) \in \operatorname{conv}\{(g^j,f_j^k,s_j^k) : j=1,\dots,k\}, \tag{2.33}$$

since a convex combination of convex combinations is again a convex combination. Then, having found x^{k+1}, we can update the linearization values by setting

$$f_j^{k+1} = f_j^k + \langle g^j,x^{k+1}-x^k\rangle \quad \text{for } j\in J^k,$$

$$f_p^{k+1} = \tilde{f}_p^k + \langle p^k,x^{k+1}-x^k\rangle.$$

On the other hand, since

$$s_j^{k+1} = s_j^k + |x^{k+1}-x^k| \text{ for } j\in J^k,$$

we shall set

$$s_p^{k+1} = \tilde{s}_p^k + |x^{k+1}-x^k|.$$

It is easy to check (see Lemma 4.1) that the above updating formulae and relation (2.33) yield

$$(p^k,f_p^{k+1},s_p^{k+1}) \in \operatorname{conv}\{(g^j,f_j^{k+1},s_j^{k+1}): j=1,\dots,k\}.$$

Comparing the above relation with (2.26) we conclude that we have completed the recursion without using the subgradients (g^j,f_j^k,s_j^k) for $j\in\{1,\dots,k\}\setminus J^k$. Consequently, these subgradients need not be stored.

In order to be able to use the above notation for k=1, we shall initialize the method by setting

$$y^1=x^1,\ p^0=g^1=g_f(y^1),\ f_p^1=f_1^1=f(y^1),\ s_p^1=s_1^1=0 \text{ and } J^1=\{1\}.$$

In the method with subgradient aggregation the sets $\{J^k\}$ can be chosen as in Chapter 2. Thus let $M_g\geq 2$ denote a user-supplied, fixed upper bound on the number of subgradients (including the aggregate subgradient) that the algorithm may use for search direction finding at any iteration. If the sets $\{J^k\}$ are selected recursively so that

$$k+1 \in J^{k+1}\subset J^k\cup\{k+1\},$$

$$|J^{k+1}|\leq M_g-1,$$

e.g.

$$J^{k+1} = \{k+1\} \cup \hat{J}^k,$$

$$\hat{J}^k \subset \{j \in J^k : \lambda_j^k > 0\},$$

$$|\hat{J}^k| \leq M_g - 1$$

as in Remark 2.2.4, then one can control the size of subproblem (2.30), and hence the core requirement and work per iteration.

The subgradient selection and aggregation strategy described above has a certain theoretical drawback that precludes obtaining global convergence results. Therefore we shall modify this strategy as follows.

Convergence analysis of existing subgradient algorithms usually requires an assumption of uniform boundedness of all calculated subgradients, see (Dixon and Gaviano,1980; Goldstein,1977; Kiwiel,1981b; Mifflin,1977b and 1982). We shall dispense with this assumption by modifying the search direction finding subproblem (2.30) at some iterations. To this end the algorithm will calculate a variable a^k satisfying

$$a^k \geq \max\{|x^k - y^j| : 1 \leq j \leq k-1 \text{ and } \tilde{\lambda}_j^{k-1} \neq 0\},$$

where the multipliers $\tilde{\lambda}_j^{k-1}$ satisfy (2.26). Simple recursive rules for computing a^k, which do not require the knowledge of $\tilde{\lambda}_j^{k-1}$, will be given in the next section. Thus a^k estimates the radius of the ball around x^k from which the past subgradient information was collected to form the aggregate subgradient (p^{k-1}, f_p^k, s_p^k). Therefore, we shall call a^k the <u>locality radius</u> of the aggregate subgradient. Whenever the value of a^k exceeds a fixed, large threshold value $\bar{a}>0$, the aggregate subgradient will be discarded at the k-th iteration and the set J^k will be reduced by deleting the j-s with large distance measures $s_j^k > \bar{a}/2$. In this case we shall say that a <u>distance resetting</u> occurs and set the reset indicator $r_a^k=1$; otherwise $r_a^k=0$. The aggregate subgradient is discarded by deleting the last constraint of subproblem (2.30) and setting the corresponding Lagrange multiplier λ_p^k to zero if $r_a^k=1$. This modification ensures that the algorithm will use only local subgradient information, i.e. only subgradients $g^j = g_f(y^j)$ with $|x^k - y^j| \leq \bar{a}$, for all k. In view of the local boundedness of ∂f, this suffices for locally uniform boundedness of such subgradients g^j.

Having described the subgradient aggregation strategy, we may now return to the method with subgradient selection. As in Chapter 2, this method solves subproblem (2.20) to find the k-th search direction d^k and a set $\hat{J}^k$ satisfying

$$\hat{J}^k = \{j \in J^k : \lambda_j^k > 0\},$$

$$|\hat{J}^k| \le N+1,$$

where λ_j^k, $j \in J^k$, are Lagrange multipliers of (2.20), which can be computed as shown in Lemma 2.2.1 (see also Remark 2.5.2). Then the aggregate subgradient $(p^k, \tilde{f}_p^k, \tilde{s}_p^k)$ is calculated according to (2.32), but with $\lambda_p^k = 0$. The next subgradient index set J^{k+1} is of the form

$$J^{k+1} = \hat{J}^k \cup \{k+1\},$$

which is analogous to the one used in Chapter 2. However, if at the k-th iteration the locality radius a^k is too large, i.e. $a^k > \bar{a}$, then the set J^k should be reduced to achieve

$$a^k = \max\{s_j^k : j \in J^k\} \le \bar{a}/2,$$

so that only local subgradients indexed by $j \in J^k$ are used for the k-th search direction finding.

Remark 2.1. As noted above, the use of distance measures s_j^k for estimating $|x^k - y^j|$ enables us to resign from storing trial points y^j. Still, for theoretical reasons, one may consider the following version of the method with subgradient selection. At the k-th iteration, let $(d^k, \hat{v}^k)$ denote the solution to the following quadratic programming problem

$$\begin{aligned} &\underset{(d,v) \in R^{N+1}}{\text{minimize}} \quad \tfrac{1}{2}|d|^2 + \hat{v}, \\ &\text{subject to} \quad -\alpha(x^k, y^j) + \langle g^j, d \rangle \le \hat{v}, \quad j \in J^k, \end{aligned} \tag{2.34}$$

and let λ_j^k, $j \in J^k$, denote the corresponding Lagrange multipliers satisfying $|\hat{J}^k| \le N+1$ for $\hat{J}^k = \{j \in J^k : \lambda_j^k > 0\}$. If we choose

$$J^{k+1} = \hat{J}^k \cup \{k+1\} \text{ for all } k,$$

then this version will need additionally to store at most N+2 points $\{y^j\}_{j \in J^k}$ for calculating the locality measures $\alpha(x^k, y^j)$, for all k. In this case the locality radius a^{k+1} can be computed directly by setting

$$a^{k+1} = \max\{|x^{k+1} - y^j| : j \in J^{k+1}\},$$

and the set J^{k+1} should be reduced, if necessary, so that $a^{k+1} \le \bar{a}$. The subsequent convergence results remain valid for this version of the method. However, we do not think that this version should be more efficient in practice, since s_j^k is not, usually, much larger than $|x^k-y^j|$, and the distance terms in the definitions of the locality measures $\alpha(x^k,y^j)$ and α_j^k are rather arbitrary, anyway.

We end this section by commenting on the relations of the above described methods with other algorithms. As shown above, the search direction finding subproblems (2.20) and (2.30) generalize the subproblems used by the methods of Chapter 2, and so also the subproblems of Pshenichny's method of linearizations for minimax problems and the classical method of steepest descent. At the same time, subproblems (2.20) and (2.34) are reduced versions of the Mifflin (1982) subproblem, which is of the form (2.34), but with $J^k = \{1,...,k\}$. Also they can be related to the "conceptual" search direction finding subproblem described in Lemma 1.2.13; see Remark 2.2.6.

3. The Algorithm with Subgradient Aggregation

We now state an algorithmic procedure for solving the problem considered. Ways of implementing each step of the method are discussed below.

Algorithm 3.1.

Step 0 (Initialization). Select the starting point $x^1 \in R^N$ and a final accuracy parameter $\varepsilon_s \geq 0$. Choose fixed positive line search parameters m_L, m_R, $\bar{a}$ and $\bar{t}$, $\bar{t} \leq 1$ and $0<m_L<m_R<1$, and a distance measure parameter $\gamma>0$ ($\gamma=0$ if f is convex). Set

$$y^1=x^1,\ p^0=g^1=g_f(y^1),\ f_p^1=f_1^1=f(y^1),\ s_p^1=s_1^1=0,\ J^1=\{1\}.$$

Set $a^1=0$ and the reset indicator $r_a^1=1$. Set the counter $k=1$.

Step 1 (Direction finding). Find the solution $(d^k,\hat{v}^k)$ to the following k-th quadratic programming problem

$$\begin{aligned} &\underset{(d,\hat{v})\in R^{N+1}}{\text{minimize}} && \tfrac{1}{2}|d|^2 + \hat{v}, \\ &\text{subject to} && -\alpha_j^k + \langle g^j,d\rangle \leq \hat{v},\ j\in J^k, \\ & && -\alpha_p^k + \langle p^{k-1},d\rangle \leq \hat{v} \quad \text{if } r_a^k=0, \end{aligned} \tag{3.1}$$

where

$$\alpha_j^k = \max\{|f(x^k) - f_j^k|, \gamma(s_j^k)^2\} \quad \text{for } j \in J^k, \tag{3.2}$$

$$\alpha_p^k = \max\{|f(x^k)|- f_p^k , \gamma(s_p^k)^2\}. \tag{3.3}$$

Compute Lagrange multipliers λ_j^k, $j \in J^k$, and λ_p^k of (3.1), setting $\lambda_p^k=0$ if $r_a^k=1$. Set

$$(p^k, \tilde{f}_p^k, \tilde{s}_p^k) = \sum_{j \in J^k} \lambda_j^k (g^j, f_j^k, s_j^k) + \lambda_p^k (p^{k-1}, f_p^k, s_p^k), \tag{3.4}$$

$$\tilde{\alpha}_p^k = \max\{|f(x^k) - \tilde{f}_p^k|, \gamma(\tilde{s}_p^k)^2 , \tag{3.5}$$

$$v^k = -\{|p^k|^2 + \tilde{\alpha}_p^k\}. \tag{3.6}$$

If $\lambda_p^k=0$ set

$$a^k = \max\{s_j^k: j \in J^k\}. \tag{3.7}$$

<u>Step 2 (Stopping criterion)</u>. Set

$$w^k = \tfrac{1}{2}|p^k|^2 + \tilde{\alpha}_p^k. \tag{3.8}$$

If $w^k \le \varepsilon_s$ then terminate. Otherwise, go to Step 3.

<u>Step 3 (Line search)</u>. By a line search procedure as given below, find two stepsizes t_L^k and t_R^k such that $0 \le t_L^k \le t_R^k$ and such that the two corresponding points defined by

$$x^{k+1} = x^k + t_L^k d^k \quad \text{and} \quad y^{k+1} = x^k + t_R^k d^k$$

satisfy $t_L^k \le 1$ and

$$f(x^{k+1}) \le f(x^k) + m_L t_L^k v^k, \tag{3.9}$$

$$t_R^k = t_L^k \qquad \text{if } t_L^k \ge \bar{t}, \tag{3.10a}$$

$$-\alpha(x^{k+1}, y^{k+1}) + \langle g_f(y^{k+1}), d^k \rangle \ge m_R v^k \qquad \text{if } t_L^k < \bar{t}, \tag{3.10b}$$

$$|y^{k+1} - x^{k+1}| \le \bar{a}/2. \tag{3.11}$$

<u>Step 4 (Subgradient updating)</u>. Select a set $\hat{J}^k \subset J^k$ and set

$$J^{k+1} = \hat{J}^k \cup \{k+1\}. \tag{3.12}$$

Set $g^{k+1} = g_f(y^{k+1})$ and

$$f_j^{k+1} = f_j^k + \langle g^j, x^{k+1} - x^k \rangle \quad \text{for } j \in \hat{J}^k, \tag{3.13a}$$

$$f_{k+1}^{k+1} = f(y^{k+1}) + \langle g^{k+1}, x^{k+1}-y^{k+1}\rangle, \tag{3.13b}$$

$$f_p^{k+1} = \tilde{f}_p^k + \langle p^k, x^{k+1}-x^k\rangle, \tag{3.13c}$$

$$s_j^{k+1} = s_j^k + |x^{k+1}-x^k| \quad \text{for } j \in \hat{J}^k, \tag{3.14a}$$

$$s_{k+1}^{k+1} = |x^{k+1}-y^{k+1}|, \tag{3.14b}$$

$$s_p^{k+1} = \tilde{s}_p^k + |x^{k+1}-x^k|. \tag{3.14c}$$

Step 5 (Distance resetting test). Set

$$a^{k+1} = \max\{a^k + |x^{k+1}-x^k|,\ s_{k+1}^{k+1}\}. \tag{3.15}$$

If $a^{k+1} \le \bar{a}$ then set $r_a^{k+1}=0$ and go to Step 7. Otherwise, set $r_a^{k+1}=1$ and go to Step 6.

Step 6 (Distance resetting). Keep deleting from J^{k+1} the smallest indices until the reset value of a^{k+1} satisfies

$$a^{k+1} = \max\{s_j^{k+1}: j \in J^{k+1}\} \le \bar{a}/2. \tag{3.16}$$

Step 7. Increase k by 1 and go to Step 1.

A few remarks on the algorithm are in order.

By Lemma 2.2.1, the k-th subproblem dual of (3.1) is to find values of multipliers λ_j^k, $j \in J^k$, and λ_p^k to

$$\begin{aligned} &\underset{\lambda}{\text{minimize}}\ \tfrac{1}{2}\Big|\sum_{j\in J^k} \lambda_j g^j + \lambda_p p^{k-1}\Big|^2 + \sum_{j\in J^k} \lambda_j \alpha_j^k + \lambda_p \alpha_p^k, \\ &\text{subject to}\ \ \lambda_j \ge 0,\ j \in J^k,\ \lambda_p \ge 0,\ \sum_{j\in J^k} \lambda_j + \lambda_p = 1. \\ &\qquad\qquad \lambda_p = 0 \ \text{if}\ r_a^k = 1. \end{aligned} \tag{3.17}$$

Any solution of (3.17) is a Lagrange multiplier vector for (3.1) and it yields the unique solution $(d^k, \hat{v}^k)$ of (3.1) as follows

$$d^k = -p^k, \tag{3.18}$$

$$\hat{v}^k = -\{|p^k|^2 + \sum_{j\in J^k} \lambda_j^k \alpha_j^k + \lambda_p^k \alpha_p^k\}, \tag{3.19}$$

where p^k is given by (3.4). Moreover, any Lagrange multipliers of (3.1) also solve (3.17). In particular, they form a convex combination:

$$\lambda_j^k \geq 0,\ j \in J^k,\ \lambda_p^k \geq 0, \quad \sum_{j \in J^k} \lambda_j^k + \lambda_p^k = 1,\ \lambda_p^k = 0 \ \text{ if } \ r_a^k = 1. \tag{3.20}$$

Thus one may equivalently solve the dual search direction finding subproblem (3.17) in Step 1 of the algorithm.

The stopping criterion in Step 2 admits of the following interpretation. The value of the locality measure $\tilde{\alpha}_p^k$ given by (3.5) indicates how far p^k is from $\partial f(x^k)$ (see the discussion of (2.27) in Section 2). A small value of w^k indicates both that $|p^k|$ is small and that p^k is close to $\partial f(x^k)$, because the value of the locality measure $\tilde{\alpha}_p^k$ is small. Thus the null vector is close to $\partial f(x^k)$, i.e. x^k is approximately stationary. In general, w^k may be thought of as a <u>measure of stationarity</u> of x^k. If f is convex and $\bar{x}$ is a minimum point of f, then we have

$$\min\{f(x): x \in R^N\} \geq f(x^k) - (2\varepsilon_s)^{1/2}\,|\bar{x}-x^k| - \varepsilon_s \tag{3.21}$$

upon termination at Step 2, see Remark 2.3.3 and results in the next section. In the nonconvex case our stopping criterion is a generalization of the standard criterion of a small value of the gradient of f.

The line search rules of Step 3 modify Mifflin's rules (2.9) and (2.10). In the next section we prove that the line search is always entered with

$$\hat{v}^k = \hat{f}_a^k(x^k+d^k) - f(x^k) \leq v^k < 0. \tag{3.22}$$

Hence the criterion (3.9) guarantees that x^{k+1} has a significantly smaller objective value than x^k if $x^{k+1} \neq x^k$. This prevents the algorithm from taking infinitely many serious steps ($t_L^k>0$) with no significant improvement in the objective value, which could impair convergence. The parameter $\bar{t}>0$, e.g. $\bar{t}=0.01$, is introduced to decrease the number of function and subgradient evaluations at line searches (observe that our requirements (3.10) are less severe than (2.9b)). Recall that in the line search rules for convex minimization described in Section 2.7 we had serious steps with $t_L^k \geq \bar{t}$ and null steps with $t_L^k=0$. Here the parameter $\bar{t}$ distinguishes <u>"long" serious steps</u> with $t_L^k \geq \bar{t}$, and <u>"short" serious steps</u> with $0 < t_L^k < \bar{t}$, for which (3.10b) is satisfied. It will be seen that as far as convergence analysis is concerned, short serious steps are essentially equivalent to null steps. If $t_L^k \geq \bar{t}$, i.e. if a significant decrease of the objective value occurs, then there is no need for detecting discontinuities in the gradient of f, so the algorithm sets $g^{k+1} = g_f(x^{k+1})$. On the other hand, if $t_L^k < \bar{t}$, which indicates that the algorithm is blocked at x^k due to the nondifferentiability of f, then the criterion (3.10b) ensures that the new subgradient

$g^{k+1} = g_f(y^{k+1})$, with y^{k+1} and x^k lying on the opposite sides of a discontinuity of the gradient of f, will force a significant modification of the next search direction finding subproblem. The criterion (3.11), which is related to the distance resetting test, prevents the algorithm from collecting irrelevant subgradient information.

Clearly, the line search rules (3.9) - (3.11) are so general that one can devise many procedures for implementing Step 3, see (Lemarechal, 1981; Mifflin, 1977b and 1982; Wierzbicki, 1982; Wolfe, 1978). For completeness, we give below a procedure for finding stepsizes $t_L=t_L^k$ and $t_R=t_R^k$, which is based on the ideas of Mifflin (1977b and 1982). In this procedure ζ is a fixed parameter satisfying $\zeta \in (0,0.5)$, $x=x^k$, $d=d^k$ and $v=v^k$.

Line Search Procedure 3.2.

(i) Set $t_L=0$ and $t=t_U=1$. Set $i=1$.

(ii) If $f(x+td) \le f(x) + m_L tv$ set $t_L=t$; otherwise set $t_U=t$.

(iii) If $t_L \ge \bar{t}$ set $t_R=t_L$ and return.

(iv) $-\alpha(x+t_L d, x+td) + \langle g_f(x+td), d\rangle \ge m_R v$ and $t_L < \bar{t}$ and $(t-t_L)|d| < \bar{a}/2$, set $t_R=t$ and return.

(v) Choose $t \in [t_L + \zeta(t_U-t_L), t_U-\zeta(t_U-t_L)]$ by some interpolation procedure.

(vi) Increase i by 1 and go to (ii).

Convergence of the above procedure can be established as in (Mifflin, 1977b, 1982) if f satisfies the following "semismoothness" hypothesis (see Bihain, 1984) :

for any $x \in R^N$, $d \in R^N$ and sequences $\{\bar{g}^i\} \in R^N$ and $\{t^i\} \subset R_+$ satisfying

$$\bar{g}^i \in \partial f(x+t^i d) \text{ and } t^i \downarrow 0, \text{ one has} \tag{3.23}$$

$$\limsup_{i\to\infty} \langle \bar{g}^i, d\rangle \ge \liminf_{i\to\infty} [f(x+t^i d) - f(x)]/t^i.$$

Lemma 3.3. If f has the property (3.23) then Line Search Procedure 3.2 terminates with $t_L^k=t_L$ and $t_R^k=t_R$ satisfying (3.9) - (3.11).

Proof. Assume, for contradiction purposes, that the search does not terminate. We recall that the line search is entered with $0<m_L<m_R<1$,

$\bar{a}>0$, $\bar{t}>0$ and $v<0$. Let t^i, $\tilde{t}_L^i$ and t_U^i denote the values taken on by t, t_L and t_U, respectively, after the i-th execution of step (ii) of the procedure, so that $t^i \in \{\tilde{t}_L^i, t_U^i\}$, for all i. Since $\zeta \in (0,0.5)$, $\tilde{t}_L^i \le \tilde{t}_L^{i+1} < t_U^{i+1} \le t_U^i$ and $t_U^{i+1} - \tilde{t}_L^{i+1} \le (1-\zeta)(t_U^i - \tilde{t}_L^i)$ for all i, there exists $\tilde{t}>0$ satisfying $\tilde{t}_L^i \uparrow \tilde{t}$ and $t_U^i \downarrow \tilde{t}$. Also $\tilde{t} \le \bar{t}$, since we must have $\tilde{t}_L^i < \bar{t}$ for all i. Let

$$TL = \{t \ge 0: f(x+td) \le f(x) + m_L tv\}.$$

Since $\{\tilde{t}_L^i\} \subset TL$, $\tilde{t}_L^i \uparrow \tilde{t}$ and f is continuous, we have $\tilde{t} \in TL$, i.e.

$$f(x+\tilde{t}d) - f(x) \le m_L \tilde{t} v. \tag{3.24a}$$

Since $t_U^i \downarrow \tilde{t}$, $\tilde{t} \in LT$ and $t_U^i \notin LT$ if $t_U^i = t^i$, there exists an infinite set $I \subset \{1,2,...\}$ such that $t_U^i = t^i > \tilde{t}$ and

$$f(x+t^i d) - f(x) > m_L t^i v \quad \text{for all } i \in I. \tag{3.24b}$$

By (3.24), we have

$$[f(x+t^i d) - f(x+\tilde{t}d)]/(t^i - \tilde{t}) > m_L v \quad \text{for all } i \in I,$$

hence

$$\liminf_{i\to\infty, i \in I} [f(x+\tilde{t}d+(t^i-\tilde{t})d) - f(x+\tilde{t}d)]/(t^i-\tilde{t}) \ge m_L v. \tag{3.25}$$

Next, for sufficiently large i we have $\tilde{t}_L^i < \bar{t}$ and $(t^i - \tilde{t}_L^i)|d| < \bar{a}/2$, because $\tilde{t}_L^i \uparrow \tilde{t}$ and $t^i \to \tilde{t}$. Therefore

$$-\alpha(x+\tilde{t}_L^i d, x+t^i d) + \langle \bar{g}^i, d \rangle < m_R v \quad \text{for all large } i,$$

where $\bar{g}^i = g_f(x+t^i d)$ for all i. But

$$\alpha(x+\tilde{t}_L^i d, x+t^i d) = \max\{|f(x+\tilde{t}_L^i d) - f(x+t^i d) - (\tilde{t}_L^i - t^i)\langle \bar{g}^i, d \rangle|,$$
$$\gamma(t^i - \tilde{t}_L^i)^2 |d|^2\} \to 0,$$

because $\tilde{t}_L^i \uparrow \tilde{t}$, $t^i \to \tilde{t}$, f is continuous and the subgradient mapping g_f is locally bounded (see Lemma 1.2.2). Therefore

$$\limsup_{i\to\infty} \langle \bar{g}^i, d \rangle \le m_R v. \tag{3.26}$$

Since $0<m_L<m_R<1$ and $v<0$, we have $m_R v < m_L v$; hence (3.25) and (3.26) contradict the property (3.23) applied to the subsequences indexed by $i \in I$. Therefore the search teminates. It is easy to show that (3.9) - (3.11) hold at termination. □

Remark 3.4. Following Mifflin (1977b), we say that f is weakly upper semismooth if in addition to (3.23) it satisfies the condition

$$\liminf_{i\to\infty} \langle \bar{g}^i, d\rangle \geq \limsup_{i\to\infty} [f(x+t^i d) - f(x)]/t^i.$$

In particular, every convex function is weakly upper semismooth, since

$$f(x) \geq f(x+t^i d) + \langle g_f(x+t^i d), x-(x+t^i d)\rangle$$

$$\geq f(x+t^i d) - t^i \langle \bar{g}^i, d\rangle$$

and hence

$$\langle \bar{g}^i, d\rangle \geq [f(x+t^i d) - f(x)]/t^i$$

if f is convex and $\bar{g}^i = g_f(x+t^i d) \in \partial f(x+t^i d)$. Many important classes of weakly upper semismooth functions are well described in (Lemarechal, 1981; Mifflin, 1977a and 1977b). In general, it may be difficult to verify whether (3.23) is satisfied in any specific situation. We believe, however, that (3.23) is likely to hold for most locally Lipschitzian functions that arise in applications. Our computational experience indicates that the above line search procedure always terminates, provided that function and subgradient evaluations are not distorted too much by gross numerical (or programming) errors.

Remark 3.5. One may choose trial stepsizes t in step(v) of Line Search Procedure 3.2 as follows. If on entering step (v) of the procedure we have $t_L=0$, which means that $t=t_U>0$ and

$$f(x+t_U d) > f(x) + m_L t_U v, \tag{3.27}$$

then one may set

$$t = \max\{\zeta t_U, \hat{t}\}, \tag{3.28a}$$

where

$$\hat{t} = 0.5v(t_U)^2/[t_U v - f(x) - f(x+t_U d)] \tag{3.28b}$$

minimizes the quadratic function h: $R\to R$ that interpolates the function $\tilde{t}\to f(x+\tilde{t}d)$ in the sense that $h(0)=f(x)$, $h(t_U)=f(x+t_U d)$ and $h'(0)=v$. It is easy to check that if (3.27) holds and $m_L \in (0,0.5)$ then

$$t \leq \max\{\zeta,\ 0.5/(1-m_L)\}t_U,$$

which ensures the necessary contraction. On the other hand, if $t_L>0$

then one may choose the next trial stepsize either by arithmetic bisection

$$t = (t_L + t_U)/2,$$

or geometric bisection

$$t = (t_L + t_U)^{1/2}.$$

It is important to observe that the algorithm never deletes the latest subgradient, i.e. we have

$$k \in J^k \quad \text{for all } k. \tag{3.29}$$

Too see this, note that $J^1=\{1\}$ and that in Step 4 the index k+1 is the largest in J^{k+1} and $s_{k+1}^{k+1} = |y^{k+1}-x^{k+1}| \le \bar{a}/2$ owing to (3.14b) and (3.11). Therefore k+1 cannot be deleted from J^{k+1} in Step 6.

4. Convergence

In this section we shall establish global convergence of Algorithm 3.1. We suppose that each execution of Line Search Procedure 3.3 is finite, e.g. that f has the additional semismoothness property (3.23). Naturally, convergence results assume that the final accuracy tolerance ε_s is set to zero. In the absence of convexity, we will content ourselves with finding stationary points for f. Our principal result states that Algorithm 3.1 either terminates at a stationary point or generates an infinite sequence $\{x^k\}$ whose accumulation points are stationary for f. When f is convex, $\{x^k\}$ is a minimizing sequence, which converges to a minimum point of f whenever f attains its infimum.

For convenience, we precede the main result by several lemmas. Besides being of interest in their own right, they can be used for establishing global convergence of the algorithms described in Chapter 2 that employ the general line search rules (2.6.2).

We shall need the following notation connected with the resets of the algorithm. Observe that the relation $r_a^k=1$ indicates that the algorithm is reset at the k-th iteration. Also $\lambda_p^k=0$ implies that the aggregate subgradient is discarded at the k-th iteration. Let

$$k_r(k) = \max\{j: j \le k \text{ and } r_a^j=1\}, \tag{4.1a}$$

$$J_r^k = J^{k_r(k)} \cup \{j: k_r(k) < j \le k\}, \tag{4.1b}$$

$$k_p(k) = \max\{j:\ j \le k \text{ and } \lambda_p^j = 0\}, \tag{4.2a}$$

$$\hat{J}_p^k = J^{k_p(k)} \cup \{j:\ k_p(k) < j \le k\}, \tag{4.2b}$$

for all k.

The following lemma shows that the aggregate subgradient is a convex combination of the subgradients retained at the latest reset and the subgradients calculated after the latest reset.

Lemma 4.1. Suppose $k \ge 1$ is such that Algorithm 3.1 did not stop before the k-th iteration. Then there exist numbers $\tilde{\lambda}_j^k$, $j \in \hat{J}_p^k$, satisfying

$$(p^k, \tilde{f}_p^k, \tilde{s}_p^k) = \sum_{j \in \hat{J}_p^k} \tilde{\lambda}_j^k (g^j, f_j^k, s_j^k), \tag{4.3a}$$

$$\tilde{\lambda}_j^k \ge 0,\ j \in \hat{J}_p^k, \quad \sum_{j \in \hat{J}_p^k} \tilde{\lambda}_j^k = 1. \tag{4.3b}$$

Moreover,

$$a^k = \max\{s_j^k:\ j \in \hat{J}_p^k\}, \tag{4.4}$$

$$|x^k - y^j| \le a^k \le \bar{a} \quad \text{for all } j \in \hat{J}_p^k . \tag{4.5}$$

Proof. (i) It follows from (4.2) and the rules of the algorithm that we always have $J^{k+1} \subset J^k \cup \{k+1\}$ and $J^k \subset \hat{J}_p^k$. Therefore, in view of (3.4) and (3.20), we can define additional multipliers

$$\lambda_j^k = 0 \quad \text{for} \quad j \in \hat{J}_p^k \setminus J^k$$

so that

$$(p^k, \tilde{f}_p^k, \tilde{s}_p^k) = \sum_{j \in \hat{J}_p^k} \lambda_j^k (g^j, f_j^k, s_j^k) + \lambda_p^k (p^{k-1}, f_p^k, s_p^k), \tag{4.6a}$$

$$\lambda_j^k \ge 0,\ j \in \hat{J}_p^k,\ \lambda_p^k \ge 0, \quad \sum_{j \in \hat{J}_p^k} \lambda_j^k + \lambda_p^k = 1,\ \lambda_p^k = 0 \quad \text{if } r_a^k = 1, \tag{4.6b}$$

for any k. Suppose that $\lambda_p^k = 0$ for some $k \ge 1$. Then $\hat{J}_p^k = J^k$ by (4.2), and (4.3) follows from (4.6) if one sets $\tilde{\lambda}_j^k = \lambda_j^k$ for all $j \in \hat{J}_p^k = J^k$. Also (4.4) is implied by (3.7) if $\lambda_p^k = 0$. Observe that $\lambda_p^1 = 0$ since $r_a^1 = 1$. Hence to prove that relations (4.3) - (4.4) are valid for any k, it suffices to show that if they hold for some fixed k and $\lambda_p^{k+1} > 0$, then they are true also for k increased by 1. Therefore, suppose that (4.3) and

(4.4) are satisfied for some k=n and $\lambda_p^{k+1} > 0$. By (4.2), we have $\hat{J}_p^{k+1} = \hat{J}_p^k \cup \{k+1\}$. Let

$$\tilde{\lambda}_j^{k+1} = \lambda_j^{k+1} + \lambda_p^{k+1}\,\tilde{\lambda}_j^k \quad \text{for } j \in \hat{J}_p^k, \quad \tilde{\lambda}_{k+1}^{k+1} = \lambda_{k+1}^{k+1}. \tag{4.7}$$

From (4.3b), (4.6b) and (4.7),

$$\tilde{\lambda}_j^{k+1} \geq 0 \quad \text{for all } j \in \hat{J}_p^{k+1},$$

$$\sum_{j \in \hat{J}_p^{k+1}} \tilde{\lambda}_j^{k+1} = \tilde{\lambda}_{k+1}^{k+1} + \sum_{j \in \hat{J}_p^k} \tilde{\lambda}_j^{k+1} = \lambda_{k+1}^{k+1} + \sum_{j \in J_p^k} (\lambda_j^{k+1} + \lambda_p^{k+1}\tilde{\lambda}_j^k) =$$

$$= \lambda_{k+1}^{k+1} + \sum_{j \in \hat{J}_p^k} \lambda_j^{k+1} + \lambda_p^{k+1} \sum_{j \in \hat{J}_p^k} \tilde{\lambda}_j^k =$$

$$= \sum_{j \in \hat{J}_p^{k+1}} \lambda_j^{k+1} + \lambda_p^{k+1} = 1,$$

which yields (4.3b) for k increased by 1. From (4.6), (4.7), (4.3) and the fact that $\hat{J}_p^{k+1} = \hat{J}_p^k \cup \{k+1\}$ we obtain

$$p^{k+1} = \sum_{j \in \hat{J}_p^{k+1}} \lambda_j^{k+1} g^j + \lambda_p^{k+1} p^k =$$

$$= \lambda_{k+1}^{k+1} g^{k+1} + \sum_{j \in \hat{J}_p^k} (\lambda_j^{k+1} + \lambda_p^{k+1}\,\tilde{\lambda}_j^k) g^j = \sum_{j \in \hat{J}_p^{k+1}} \tilde{\lambda}_j^{k+1} g^j,$$

and

$$\tilde{f}_p^{k+1} = \sum_{j \in \hat{J}^{k+1}} \lambda_j^{k+1} f_j^{k+1} + \lambda_p^{k+1} f_p^{k+1} =$$

$$= \sum_{j \in \hat{J}_p^{k+1}} \lambda_j^{k+1} f_j^{k+1} + \lambda_p^{k+1} [\tilde{f}_p^k + \langle p^k, x^{k+1} - x^k \rangle] =$$

$$= \sum_{j \in \hat{J}_p^{k+1}} \lambda_j^{k+1} f_j^{k+1} + \lambda_p^{k+1} \sum_{j \in J_p^k} \tilde{\lambda}_j^k [f_j^k + \langle g^j, x^{k+1} - x^k \rangle] =$$

$$= \lambda_{k+1}^{k+1} f_{k+1}^{k+1} + \sum_{j \in \hat{J}_p^k} (\lambda_j^{k+1} + \lambda_p^k\,\tilde{\lambda}_j^k)\, f_j^{k+1} = \sum_{j \in \hat{J}_p^{k+1}} \tilde{\lambda}_j^{k+1} f_j^{k+1}$$

from (3.13), and

$$\tilde{s}_p^{k+1} = \sum_{j\in \hat{J}_p^{k+1}} \lambda_j^{k+1} s_j^{k+1} + \lambda_p^{k+1} s_p^{k+1} =$$

$$= \sum_{j\in \hat{J}_p^{k+1}} \lambda_j^{k+1} s_j^{k+1} + \lambda_p^{k+1}(\tilde{s}_p^k + |x^{k+1}-x^k|) =$$

$$= \sum_{j\in \hat{J}_p^{k+1}} \lambda_j^{k+1} s_j^{k+1} + \lambda_p^{k+1} \sum_{j\in \hat{J}_p^k} \tilde{\lambda}_j^k (s_j^k + |x^{k+1}-x^k|) =$$

$$= \lambda_{k+1}^{k+1} s_{k+1}^{k+1} + \sum_{j\in \hat{J}_p^k} (\lambda_j^{k+1} + \lambda_p^{k+1} \tilde{\lambda}_j^k) s_j^{k+1} = \sum_{j\in \hat{J}^{k+1}} \tilde{\lambda}_j^{k+1} s_j^{k+1}$$

from (3.14). This yields (4.3a) for k=n+1. Next, since $\lambda_p^{k+1}>0$ by assumption, the rules of the algorithm imply that $r_a^{k+1}=0$, and so a^{k+1} is computed by (3.15), i.e.

$$a^{k+1} = \max\{a^k + |x^{k+1}-x^k|,\ s_{k+1}^{k+1}\} \quad \text{if } \lambda_p^{k+1}>0.$$

Combining this with (4.4) and the fact that $s_j^{k+1}=s_j^k+|x^{k+1}-x^k|$ for all $j\in \hat{J}_p^k$ and that $\hat{J}_p^{k+1}=\hat{J}_p^k \cup \{k+1\}$, we obtain

$$\max\{s_j^{k+1}: j\in \hat{J}_p^{k+1}\} = \max\{\max\{s_j^{k+1}: j\in \hat{J}_p^k\},\ s_{k+1}^{k+1}\} =$$

$$= \max\{\max\{s_j^k + |x^{k+1}-x^k| : j\in \hat{J}_p^k\},\ s_{k+1}^{k+1}\} =$$

$$= \max\{a^k + |x^{k+1}-x^k|,\ s_{k+1}^{k+1}\} = a^{k+1},$$

which shows that (4.4) holds for k=n+1.

(ii) Since $|x^k-y^j|\le s_j^k$ for all $j\le k$, and $a^k\le \bar{a}$ by the rules of Step 5 and Step 6, (4.5) follows from (4.4). □

In the convex case we have the following additional result.

<u>Lemma 4.2</u>. Suppose that $k\geq 1$ is such that Algorithm 3.1 did not stop before the k-th iteration, and that f is convex. Then

$$p^k \in \partial_\varepsilon f(x^k) \quad \text{for } \varepsilon = f(x^k) - \tilde{f}_p^k = \tilde{\alpha}_p^k \ge 0. \tag{4.8}$$

<u>Proof</u>. As in the proof of Lemma 2.4.2, use (4.3) and the fact that $\tilde{\alpha}_p^k = |f(x^k)-\tilde{f}_p^k|$ if f is convex, since $\gamma=0$ in the convex case. □

Our next result states that in fact the aggregate subgradient can be expressed as a convex combination of N+3 (not necessarily different) past subgradients calculated at points whose distances from the current point do not exceed the threshold value $\bar{a}$.

Lemma 4.3. Suppose $k \geq 1$ is such that Algorithm 3.1 did not stop before the k-th iteration, and let M=N+3. Then there exist numbers $\hat{\lambda}_i^k$ and vectors $(y^{k,i}, f^{k,i}, s^{k,i}) \in R^N \times R \times R$, $i=1,\ldots,M$, satisfying

$$(p^k, \tilde{f}_p^k, \tilde{s}_p^k) = \sum_{i=1}^{M} \tilde{\lambda}_i^k \, (g_f(y^{k,i}), f^{k,i}, s^{k,i}), \tag{4.9a}$$

$$\hat{\lambda}_i^k \geq 0, \quad i=1,\ldots,M, \quad \sum_{i=1}^{M} \hat{\lambda}_i^k = 1, \tag{4.9b}$$

$$(g_f(y^{k,i}), f^{k,i}, s^{k,i}) \in \{(g_f(y^j), f_j^k, s_j^k) : j \in \hat{J}_p^k\}, \quad i=1,\ldots,M, \tag{4.9c}$$

$$|y^{k,i} - x^k| \leq s^{k,i}, \quad i=1,\ldots,M, \tag{4.9d}$$

$$\max\{s^{k,i} : i=1,\ldots,M\} \leq a^k \leq \bar{a}. \tag{4.9e}$$

Proof. (4.9) follows from Lemma 4.1, Caratheodory's theorem (Lemma 1.2.1), and the fact that $g^j = g_f(y^j)$ for $1 \leq j \leq k$. □

In order to deduce stationarity results from the representation (4.9), we shall need the following lemma.

Lemma 4.4. Let $\bar{x} \in R^N$ be given and suppose that the following hypothesis is fulfilled:

there exist N-vectors $\bar{p}$, $\bar{y}^i$, $\bar{g}^i$ for $i=1,\ldots,M$, M=N+3, and numbers $\bar{f}_p$, $\bar{f}^i$, $\bar{s}^i$, satisfying

$$(\bar{p}, \bar{f}_p, \bar{s}_p) = \sum_{i=1}^{M} \bar{\lambda}_i \, (\bar{g}^i, \bar{f}^i, \bar{s}^i), \tag{4.10a}$$

$$\bar{\lambda}_i \geq 0, \quad i=1,\ldots,M, \quad \sum_{i=1}^{M} \bar{\lambda}_i = 1, \tag{4.10b}$$

$$\bar{g}^i \in \partial f(\bar{y}^i), \quad i=1,\ldots,M \tag{4.10c}$$

$$\bar{f}^i = f(\bar{y}^i) + \langle \bar{g}^i, \bar{x} - \bar{y}^i \rangle, \quad i=1,\ldots,M, \tag{4.10d}$$

$$|\bar{y}^i - \bar{x}| \leq \bar{s}^i, \quad i=1,\ldots,M, \tag{4.10e}$$

$$f(\bar{x}) = \bar{f}_p, \tag{4.10f}$$

$$\gamma \bar{s}_p = 0. \tag{4.10g}$$

(Recall that $\gamma=0$ only if f is convex; otherwise $\gamma>0$.) Then $\bar{p} \in \partial f(\bar{x})$.

Proof. (i) First, suppose that $\gamma>0$. Let $I=\{i: \bar{\lambda}_i \neq 0\}$. By (4.10g), $\bar{s}_p=0$, hence (4.10a,b) and (4.10e) imply $\bar{y}^i=\bar{x}$ for all $i \in I$, so (4.10c) yields $\bar{g}^i \in \partial f(\bar{x})$ for all $i \in I$. Thus we have $\bar{p}= \sum_{i \in I} \bar{\lambda}_i \bar{g}^i$, $\bar{\lambda}_i>0$ for $i \in I$, $\sum_{i \in I} \bar{\lambda}_i=1$ and $\bar{g}^i \in \partial f(\bar{x})$, $i \in I$, so $\bar{p} \in \partial f(\bar{x})$ by the convexity of $\partial f(\bar{x})$.

(ii) Next, suppose that $\gamma=0$. Then f is convex and (4.10c) and (4.10d) give

$$f(z) \geq f(\bar{y}^i) + \langle \bar{g}^i, z-\bar{y}^i \rangle = f(\bar{x}) + \langle \bar{g}^i, z-\bar{x} \rangle - [f(\bar{x}) - \bar{f}^i]$$

for all $z \in R^N$ and $i=1,\ldots,M$. Multiplying the above inequality by $\bar{\lambda}_i$ and summing, we obtain for each z

$$f(z) \geq f(\bar{x}) + \langle \bar{p}, z-\bar{x} \rangle - [f(\bar{x}) - \bar{f}_p] = f(\bar{x}) + \langle \bar{p}, z-\bar{x} \rangle$$

from (4.10b), (4.10a) and (4.10f). Thus $\bar{p} \in \partial f(\bar{x})$ by the definition of subdifferential in the convex case. □

First we consider the case when the method terminates.

Lemma 4.5. If Algorithm 3.1 terminates at the k-th iteration, $k \geq 1$, then the point $\bar{x}=x^k$ is stationary for f.

Proof. If the algorithm terminates at Step 2 due to $w^k \leq \varepsilon_f=0$, then, since $w^k=\frac{1}{2}|p^k|^2 + \tilde{\alpha}_p^k$ and $\tilde{\alpha}_p^k \geq 0$, we have $p^k=0$ and $\tilde{\alpha}_p^k= \max\{|f(x^k)-\tilde{f}_p^k|, \gamma(\tilde{s}_p^k)^2\}=0$, hence $f(x^k) = \tilde{f}_p^k$ and $\gamma \tilde{s}_p^k=0$. Combining this with (4.9a) - (4.9d), we see that the assumption of Lemma 4.4 is fulfilled by $\bar{x}=x^k$, $\bar{p}=p^k$, $\bar{f}_p=\tilde{f}_p^k$, $\bar{s}_p=\tilde{s}_p^k$, etc. Therefore $0=p^k \in \partial f(\bar{x})$. □

From now on we suppose that the algorithm calculates an infinite sequence $\{x^k\}$, i.e. $w^k>0$ for all k.

The following lemma states useful asymptotic properties of the aggregate subgradients.

Lemma 4.6. Suppose that there exist a point $\bar{x} \in R^N$ and an infinite set $K \subset \{1,2,...\}$ satisfying $x^k \xrightarrow{K} \bar{x}$. Then there exists an infinite set $\bar{K} \subset K$ such that the hypothesis (4.10a) - (4.10e) is fulfilled at $\bar{x}$ and

$$(p^k, \tilde{f}_p^k, \tilde{s}_p^k) \xrightarrow{\bar{K}} (\bar{p}, \bar{f}_p, \bar{s}_p). \tag{4.11}$$

If additionally $\tilde{\alpha}_p^k \xrightarrow{K} 0$, then $\bar{p} \in \partial f(\bar{x})$.

Proof. (i) From (4.9d,e), the fact that $\bar{a}<\infty$ and the assumption that $x^k \xrightarrow{K} \bar{x}$, we deduce the existence of points $\bar{y}^i$, i=1,...,M, and an infinite set $K^1 \subset K$, satisfying

$$y^{k,i} \xrightarrow{K^1} \bar{y}^i \quad \text{for } i=1,...,m. \tag{4.12a}$$

By (4.12a), (4.9c), and the local boundedness and uppersemicontinuity of ∂f (see Lemma 1.2.2), there exist N-vectors $\bar{g}^i$ and numbers $\bar{f}^i$, i=1,...,M, and an infinite set $K^2 \subset K^1$, satisfying

$$g_f(y^{k,i}) \xrightarrow{K^2} \bar{g}^i \in \partial f(\bar{y}^i) \quad \text{for } i=1,...,M, \tag{4.12b}$$

$$f^{k,i} \xrightarrow{K^2} \bar{f}^i = f(\bar{y}^i) + \langle \bar{g}^i, \bar{x}-\bar{y}^i \rangle \quad \text{for } i=1,...,M, \tag{3.12c}$$

since $f^{k,i} = f(y^{k,i}) + \langle g_f(y^{k,i}), x^k - y^{k,i} \rangle$ for i=1,...,M and all k. In view of (4.9b,e) there exist numbers $\bar{\lambda}_i$ and $\bar{s}^i$, i=1,...,M, and an infinite set $\bar{K} \subset K^2$ such that

$$\hat{\lambda}_i^k \xrightarrow{\bar{K}} \bar{\lambda}_i \quad \text{for } i=1,...,M, \tag{4.12d}$$

$$s^{k,i} \xrightarrow{\bar{K}} \bar{s}^i \quad \text{for } i=1,...,M. \tag{4.12e}$$

Letting $k \in \bar{K}$ approach infinity in (4.9a), (4.9b) and (4.9d), and using (4.12), we obtain (4.10a) - (4.10e) and (4.11).
(ii) Suppose that $\tilde{\alpha}_p^k = \max\{|f(x^k) - \tilde{f}_p^k|, \gamma(\tilde{s}_p^k)^2\} \xrightarrow{K} 0$. Then $\tilde{f}_p^k \xrightarrow{\bar{K}} \bar{f}_p = f(\bar{x})$ and $\gamma \tilde{s}_p^k \xrightarrow{\bar{K}} \gamma \bar{s}_p = 0$, so $\bar{p} \in \partial f(\bar{x})$ by Lemma 4.4. □

Our next result describes a crucial property of the stationarity measure w^k of the current point w^k.

Lemma 4.7. Suppose that for some point $\bar{x} \in R^N$ we have

$$\liminf_{k \to \infty} \max \{w^k, |\bar{x} - x^k|\} = 0, \tag{4.13}$$

or equivalently

there exists an infinite set $K \subset \{1,2,\ldots\}$ such that $x^k \xrightarrow{K} \bar{x}$
and $w^k \xrightarrow{K} 0$. (4.14)

Then $0 \in \partial f(\bar{x})$.

Proof. The equivalence of (4.13) and (4.14) follows from the fact that w^k is nonnegative for all k, since we always have $w^k = \frac{1}{2}|p^k|^2 + \tilde{\alpha}_p^k$ and $\tilde{\alpha}_p^k \geq 0$. Thus (4.14) implies $p^k \xrightarrow{K} 0$ and $\tilde{\alpha}_p^k \xrightarrow{K} 0$, so Lemma 4.6 yields the desired conclusion. $\square$

The above lemma enables us to reduce further convergence analysis to checking if w^k approaches zero around any accumulation point of $\{x^k\}$. To this end, as in Chapter 2, we shall now relate the stationarity measures with the optimal values of the dual search direction finding subproblems.

Let $\hat{w}^k$ denote the optimal value of the k-th dual search direction finding subproblem (3.17), for all k. By (3.4) and the fact that the Lagrange multipliers of (3.1) solve (3.17), we always have

$$\hat{w}^k = \frac{1}{2}|p^k|^2 + \hat{\alpha}_p^k, \tag{4,14a}$$

where

$$\hat{\alpha}_p^k = \sum_{j \in J^k} \lambda_j^k \alpha_j^k + \lambda_p^k \alpha_p^k. \tag{4.15b}$$

A useful relation between w^k and $\hat{w}^k$ is established in the following lemma.

Lemma 4.8. (i) At the k-th iteration of Algorithm 3.1, one has

$$0 \leq \alpha_p^k \leq \hat{\alpha}_p^k, \tag{4.16}$$

$$0 \leq w^k \leq \hat{w}^k, \tag{4.17}$$

$$v^k \leq -w^k \leq 0, \tag{4.18}$$

$$\hat{v}^k \leq v^k. \tag{4.19}$$

(ii) If f is convex then $\tilde{\alpha}_p^k = \hat{\alpha}_p^k$, $w^k = \hat{w}^k$ and $\hat{v}^k = v^k$, for all k.

Proof.(i) By (3.4) and (3.20),

$$|f(x^k) - \tilde{f}_p^k| = |\sum_{j\in J^k} \lambda_j^k[f(x^k)-f_j^k] + \lambda_p^k[f(x^k)-f_p^k]| <$$

$$\leq \sum_{j\in J^k} \lambda_j^k|f(x^k)-f_j^k| + \lambda_p^k|f(x^k)-f_p^k|, \qquad (4.20a)$$

and, since the function $t\to\gamma t^2$ is convex ($\gamma\geq 0$),

$$\gamma(\tilde{s}_p^k)^2 = \gamma(\sum_{j\in J^k} \lambda_j^k s_j^k + \lambda_p^k s_p^k)^2 \leq$$

$$\leq \sum_{j\in J^k} \lambda_j^k\gamma(s_j^k)^2 + \lambda_p^k\gamma(s_p^k)^2, \qquad (4.20b)$$

for all k. Since the Lagrange multipliers λ_j^k and λ_p^k are nonnegative, we obtain from (4.20)

$$\tilde{\alpha}_p^k = \max\{|f(x^k) - \tilde{f}_p^k|, \gamma(\tilde{s}_p^k)^2\} \leq$$

$$\leq \sum_{j\in J^k} \lambda_j^k \max\{|f(x^k)-f_j^k|, \gamma(s_j^k)^2\} + \lambda_p^k \max\{|f(x^k)-f_p^k|, \gamma(s_p^k)^2\} =$$

$$= \sum_{j\in J^k} \lambda_j^k \alpha_j^k + \lambda_p^k \alpha_p^k = \hat{\alpha}_p^k,$$

which yields (4.16). Next, (4.17) - (4.19) follow immediately from (4.16), (3.6), (3.8), (3.19) and (4.15).
(ii) If f is convex then we have $f(x^k)-f_j^k\geq 0$ for all $j\in J^k$, $f(x^k)-f_p^k\geq 0$ and $f(x^k)-\tilde{f}_p^k\geq 0$, hence equality holds in (2.20a). Therefore, since $\gamma=0$ in the convex case, we have $\tilde{\alpha}_p^k = \hat{\alpha}_p^k$, ant the preceding argument yields $w^k=\hat{w}^k$ and $v^k=\hat{v}^k$. □

We conclude from the above lemma that in the convex case the variables involved in line searches and the search direction finding subproblems satisfy relations analogous to those developed for the algorithms in Chapter 2.

Returning to relation (3.22), we see that (3.22) follows from (4.18), (4.19) and the fact that w^k is always positive at line searches. Note that for nonconvex f our estimate v^k of the derivative of f at x^k in the direction d^k can be less optimistic than the primal estimate $\hat{v}^k$, since $\hat{v}^k\leq v^k$ for all k. Thus v^k is always negative, hence the criterion (3.9) with $m_L>0$ and $t_L^k\geq 0$ ensures that the sequence $\{f(x^k)\}$ is nonincreasing and $f(x^{k+1})<f(x^k)$ if $x^{k+1}\neq x^k$.

Consider the following condition for some fixed point $\bar{x}\in R^N$:

there exists an infinite set $K\subset\{1,2,...\}$ satisfying $x^k \xrightarrow{K} \bar{x}$. (4.21)

Our aim is to show that $w^k \xrightarrow{\bar{K}} 0$ for some set $\bar{K} \subset K$. In Chapter 2 this was done by considering first the case of an infinite number of serious steps, where the line search criterion (3.9) and the fact that $t_L^k \geq \bar{t} > 0$ for each serious step yielded the desired conclusion. Since in Algorithm 3.1 one can have arbitrarily small serious stepsizes $t_L^k > 0$, the same argument cannot be applied here. However, we can still analyze "long" serious steps as follows.

<u>Lemma 4.9</u>. (i) If (4.21) holds then

$$f(x^k) \downarrow f(\bar{x}) \quad \text{as } k \to \infty, \tag{4.22}$$

$$t_L^k v^k \to 0 \quad \text{as } k \to \infty. \tag{4.23}$$

(ii) If (4.21) is fulfilled and there exist a number $\hat{t} > 0$ and an infinite set $\bar{K} \subset K$ such that $t_L^k \geq \hat{t}$ for all $k \in \bar{K}$, then (4.14) holds.

<u>Proof</u>. (i) (4.21) and the continuity of f yield $f(x^k) \xrightarrow{K} f(\bar{x})$, so (4.22) follows from the monotonicity of $\{f(x^k)\}$. Since we always have $v^k < 0$, $m_L \in (0,1)$ and $t_L^k \geq 0$, we obtain from (3.9)

$$0 \leq -t_L^k v^k \leq [f(x^k) - f(x^{k+1})]/m_L \quad \text{for all } k,$$

which yields (4.23) in virtue of (4.22).
(ii) If $x^k \xrightarrow{\bar{K}} \bar{x}$ and $t_L^k \geq \hat{t} > 0$ for all $k \in \bar{K}$, then (4.23) yields $v^k \xrightarrow{\bar{K}} 0$, and $w^k \xrightarrow{\bar{K}} 0$ by (4.18). Thus $x^k \xrightarrow{\bar{K}} \bar{x}$ and $w^k \xrightarrow{\bar{K}} 0$, which implies (4.14). □

<u>Corollary 4.10</u>. Suppose (4.21) is satisfied, but (4.13) does not hold, i.e.

$$\liminf_{k \to \infty} \max\{w^k, |\bar{x} - x^k|\} \geq \bar{\varepsilon} > 0 \tag{4.24}$$

for some $\bar{\varepsilon}$. Then $t_L^k \xrightarrow{K} 0$.

<u>Proof</u>. Since we always have $t_L^k > 0$, the desired conclusion follows from Lemma 4.9 and the equivalence of (4.13) and (4.14). □

In view of the above results, it remains to consider the case of arbitrarily short serious and null steps, i.e. $t_L^k \xrightarrow{K} 0$. Recall that in

Observe that for a null step $t_L^k=0$ the above lemma reduces to Lemma 2.4.11, since then $\alpha_p^k=\tilde{\alpha}_p^{k-1}$. In this case w^k is a fraction of w^{k-1}. For short serious steps the rate of decrease of w^k depends on the value of $|\alpha_p^k-\tilde{\alpha}_p^{k-1}|$ and the following properties of the function ϕ_C. Note that in fact ϕ_C depends on the value of $m_R\in(0,1)$, which is fixed in our analysis.

Lemma 4.12. For any $\varepsilon_w>0$ and $C>0$ there exist numbers $\varepsilon_\alpha=\varepsilon_\alpha(\varepsilon_w,C)>0$ and $\bar{N}=\bar{N}(\varepsilon_w,C)\geq 1$ such that for any sequence of numbers $\{t^i\}$ satisfying

$$0\leq t^{i+1}\leq\phi_C(t^i)+\varepsilon_\alpha \text{ for } i\geq 1,\ 0\leq t^1\leq 4C^2, \tag{4.29}$$

one has $t^i<\varepsilon_w$ for all $i\geq\bar{N}$.

Proof. For any $\varepsilon_\alpha>0$ define the number $t(\varepsilon_\alpha)$ by $t(\varepsilon_\alpha)=\phi_C(t(\varepsilon_\alpha))+\varepsilon_\alpha$ and observe that $\phi_C(t)+\varepsilon_\alpha<t$ for any $t>t(\varepsilon_\alpha)$. Then it is easy to show that $\limsup_{i\to\infty} t^i\leq t(\varepsilon_\alpha)$ for any sequence $\{t^i\}$ satisfying (4.29), because the function $\phi_C(\cdot)+\varepsilon_\alpha$ is continuous. Define the sequence $\bar{t}^1=4C^2$, $\bar{t}^{i+1}=\phi_C(t^i)+\varepsilon_\alpha$ for $i\geq 1$. Clearly, $\limsup_{i\to\infty}\bar{t}^i\leq t(\varepsilon_\alpha)$ and for any sequence $\{t^i\}$ satisfying (4.28) we have $t^i\leq\bar{t}^i$ if $\bar{t}^i\geq t(\varepsilon_\alpha)$, for all i. Then the desired conclusion follows from the fact that $t(\varepsilon_\alpha)\downarrow 0$ as $\varepsilon_\alpha\downarrow 0$. □

We conclude from Lemma 4.11 and Lemma 4.12 that w^k will become arbitrarily small, i.e. $w^k<\varepsilon_w$ for any fixed $\varepsilon_w>0$, provided that for sufficiently many $\bar{N}=\bar{N}(\varepsilon_w,C)$ consecutive iterations a local bound of the form (4.27) is valid, we have sufficiently small $|\alpha_p^k-\tilde{\alpha}_p^{k-1}|\leq\varepsilon_\alpha$ and $t_L^{k-1}<\bar{t}$, and no reset occurs. These properties will be established by the following four lemmas.

A locally uniform bound of the form (4.27) will result from the following lemma, which gives the reason for the line search rule (3.11).

Lemma 4.13.(i) For each $k\geq 1$

$$\max\{|p^k|,\ \tilde{\alpha}_p^k\}\leq\max\{\tfrac{1}{2}|g^k|^2+\alpha_k^k,\ (|g^k|^2+2\alpha_k^k)^{1/2}\}. \tag{4.30}$$

(ii) Suppose $\bar{x}\in R^N$, $B=\{y\in R^N: |\bar{x}-y|\leq 2\bar{a}\}$, where $\bar{a}>0$ is the line search parameter involved in (3.11), and let

$$C_g=\sup\{|g_f(y)|:\ y\in B\}, \tag{4.31a}$$

Chapter 2 this case was equivalent to having $t_L^k=0$ for all sufficiently large k, and was analyzed by showing that the optimal value of the dual search direction finding subproblem decreases after a null step owing to line search requirements of the form (3.10b). Proceeding along similar lines, we shall now show that the stationarity measure w^k decreases whenever the algorithm cannot obtain a significant improvement in the objective value, i.e. after a null step or a short serious step.

Lemma 4.11. Suppose that $t_L^{k-1}<\bar{t}$ and $r_a^k=0$ for some $k>1$. Then

$$w^k\le\hat{w}^k\le\phi_C(w^{k-1})+|\alpha_p^k-\tilde{\alpha}_p^{k-1}|, \tag{4.25}$$

where the function ϕ_C is defined (for the fixed value of the line search parameter $m_R\in(0,1)$) by

$$\phi_C(t) = t - (1-m_R)^2t^2/(8C^2), \tag{4.26}$$

and C is any number satisfying

$$\max\{|p^{k-1}|,\ |g^k|,\ \tilde{\alpha}_p^{k-1},1\}\ \le C. \tag{4.27}$$

Proof. (i) Observe that $k>1$, $t_L^{k-1}<\bar{t}$ and the line search rule (3.10b) yield $-\alpha(x^k,y^k) + \langle g_f,(y^k),d^{k-1}\rangle \ge m_R v^{k-1}$, so

$$-\alpha_k^k + \langle g^k,d^{k-1}\rangle \ge m_R v^{k-1}. \tag{4.28}$$

(ii) Define the multipliers

$$\lambda_k(\nu) = \nu,\ \lambda_j(\nu) = 0 \quad \text{for} \quad j\in J^k\setminus\{k\},\quad \lambda_p(\nu) = 1-\nu$$

for each $\nu\in[0,1]$. Since $r_a^k=0$ by assumption and $k\in J^k$ by (3.29), the above multipliers are feasible for the k-th dual subproblem (3.17). Therefore, reasoning as in the proof of Lemma 2.4.11, we obtain

$$\begin{aligned}\hat{w}^k&\le\min\{\tfrac{1}{2}|(1-\nu)p^{k-1}+\nu g^k|^2+(1-\nu)\alpha_p^k+\nu\alpha_k^k:\ \nu\in[0,1]\}\le\\ &\le\min\{\tfrac{1}{2}|(1-\nu)p^{k-1}+\nu g^k|^2+(1-\nu)\tilde{\alpha}_p^{k-1}+\nu\alpha_k^k:\ \nu\in[0,1]\}+\\ &\qquad+|\alpha_p^k-\tilde{\alpha}_p^{k-1}|.\end{aligned}$$

Then one may use (4.28) and the various definitions of the algorithm to obtain the desired conclusion by invoking Lemma 2.4.10 as in the proof of Lemma 2.4.11. □

$$C_\alpha = \sup\{\alpha(x,y) : x \in B,\ y \in B\}, \tag{4.31b}$$

$$C = \max\{\tfrac{1}{2} C_g^2 + C_\alpha,\ (C_g^2 + 2C_\alpha)^{1/2}, 1\}. \tag{4.31c}$$

Then C is finite and

$$\max\{|p^k|, \tilde{\alpha}_p^k, |g^k|, 1\} \le C \quad \text{if} \quad |x^k - \bar{x}| \le \bar{a}. \tag{4.32}$$

Proof. (i) Let $k>1$ be fixed and define the multipliers

$$\lambda_k = 1,\ \lambda_j = 0 \quad \text{for} \quad j \in J^k \setminus \{k\},\ \lambda_p = 0 .$$

Since $k \in J^k$, the above multipliers are feasible for the k-th dual subproblem (3.17). Therefore the optimal value $\hat{w}^k$ of (3.17) satisfies $\hat{w}^k \le \frac{1}{2}|g^k|^2 + \alpha_k^k$, hence $\frac{1}{2}|p^k|^2 + \tilde{\alpha}_p^k = w^k \le \frac{1}{2}|g^k|^2 + \alpha_k^k$, and (4.30) follows.
(ii) We deduce from the local boundedness of ∂f (Lemma 1.2.1) that the mappings $g_f(\cdot)$ and $\alpha(\cdot,\cdot)$ are bounded on the bounded sets B and $B \times B$, respectively. Therefore, the constants defined by (4.31) are finite. If $|\bar{x} - x^k| \le \bar{a}$ then $|\bar{x} - y^k| \le |\bar{x} - x^k| + |x^k - y^k| \le 2\bar{a}$, because $|y^k - x^k| \le \bar{a}$ by (3.11). Thus we have $x^k \in B$, $y^k \in B$, $g^k = g_f(y^k)$ and $\alpha_k^k = \alpha(x^k, y^k)$, hence (4.32) follows from (4.30) and (4.31). □

Our next result will provide bounds on the term $|\alpha_p^k - \tilde{\alpha}_p^{k-1}|$ involved in (4.25).

Lemma 4.14. Suppose (4.21) holds. Then

$$|f(x^{k+1}) - f_p^{k+1}| - |f(x^k) - \tilde{f}_p^k| \longrightarrow 0 \text{ as } k \to \infty. \tag{4.33}$$

Proof. For any k

$$\begin{aligned} ||f(x^{k+1}) - f_p^{k+1}| - |f(x^k) - \tilde{f}_p^k|| &\le |f(x^{k+1}) - f_p^{k+1} - f(x^k) + \tilde{f}_p^k| \le \\ &\le |f(x^{k+1}) - f(x^k)| + |f_p^{k+1} - \tilde{f}_p^k| \le \\ &\le |f(x^{k+1}) - f(x^k)| + |\langle p^k, x^{k+1} - x^k \rangle| \end{aligned} \tag{4.34a}$$

from (3.13c). Next, since $t_L^k \ge 0$ and $x^{k+1} = x^k + t_L^k d^k = x^k - t_L^k p^k$ (see (3.18) for all k, we obtain from (3.6)

$$-t_L^k v^k \ge t_L^k |p^k|^2 = \langle x^k - x^{k+1}, p^k \rangle \ge 0. \tag{4.34b}$$

Combining (4.34) with Lemma 4.9(i) we establish (4.33). □

The following lemma will imply that $t_L^{k-1} < \bar{t}$ for sufficiently many consecutive iterations.

<u>Lemma 4.15</u>. Suppose (4.21) and (4.24) hold. Then for any fixed integer $m \geq 0$ there exists a number $\hat{k}_m$ such that for any integer $n \in [0,m]$

$$x^{k+n} \to \bar{x} \quad \text{as } k \to \infty,\ k \in K, \tag{4.35a}$$

$$w^{k+n} \geq \bar{\varepsilon}/2 \quad \text{for all } k > \hat{k}_m,\ k \in K, \tag{4.35b}$$

$$t_L^{k+n} \to 0 \quad \text{as } k \to \infty,\ k \in K. \tag{4.35c}$$

Moreover, for any numbers $\hat{k}$, $\tilde{N}$ and $\varepsilon_\alpha > 0$ there exists a number $\tilde{k} \geq \hat{k}$, $\tilde{k} \in K$, such that

$$w^k \geq \bar{\varepsilon}/2 \qquad \text{for } \tilde{k} \leq k \leq \tilde{k}+\tilde{N}, \tag{4.36a}$$

$$\max\{|p^{k-1}|, |g^k|, \tilde{\alpha}_p^{k-1}, 1\} \leq C \qquad \text{for } \tilde{k} \leq k \leq \tilde{k}+\tilde{N}, \tag{4.36b}$$

$$|\alpha_p^k - \tilde{\alpha}_p^{k-1}| \leq \varepsilon_\alpha \qquad \text{for } \tilde{k} \leq k \leq \tilde{k}+\tilde{N}, \tag{4.36c}$$

$$t_L^k < \bar{t} \qquad \text{for } \tilde{k} \leq k \leq \tilde{k}+\tilde{N}, \tag{4.36d}$$

where C is the constant defined in Lemma 4.13.

<u>Proof</u>. (i) We shall first establish (4.35). For $m=n=0$, (4.35a) follows from our assumption (4.21). Suppose that (4.35a) holds for some fixed $m=n>0$. From (4.35a) and (4.24) we deduce the existence of a number $\tilde{k}_n$ such that

$$w^{k+n} \geq \bar{\varepsilon}/2 \quad \text{for all } k \geq \tilde{k}_n,\ k \in K. \tag{4.37a}$$

(4.35c) follows from (4.35a), (4.24) and Corollary 4.10. Using (4.35a) and Lemma 4.13, we deduce that $|p^{k+n}| \leq C$ for all $k \in K$. Then

$$|x^{k+n} - x^{k+n}| = t_L^{k+n}|d^{k+n}| = t_L^{k+n}|p^{k+n}| \leq C t_L^{k+n} \tag{4.37b}$$

for all $k \in K$, since we always have $d^k = -p^k$ by (3.18). (4.37b) and (4.35c) yield $|x^{k+n+1} - x^{k+n}| \to 0$ as $k \to \infty$, $k \in K$, hence (4.35a) implies $x^{k+n+1} \to \bar{x}$ as $k \to \infty$, $k \in K$. Thus (4.35a) holds for n increased by 1. Therefore one can repeat the above reasoning for all $n \in [0,m]$. Setting

$\hat{k}_m = \max\{\tilde{k}_n : n\in[0,m]\}$, see (4.37a), we complete the proof of (4.35).
(ii) If $\varepsilon_\alpha>0$ then (4.21) and Lemma 4.14 imply the existence of a number $k_1>\hat{k}$ satisfying

$$||f(x^{k+1})-f_p^{k+1}|-|f(x^k)-\tilde{f}_p^k|| \le \varepsilon_\alpha/2 \quad \text{for all } k\ge k_1. \tag{4.38a}$$

Next, by (3.14c) and (3.5), we always have

$$\begin{aligned}\gamma(\tilde{s}_p^k) \le \gamma(s_p^{k+1})^2 &= \gamma(\tilde{s}_p^k + |x^{k+1}-x^k|)^2 < \\ &\le \gamma(\tilde{s}_p^k)^2 + 2\gamma\tilde{s}_p^k|x^{k+1}-x^k|+ \gamma|x^{k+1}-x^k|^2 \\ &\le \gamma(\tilde{s}_p^k)^2 +2(\gamma\tilde{\alpha}_p^k)^{1/2}\ |x^{k+1}-x^k| + \gamma|x^{k+1}-x^k|^2,\end{aligned}$$

hence

$$\gamma(\tilde{s}_p^k)^2\le\gamma(s_p^{k+1})^2\le\gamma(\tilde{s}_p^k)^2 + |x^{k+1}-x^k|(2(\gamma C)^{1/2} + \gamma|x^{k+1}-x^k|) \tag{4.38b}$$

if $|x^k-\bar{x}|\le\bar{a}$ by (4.32). It follows from (3.2), (3.5) and (4.38) that

$$|\alpha_p^{k+1}-\tilde{\alpha}_p^k| \le\varepsilon_\alpha/2 + |x^{k+1}-x^k|(2(\gamma C)^{1/2} +\gamma|x^{k+1}-x^k|) \tag{4.39}$$

for any $k\ge k_1$ such that $|\bar{x}-x^k|\le\bar{a}$. Using (4.32) and (4.39), we deduce from the first part of the lemma the existence of $\tilde{k}\ge k_1$ satisfying (4.36). □

We can now prove the principal result of this section.

<u>Lemma 4.16</u>. Suppose that (4.21) holds. Then (4.13) is satisfied.

<u>Proof</u>. For purposes of a proof by contradiction, assume that (4.13) does not hold, i.e. (4.24) is satisfied for some $\bar{\varepsilon}>0$.
(i) Let $\varepsilon_w=\bar{\varepsilon}/2>0$ and choose $\varepsilon_\alpha=\varepsilon_\alpha(\varepsilon_w,C)$ and $\bar{N}=\bar{N}(\varepsilon_w,C)<+\infty$, $\bar{N}\ge1$, as specified in Lemma 4.12, where C is the constant defined in Lemma 4.13.
(ii) Let $\tilde{N}=10\bar{N}(\varepsilon_w,C)$. Using Lemma 4.15 and the fact that $\bar{a}>0$ by assumption, we can choose $\tilde{k}$ satisfying (4.36) and

$$\sum_{k=\tilde{k}}^{\tilde{k}+\tilde{N}} |x^{k+1}-x^k| \le \bar{a}/4. \tag{4.40}$$

(iii) Suppose that there exists a number $\hat{k}$ satisfying

$$\tilde{k} \le \hat{k} \le \tilde{k} + \tilde{N} - 2\bar{N}, \tag{4.41a}$$

$$r_a^k = 0 \quad \text{for all } k \in [\hat{k}, \hat{k}+\bar{N}]. \tag{4.41b}$$

Then (4.36b) - (4.36d), (4.41b), Lemma 4.11, Lemma 4.12 and our choice of ε_α and $\bar{N}$ imply $w^k < \varepsilon_w = \bar{\varepsilon}/2$ for some $k \in [\hat{k}, \hat{k}+\bar{N}]$, which contradicts (4.36a) and (4.41a). Consequently, we have shown that for any number $\hat{k}$ satisfying (4.41a) we have

$$r_a^k = 1 \quad \text{for} \quad \text{some } k \in [\hat{k}, \hat{k}+\bar{N}].$$

(iv) Letting $\hat{k} = \tilde{k}$ we obtain from part (iii) of the proof that $r^{k_1} = 1$ for some $k_1 \in [\tilde{k}, \tilde{k}+\bar{N}]$, and the rules of Step 5 and Step 6 yield

$$a^{k_1} \le \bar{a}/2. \tag{4.42}$$

(v) Since we always have $|y^{k+1} - x^{k+1}| \le \bar{a}/2$, at Step 5

$$a^{k+1} = \max\{a^k + |x^{k+1} - x^k|,\ |y^{k+1} - x^{k+1}|\} \le$$

$$\le \max\{a^k + |x^{k+1} - x^k|,\ \bar{a}/2\}. \tag{4.43}$$

The above estimate, (4.40) and (4.42) yield $a^k \le \frac{3}{4}\bar{a} < \bar{a}$ for all $k \in (k_1, \tilde{k}+\tilde{N}]$, hence for such k no resetting due to $a^k > \bar{a}$ occurs, i.e.

$$r_a^k = 0 \quad \text{for} \quad k = k_1 + 1, \ldots, k_1 + 1 + \bar{N}. \tag{4.44}$$

(vi) Since $\hat{k} = k_1 + 1$ satisfies (4.41a), from part (iii) of the proof we deduce a contradiction with (4.44). Therefore, (4.13) must hold. □

Combining Lemma 4.16 with Lemma 4.7, we obtain

Theorem 4.17. Each accumulation point of the sequence $\{x^k\}$ generated by Algorithm 3.1 is stationary for f.

In the convex case, the above result can be strengthened as follows.

Theorem. 4.18. If f is convex then Algorithm 3.1 constructs a minimizing sequence $\{x^k\}$: $f(x^k) \downarrow \inf\{f(x): x \in R^N\}$. Moreover, if f attains its minimum value, then $\{x^k\}$ converges to a minimum point of f.

Proof. One can check that Lemma 2.4.7 and Lemma 2.4.14 hold also for Algorithm 3.1 in the convex case. Then Theorem 4.17 and the proofs of Theorem 2.4.15 and Theorem 2.4.16 yield the desired result. □

The following result substantiates the stopping criterion of the method.

Corollary 4.19. If the level set $S=\{x \in R^N : f(x) \le f(x^1)\}$ is bounded and the final accuracy tolerance ε_s is positive, then Algorithm 3.1 terminates in a finite number of iterations.

Proof. If the assertion were false, then the infinite sequence $\{x^k\} \subset S$ would have an accumulation point, say $\bar{x}$. Then Lemma 4.16 would yield (4.14), and the algorithm should stop owing to $w^k \le \varepsilon_s$ for large k. □

Remark 4.20. It is worth observing that our convergence analysis does not use explicitly the semismoothness hypothesis (3.23). In fact, it remains valid when a procedure different from Line Search Procedure 3.2 is used for finding stepsizes satisfying the line search requirements (3.9)-(3.11) at each iteration of the method. For instance, one may use expansion in Line Search Procedure 3.2 by repeating its step (ii) with a doubled t if $t=t_L$ until t becomes t_U, i.e. by increasing the initial stepsize if this yields significantly smaller objective values; see (Mifflin, 1982). However, one should ensure boundedness of $\{t_L^k\}$, since it is necessary for the proof of Theorem 4.18 in the convex case. For this reason we had $t_L^k \le 1$ for all k in Algorithm 3.1, although this bound may be larger . Also one may delete the criterion (3.10a).

5. The Algorithm with Subgradient Selection

In this section we state in detail and analyze the method for nonconvex minimization that uses subgradient selection in the way described in Section 2.

Algorithm 5.1.

Step 0 (Initialization). Select the starting point $x^1 \in R^N$ and a final accuracy parameter $\varepsilon_s \geq 0$. Choose fixed positive line search parameters $m_L, m_R, \bar{a}$ and $\bar{t}$ satisfying $0 < m_L < m_R < 1$, and a distance measure parameter $\gamma > 0$ ($\gamma = 0$ if f is convex). Set $a^1 = 0$ and

$$y^1 = x^1,\ g^1 = g_f(y^1),\ f_1^1 = f(y^1),\ s_1^1 = 0,\ J^1 = \{1\}.$$

Set the counter $k=1$.

Step 1 (Direction finding). Find the solution (d^k, v^k) to the following k-th quadratic programming problem

$$\underset{(d,\hat{v}) \in R^{N+1}}{\text{minimize}} \quad \tfrac{1}{2}|d|^2 + \hat{v}, \tag{5.1}$$

$$\text{subject to} \quad -\alpha_j^k + \langle g^j, d \rangle \leq \hat{v},\ j \in J^k,$$

where

$$\alpha_j^k = \max\{|f(x^k) - f_j^k|,\ \gamma (s_j^k)^2\} \quad \text{for} \quad j \in J^k. \tag{5.2}$$

Find Lagrange multipliers λ_j^k, $j \in J^k$, of (5.1) and a set $\hat{J}^k$ satisfying

$$\hat{J}^k = \{j \in J^k : \lambda_j^k \neq 0\}, \tag{5.3a}$$

$$|\hat{J}^k| \leq N+1. \tag{5.3b}$$

Step 2 (Stopping criterion). Set

$$\hat{\alpha}_p^k = \sum_{j \in J^k} \lambda_j^k \alpha_j^k, \tag{5.4}$$

$$\hat{w}^k = \tfrac{1}{2}|d^k|^2 + \hat{\alpha}_p^k. \tag{5.5}$$

If $\hat{w}^k \leq \varepsilon_s$ then terminate. Otherwise, go to Step 3.

Step 3 (Line search). By a line search procedure as discussed below, find two stepsize t_L^k such that $0 \leq t_L^k \leq t_R^k$ and such that the two corresponding points defined by

$$x^{k+1}=x^k+t_L^k d^k \quad \text{and} \quad y^{k+1}=x^k+t_R^k d^k$$

satisfy $t_L^k \le 1$ and

$$f(x^{k+1}) \le f(x^k)+m_L t_L^k \hat{v}^k, \tag{5.6a}$$

$$t_R^k = t_L^k \qquad \text{if} \quad t_L^k \ge \bar{t}, \tag{5.6b}$$

$$-\alpha(x^{k+1},y^{k+1})+ < g_f(y^{k+1}),d^k > \ge m_R \hat{v}^k \qquad \text{if} \quad t_L^k < \bar{t}, \tag{5.6c}$$

$$|y^{k+1}-x^{k+1}| \le \bar{a}/2 . \tag{5.6d}$$

Step 4 (Subgradient updating). Set

$$J^{k+1} = \hat{J}^k \cup \{k+1\}. \tag{5.7}$$

Set $g^{k+1}=g_f(y^{k+1})$ and

$$f_{k+1}^{k+1} = f(y^{k+1})+ < g^{k+1},x^{k+1}-y^{k+1} > , \tag{5.8a}$$

$$f_j^{k+1} = f_j^k+ < g^j,x^{k+1}-x^k > \quad \text{for} \quad j\in \hat{J}^k, \tag{5.8b}$$

$$s_{k+1}^{k+1} = |y^{k+1}-x^{k+1}|, \tag{5.8c}$$

$$s_j^{k+1} = s_j^k+|x^{k+1}-x^k| \quad \text{for} \quad j\in \hat{J}^k. \tag{5.8d}$$

Step 5 (Distance resetting test). Set

$$a^{k+1}=\max\{s_j^{k+1}: j\in J^{k+1}\}. \tag{5.9}$$

If $a^{k+1} \le \bar{a}$ then set $r_a^{k+1}=0$ and go to Step 7. Otherwise, set $r_a^{k+1}=1$ and go to Step 6.

Step 6 (Distance resetting). Keep deleting from J^{k+1} the smallest indices until the reset value of a^{k+1} satisfies

$$a^{k+1} = \max\{s_j^{k+1}: j\in J^{k+1}\} \le \bar{a}/2. \tag{5.10}$$

Step 7. Increase k by 1 and go to Step 1.

We shall now comment on relations between the above method and Algorithm 3.1.

By Lemma 2.2.1, the k-th subproblem dual of (5.1) is to find values of the multipliers λ_j^k, $j \in J^k$, to

$$\begin{aligned} &\underset{\lambda}{\text{minimize}}\ \tfrac{1}{2}\Big|\sum_{j\in J^k}\lambda_j g^j\Big|^2+\sum_{j\in J^k}\lambda_j\alpha_j^k, \\ &\text{subject to}\ \ \lambda_j\geq 0,\ j\in J^k,\ \ \sum_{j\in J^k}\lambda_j=1. \end{aligned} \tag{5.11}$$

Any solution of (5.11) is a Lagrange multiplier vector for (5.1) and it yields the unique solution $(d^k,\hat{v}^k)$ of (5.1) as follows

$$d^k = -p^k, \tag{5.12}$$

$$\hat{v}^k = -\{|p^k|^2+\sum_{j\in J^k}\lambda_j^k\alpha_j^k\}, \tag{5.13}$$

where

$$p^k = \sum_{j\in J^k}\lambda_j^k g^j. \tag{5.14}$$

Moreover, any Lagrange multipliers of (5.1) also solve (5.11). In particular, we have

$$\lambda_j^k \geq 0 \quad \text{for} \quad j\in J^k, \quad \sum_{j\in J^k}\lambda_j^k = 1. \tag{5.15}$$

Thus we see that, as far as the search direction finding is concerned, the above relations (5.11)-(5.15) can be obtained from the corresponding relations developed for Algorithm 3.1 in Section 3 by setting

$$\lambda_p^k = 0 \quad \text{for all } k. \tag{5.16}$$

This corresponds to deleting the last constraint of the k-th primal search direction finding subproblem (3.1), and thus reducing subproblem (3.1) to subproblem (5.1).

We refer the reader to Remark 2.5.2 for a discussion of the possible ways of finding the k-th Lagrange multipliers satisfying the requirement (5.3).

The stopping criterion in Step 2 can be interpreted similarly to the termination rule of Algorithm 3.1. A slight difference between the two stopping criteria arises from the fact that the values of $\hat{\alpha}_p^k$ and $\hat{w}^k$ can be larger than the values of the variables $\tilde{\alpha}_p^k$ and w^k defined by

$$(\tilde{f}_p^k,\tilde{s}_p^k) = \sum_{j\in J^k}\lambda_j^k(f_j^k,s_j^k), \tag{5.17}$$

$$\tilde{\alpha}_p^k = \max\{|f(x^k)-\tilde{f}_p^k|,\gamma(\tilde{s}_p^k)\}, \tag{5.18}$$

$$w^k = \tfrac{1}{2}|p^k|^2+\tilde{\alpha}_p^k. \tag{5.19}$$

To see this, note that, by (5.5) and (5.12), we always have

$$\hat{w}^k = \tfrac{1}{2}|p^k|^2+\hat{\alpha}_p^k, \tag{5.20}$$

hence one can use the proof of Lemma 4.8 and the convention (5.16) to show that

$$\tilde{\alpha}_p^k \le \hat{\alpha}_p^k, \tag{5.21}$$

$$w^k \le \hat{w}^k, \tag{5.22}$$

and that the above variables are nonnegative. Thus both w^k and $\hat{w}^k$ can be regarded as stationarity measures of x^k; see Section 3. We also have the optimality estimate (3.21) upon termination if f is convex.

The line search rules (5.6) differ from the rules (3.9)-(3.11) inasmuch as the value of $\hat{v}^k$ can be lower that the value of the variable

$$v^k = -\{|p^k|^2+\tilde{\alpha}_p^k\}, \tag{5.23}$$

which corresponds to the construction used in Algorithm 3.1. By (5.13), (5.4),(5.21) and (5.23), we have

$$\hat{v}^k = -\{|p^k|^2+\hat{\alpha}_p^k\}, \tag{5.24}$$

$$\hat{v}^k \le v^k. \tag{5.25}$$

Also $\hat{v}^k \le -\hat{w}^k$ from (5.19) and (5.24), so the line search is always entered with negative $\hat{v}^k$. As observed in Section 2,

$$\hat{v}^k = \hat{f}_s^k(x^k+d^k)-f(x^k)$$

is an approximation to the derivative of f at x^k in the direction d^k. Thus, except for the difference in the values of v^k and $\hat{v}^k$, the line search criteria (5.6) may be interpreted esentially as in Section 3. We may add that for implementing Step 3 of Algorithm 5.1 one can use Line Search Procedure 3.2 with v^k replaced by $\hat{v}^k$.

In algorithm 5.1 the locality radius a^{k+1} is calculated directly via (5.9) and (5.10), instead of using the recursive formulae (3.7) and (3.15). We also observe that the subgradient deletion rules ensure that

$$k \in J^k \quad \text{for all } k, \tag{5.26}$$

as in Section 3, i.e. the latest subgradient is always used for the current search direction finding.

Remark 5.2. The requirement of Algorithm 5.1 that at any iteration at most N+1 past subgradients should be retained for the next search direction finding can be modified as follows. Let $M_g \geq N+2$ denote the maximum number of subgradients that the algorithm may store. Then one may choose Lagrange multipliers λ_j^k and a set $\hat{J}^k$ subject only to the following requirement

$$\hat{J}^k \supset \{j \in J^k : \lambda_j^k \neq 0\} \quad \text{and} \quad |\hat{J}^k| \leq M_g - 1$$

and set $J^{k+1} = \hat{J}^k \cup \{k+1\}$, for all k. Such a choice is always possible if $M_g - 1 \geq N+1$, cf. Lemma 2.2.1 and Remark 2.5.2. It will be seen that such modifications do not impair the subsequent convergence results.

We now pass to convergence analysis. Global convergence of Algorithm 5.1 can be established by modifying the results of Section 4. To this end one may proceed as in Section 2.5, where the convergence of the method with subgradient selection was deduced from the results on the convergence of the corresponding method with subgradient aggregation. Therefore we shall give only an outline of required results.

In the proof of Lemma 4.1 for Algorithm 5.1, observe that no induction is needed, since

$$\hat{J}_p^k = J^k \quad \text{for all } k$$

by (4.2) and (5.16), hence we may let $\tilde{\lambda}_j^k = \lambda_j^k$ for all $j \in J^k$ to obtain (4.3) from (5.14), (5.17) and (5.15). Also relations (4.4) and (4.5) follow immediately from (5.9) and (5.10).

The proofs of Lemma 4.2 through Lemma 4.8 require no modifications if one uses the simplifying convention (5.16). Also it can be checked that the assertions of Lemma 4.7 and Lemma 4.8 remain valid upon replacing $\tilde{\alpha}_p^k$ and w^k by $\hat{\alpha}_p^k$ and $\hat{w}^k$; see (5.21)-(5.22).

In the formulation of Lemma 4.9, substitute relation (4.23) by the following

$$t_L^k \hat{v}^k \longrightarrow 0 \quad \text{as} \quad k \to \infty,$$

while in the proof one can refer to (5.6a) instead of (3.9), replace v^k by $\hat{v}^k$ and use (5.25). Of course, Corollary 4.10 remains valid, even if one replaces w^k by $\hat{w}^k$ in (4.13), (4.14) and (4.24).

Lemma 4.11 is replaced by

Lemma 5.3. Suppose that $t_L^{k-1} < \bar{t}$ and $r_a^k=0$ for some $k > 1$. Then

$$w^k \le \hat{w}^k \le \phi_C(\hat{w}^{k-1})+\Delta_\alpha^k, \tag{5.27}$$

where ϕ_C is defined by (4.26) for any C satisfying

$$\max\{|p^{k-1}|,|g^k|,\hat{\alpha}_p^{k-1},1\} \le C, \tag{5.28}$$

and

$$\Delta_\alpha^k = \Big| \sum_{j\in \hat{J}^{k-1}} \lambda_j^{k-1}\alpha_j^k - \sum_{j\in \hat{J}^{k-1}} \lambda_j^{k-1}\alpha_j^{k-1}\Big| \tag{5.29}$$

Proof. (i) If $k > 1$ and $t_L^{k-1} < \bar{t}$ then the line search rule (5.6c) yields $-\alpha(x^k,y^k)+\langle g_f(y^k),d^{k-1}\rangle \ge m_R\hat{v}^{k-1}$, so

$$-\alpha_k^k+\langle g^k,d^{k-1}\rangle \ge m_R\hat{v}^{k-1}. \tag{5.30}$$

(ii) Define the multipliers

$$\lambda_k(\nu) = \nu, \quad \lambda_j(\nu)=(1-\nu)\lambda_j^{k-1} \quad \text{for } j\in \hat{J}^{k-1} \tag{5.31}$$

for each $\nu\in [0,1]$. By (5.14), (5.15), (5.4) and (5.3a), we have

$$p^{k-1}= \sum_{j\in \hat{J}^{k-1}}\lambda_j^{k-1}g^j, \tag{5.32a}$$

$$\lambda_j^{k-1} \ge 0 \quad \text{for } j\in \hat{J}^{k-1}, \quad \sum_{j\in \hat{J}^{k-1}}\lambda_j^{k-1}=1, \tag{5.32b}$$

$$\hat{\alpha}_p^{k-1}= \sum_{j\in \hat{J}^{k-1}}\lambda_j^{k-1}\alpha_j^{k-1}. \tag{5.32c}$$

Since $r_a^k=0$ by assumption, we have $J^k=\hat{J}^{k-1}\cup\{k\}$, so (5.32b) implies that the multipliers defined by (5.31) satisfy the constraints of the k-th dual subproblem (5.11) for each $\nu\in[0,1]$. Noting that $\hat{w}^k$ is the optimal value of (5.11)(see (5.14), (5.4) and (5.20)), we deduce from (5.31) and (5.32a) that

$$\hat{w}^k \le \tfrac{1}{2}|(1-\nu)p^{k-1}+\nu g^k|^2+(1-\nu)\sum_{j\in \hat{J}^{k-1}}\lambda_j^{k-1}\alpha_j^k+\nu\alpha_k^k$$

for all $\nu\in[0,1]$, so (5.32c) and (5.29) yield

$$\hat{w}^k \le \tfrac{1}{2}|(1-\nu)p^{k-1}+\nu g^k|^2+(1-\nu)\hat{\alpha}_p^{k-1}+\nu\alpha_k^k+\Delta_\alpha^k \tag{5.33}$$

for all $\nu \in [0,1]$. Using Lemma 2.4.10 and relations (5.12), (5.20), (5.24), (5.30) and (5.28), we obtain

$$\min\{\tfrac{1}{2}|(1-\nu)p^{k-1}+\nu g^k|^2+(1-\nu)\hat{\alpha}_p^{k-1}+\nu\alpha_k^k : \nu \in [0,1]\} \le \phi_C(\hat{w}^{k-1}). \tag{5.34}$$

Combining (5.33), (5.34) and (5.22) we obtain (5.27), as required. □

We conclude from the above lemma that after a null step (or a short serious step) of Algorithm 5.1 one can expect a significant decrease of the stationarity measure $\hat{w}^k$, while in Algorithm 3.1 the same observation applies to the stationarity measure w^k, cf. (5.27) and (4.25). The rate of decrease of $\hat{w}^k$ is established by Lemma 4.12 (if $\Delta_\alpha^k \le \varepsilon_\alpha$).

As far as Lemma 4.13 is concerned, substitute $\tilde{\alpha}_p^k$ by $\hat{\alpha}_p^k$ in relations (4.30) and (4.32), and use the fact that $\hat{w}^k$ is the optimal value of the dual subproblem (5.11).

Recall that Lemma 4.14 was instrumental only in the proof of Lemma 4.15. Therefore it suffices now to consider the following substitute of Lemma 4.15.

Lemma 5.4. Suppose that (4.21) and (4.24) hold. Then the assertions of Lemma 4.15 are valid for Algorithm 5.1 if one replaces (4.36b)-(4.36c) by

$$\max\{|p^{k-1}|,|g^k|,\hat{\alpha}_p^{k-1},1\} \le C \quad \text{for} \quad \tilde{k} \le k \le \tilde{k}+\tilde{N}, \tag{5.35a}$$

$$\Delta_\alpha^k \le \varepsilon_\alpha \quad \text{for} \quad \tilde{k} \le k \le \tilde{k}+\tilde{N}. \tag{5.35b}$$

Proof. It is easily verified that we only need to prove the assertion concerning (5.35b). To this end, observe that

$$\Delta_\alpha^k = \Big|\sum_{j\in\hat{J}^{k-1}}\lambda_j^{k-1}\alpha_j^k - \sum_{j\in\hat{J}^{k-1}}\lambda_j^{k-1}\alpha_j^{k-1}\Big| = \Big|\sum_{j\in\hat{J}^{k-1}}\lambda_j^{k-1}(\alpha_j^k-\alpha_j^{k-1})\Big| \le$$

$$\le \max\{|\alpha_j^k-\alpha_j^{k-1}| : j\in J^{k-1}\}\sum_{j\in\hat{J}^{k-1}}\lambda_j^{k-1},$$

i.e.

$$\Delta_\alpha^{k+1} \le \max\{|\alpha_j^{k+1}-\alpha_j^k| : j \in J^k\} \quad \text{for all } k, \tag{5.36}$$

from (5.15) and (5.3a). Next, for all k and $j \in J^k$ we have

$$||f(x^{k+1})-f_j^{k+1}|-|f(x^k)-f_j^k|| \le |f(x^{k+1})-f_j^{k+1}-f(x^k)+f_j^k| \le$$

$$\le |f(x^{k+1})-f(x^k)| + |< g^j, x^{k+1}-x^k >| \le$$

$$\le |f(x^{k+1})-f(x^k)| + |g_f(y^j)|\,|x^{k+1}-x^k| \tag{5.37a}$$

from (5.8b), while (5.8d) yield

$$\gamma(s_j^k)^2 \le \gamma(s_j^{k+1})^2 = \gamma(s_j^k+|x^{k+1}-x^k|)^2 \le$$

$$\le \gamma(s_j^k)^2+2\gamma s_j^k|x^{k+1}-x^k|+\gamma|x^{k+1}-x^k|^2 \le$$

$$\le \gamma(s_j^k)^2+|x^{k+1}-x^k|(2\gamma\bar{a}+\gamma|x^{k+1}-x^k|) \tag{5.37b}$$

since $s_j^k \le a^k \le \bar{a}$. From (5.2) and (5.37)

$$|\alpha_j^{k+1}-\alpha_j^k| \le |f(x^{k+1})-f(x^k)|+|x^{k+1}-x^k|(|g_f(y^j)|+2\gamma\bar{a}+\gamma|x^{k+1}-x^k|) \tag{5.38}$$

for all $j \in J^k$ and $k \ge 1$. Since $|x^k-y^j| \le s_j^k \le a^k \le \bar{a}$ for all k and $j \in J^k$, from Lemma 4.13 we obtain that

$$\max\{|g_f(y^j)|: j \in J^k\} \le C \quad \text{if} \quad |\bar{x}-x^k| \le \bar{a}, \text{ for all } k,$$

and that $|f(x^{k+1})-f(x^k)| \longrightarrow 0$ as $k\to\infty$. Therefore one may complete the proof by using (5.36), (5.38) and the first assertion of Lemma 4.15. □

In the proof of Lemma 4.16, substitute (4.43) with the following relation

$$a^{k+1}=\max\{s_j^{k+1}: j \in J^{k+1}\}=\max\{s_j^{k+1}: j\in \hat{J}^k \cup \{k+1\}\}=$$

$$=\max\left[\max\{s_j^k+|x^{k+1}-x^k|: j \in \hat{J}^k\}, |y^{k+1}-x^{k+1}|\right\} \le$$

$$\le \max\{a^k+|x^{k+1}-x^k|, \bar{a}/2\},$$

which follows from the fact that $\max\{s_j^k: j\in \hat{J}^k\} \le \max\{s_j^k: j \in J^k\}=a^k$.

In this way we arrive at the following result.

Theorem 5.5. Every accumulation point of a sequence $\{x^k\}$ generated by Algorithm 5.1 is stationary for f.

Using the results of Section 2.5 one can check that in the convex case the convergence properties of Algorithm 5.1 may be expressed in the form of Theorem 4.18. Moreover, Corollary 4.19 is valid for Algorithm 5.1. This follows from the fact that the proof of Lemma 4.16 shows that one may substitute in the lemma relation (4.14) by a modification of

(4.14) in which $\hat{w}^k$ replaces w^k.

To sum up, we have extended all the convergence results of Section 4 to Algorithm 5.1.

6. Modifications of the Methods

In this section we describe several modifications of the methods discussed so far and analyze their convergence within the framework established in the preceding sections.

First, we want to demonstrate global convergence of versions of the methods for convex minimization presented in Chapter 2 that use the general line search criteria (2.7.2), which allow for arbitrarily small serious stepsizes. To this end, suppose that f is convex and consider the following method which is obtained from Algorithm 2.3.1 by modifying its line search rules.

Algorithm 6.1.

Step 0 through Step 2. These are the corresponding steps of Algorithm 2.3.1. Step 3 (Line search). Find two stepsizes t_L^k and t_R^k such that $0 \le t_L^k \le t_R^k$ and such that the two corresponding points defined by

$$x^{k+1}=x^k+t_L^k d^k \quad \text{and} \quad y^{k+1}=x^k+t_R^k d^k$$

satisfy

$$f(x^{k+1}) \le f(x^k)+m_L t_L^k v^k, \tag{6.1a}$$

$$-\alpha(x^{k+1},y^{k+1})+\langle g_f(y^{k+1}),d^k\rangle \ge m_R v^k \quad \text{if} \quad t_L^k < \bar{t}, \tag{6.1b}$$

$$|y^{k+1}-x^{k+1}| \le \bar{a}, \tag{6.1c}$$

$$t_R^k \le \tilde{t}. \tag{6.1d}$$

Step 4. The same as in Algorithm 2.3.1.

We suppose that in (6.1) $\bar{a}, m_L, m_R, \bar{t}$ and $\tilde{t}$ are fixed positive parameters satisfying $0 < m_L < m_R < 1$ and $\bar{t} \le 1 \le \tilde{t}$, and

$$\alpha(x,y)=f(x)-f(y)-\langle g_f(y),x-y\rangle .$$

It is easy to verify that Line Search Procedure 3.2 can be employed for finding stepsizes t_L^k and t_R^k satisfying (6.1).

One may observe that, except for Step 3, Algorithm 6.1 is obtained from Algorithm 3.1 by deleting in the latter method Step 5 and Step 6, and setting $\gamma=0$ and $r_a^k=0$ for all k. In other words, the constructions involving distance measures s_j^k and the resetting strategy are not necessary in the convex case.

To show that Algorithm 6.1 is globally convergent in the sense of Theorem 4.18 and Corollary 4.19, one may proceed as follows. In view of the results of Section 2.4, it suffices to establish Lemma 4.16 for Algorithm 6.1. To this end, observe that Lemma 4.9 through Lemma 4.15 remain true for Algorithm 6.1, since we have $\gamma=0$, $r_a^k=0$ and $|y^{k+1}-x^{k+1}|\leq \bar{a}$ for all k. Therefore we may use parts (i)-(iii) of the proof of Lemma 4.16, deducing in part (iii) a contradiction, since we now have $r_a^k=0$ for all k. Therefore Lemma 4.16 and the subsequent results hold for Algorithm 6.1.

Let us now consider the method with subgradient selection for convex minimization which is obtained by using the line search criteria (6.1) in Algorithm 2.5.1. This method can be also derived from Algorithm 5.1 by replacing (5.6) with (6.1) and deleting Step 5 and Step 6. To establish global convergence of the method, one may reason as above, but this time using the results of Section 2.5 and Section 5.

To sum up, we have shown that one may employ the general line search criteria (6.1) in the methods for convex minimization from Chapter 2 without impairing the global convergence results.

Next, consider the version of the subgradient selection method for nonconvex minimization that uses the measures $\alpha(x^k,y^j)$ instead of α_j^k, see Remark 2.1. The method results from replacing in Algorithm 5.1 the variables α_j^k by $\alpha(x^k,y^j)$, and calculating a^{k+1} by

$$a^{k+1}=\max\{|x^{k+1}-y^j|:j\in J^{k+1}\} \tag{6.2}$$

instead of (5.9) and (5.10). For verifying that the global convergence results of Section 5 cover this version of the method, modify (5.17) as follows

$$(\tilde{f}_p^k,\tilde{s}_p^k)=\sum_{j\in J^k}\lambda_j^k(f_j^k,|x^k-y^j|) \tag{6.3}$$

and use the fact that we always have

$$|x^{k+1}-y^j|\leq|x^k-y^j|+|x^{k+1}-x^k|. \tag{6.4}$$

In effect, this version can be analyzed by assuming that the variables s_j^k are substituted by $|x^k-y^j|$ everywhere in Algorithm 5.1 (including relation (5.2) defining α_j^k).

For the sake of completeness of the theory, let us now consider a method that uses all the past subgradients for search direction finding at each iteration. Of course, such a strategy of total subgradient accumulation is only of theoretical interest, since it requires infinite storage and an increasing amount of work per iteration. Our method with subgradient accumulation is obtained from Algorithm 5.1 by replacing everywhere α_j^k by $\alpha(x^k,y^j)$, deleting Step 5 and Step 6, and setting

$$J^k = \{1,\dots,k\} \quad \text{for all } k.$$

Thus we always have

$$J^{k+1} = J^k \cup \{k+1\},$$

hence there is need for selecting Lagrange multipliers of (5.1) to meet the requirement (5.3). As far as our techniques of convergence analysis are concerned, the principal difference between this method and the corresponding algorithm with subgradient selection described above consists in the fact that the locality radii defined by

$$a^k = \max\{|x^k-y^j|: j=1,\dots,k\}$$

need not be bounded, since the method has no subgradient deletion rules. In this context we recall that the proofs in Section 4 and Section 5 rely heavily on the local boundedness arguments, which depend on the boundedness of a^k. However, if one makes an additional assumption that there exists a constant $\tilde{a}$ such that $a^k \le \tilde{a}$ for all k, then one can establish global convergence of the method with subgradient accumulation by using (6.3) and (6.4) as above. For this reason, consider the following assumption on f and the starting point x^1:

$$\text{the set } S=\{x \in R^N: f(x) \le f(x^1)\} \text{ is bounded.} \tag{6.5}$$

Then there exists $\tilde{a} > \bar{a}$ such that

$$\sup\{|x-\tilde{x}|: x \in S,\ \tilde{x} \in S\} \le \tilde{a}/2,$$

so that we always have

$$|x^k-y^j| \le |x^k-x^j|+|x^j-y^j| \le \tilde{a}/2+\bar{a}/2 \le \tilde{a},$$

since $\{x^k\} \subset S$ by the monotonicity of $\{f(x^k)\}$, while $|x^j-y^j| \le \bar{a}/2$ owing to the line search requirement (5.6d). It follows that $a^k \le \tilde{a}$ for all k. In effect, we have shown that the above-described method with subgradient accumulation is convergent in the sense of Theorem

5.5, Theorem 4.18 and Corollary 4.19 under the additional assumption (6.5).

We shall now present a simple modification of the methods of this chapter that we have found useful in calculations. This modification, which amounts to calculating and using more subgradients for search direction finding, can be motivated as follows. The methods described so far evaluate subgradients $g^j = g_f(y^j)$ at trial points y^j which can, in general, be different from the points x^j. This is due to the use of the two-point line search for detecting discontinuities in the gradient of f. However, the lack of subgradient information associated with the points x^j may unnecessarily slow down convergence. For instance, consider the k-th iteration of Algorithm 3.1. Recall that for line searches the variable v^k is regarded as an approximation to the directional derivative of f at x^k in the direction d^k. For this interpretation to be valid, we should try to achieve the relation

$$\max\{ \langle \tilde{g}, d^k \rangle : \tilde{g} \in \partial f(x^k) \} \le v^k, \tag{6.6}$$

since the left side of (6.6) is equal to that directional derivative (see Section 1.2). For instance, if f is smooth at x^k, so that the gradient $\nabla f(x^k)$ exists and $g_f(x^k) = \nabla f(x^k)$, then we should have

$$\langle g_f(x^k), d^k \rangle \le v^k, \tag{6.7}$$

which would yield $f'(x^k; d^k) = \langle \nabla f(x^k), d^k \rangle \le v^k$. But if $y^k \ne x^k$ and $g_f(x^k)$ is not evaluated then we cannot even verify (6.7), let alone ensuring that it holds. A simple way out of this difficulty is to calculate $g_f(x^k)$ and use it for the k-th search direction finding by appending the following additional constraint

$$\langle g_f(x^k), d \rangle \le \hat{v} \tag{6.8}$$

to the k-th primal subproblem (3.1). Then we shall have

$$\langle g_f(x^k), d^k \rangle \le \hat{v}^k, \tag{6.9}$$

which will yield (6.7), since $\hat{v}^k \le v^k$. Noting that (6.8) can be formulated as

$$-\alpha(x^k, x^k) + \langle g_f(x^k), d \rangle \le \hat{v}, \tag{6.10}$$

we conclude that $g_f(x^k)$ is treated here as any other subgradient $g^j = g_f(y^j)$.

Once the subgradient $g_f(x^j)$ is evaluated, it can be employed for search direction finding at x^k for $k > j$. Thus, for ease of subsequent notation, let

$$y^{-j}=x^j \quad \text{for} \quad j=1,2,\ldots,k \quad \text{and all } k, \tag{6.11}$$

and

$$g^j=g_f(y^j) \quad \text{for} \quad j=\pm 1,\pm 2,\ldots, \tag{6.12}$$

i.e. $g^j=g_f(x^{|j|})$ if $j<0$, and $g^j=g_f(y^j)$ if $j>0$, as before. Then at the k-th iteration the past points y^j, $j=\pm 1,\pm 2,\ldots,\pm k$, are characterized by the linearizations

$$f_j(x)=f(y^j)+\langle g^j,x-y^j\rangle=f_j^k+\langle g^j,x-x^k\rangle \quad \text{for all } x$$

and the following upper estimates of $|x^k-y^j|$

$$s_j^k=|x^{|j|}-y^j|+\sum_{i=|j|}^{k-1}|x^{i+1}-x^i| \quad \text{for} \quad |j|<k,\ s_{\pm k}^k=|x^k-y^{\pm k}|, \tag{6.13}$$

from which we can calculate the subgradient locality measures

$$\alpha_j^k=\max\{|f(x^k)-f_j^k|,\ \gamma(s_j^k)\} \quad \text{for} \quad 1\le|j|\le k.$$

To sum up, the points x^j are treated exactly as the points y^j, $j\ge 1$, used before; see (2.11) and (2.14). Therefore we may now choose sets

$$J^k\subset\{j:\ j=\pm 1,\ldots,\pm k\} \quad \text{for all } k.$$

Noting that $\alpha_{-k}^k=0$, since $f_{-k}^k=f(x^k)$ and $s_{-k}^k=0$, we see that the k-th subproblem (3.1) will have (6.8) among its constraints if

$$-k\in J^k.$$

Together with the previously motivated condition $k\in J^k$, this leads to the requirement

$$\pm k\in J^k \quad \text{for all } k, \tag{6.14}$$

which can be met by selecting in Step 4 of Algorithm 3.1 a (possibly empty) set $\hat{J}^k\subset J^k$ and setting

$$J^{k+1}=\hat{J}^k\cup\{k+1,-(k+1)\} \tag{6.15}$$

with $g^{-(k+1)}=g_f(x^{k+1})$ and

$$f_{-(k+1)}^{k+1}=f(x^{k+1}), \tag{6.16a}$$

$$s_{-(k+1)}^{k+1}=0, \tag{6.16b}$$

instead of using (3.12).

Of course, Algorithm 5.1 may also be modified in this way. To this end, replace the requirement (5.7) by (6.15), and use (6.16) in addition to (5.8). Moreover, similar modifications may be introduced in all the versions of the methods considered in this section.

It is elementary to check that the preceding results on the global convergence of the methods are not influenced by the additional use of subgradients $g_f(x^j)$. This exercise is left to the reader, who may, for instance, define additional multipliers $\tilde{\lambda}_j^k$ involved in (4.3) for j belonging to the set

$$\hat{J}_p^k = J^{k_p(k)} \cup \{j: k_p(k) < |j| \le k\},$$

etc.

To sum up, the above described modification has no impact on the theory of our methods, but may increase their computational efficiency by decreasing the number of null steps (or short serious steps), thus leading to faster convergence. Of course, this advantage should be weighed againts the work involved in evaluating additional subgradients. We may add that in many applications it is relatively easy to calculate $g_f(x^{k+1})$ once f has been evaluated at x^{k+1}.

It is worthwile to observe that, although the notation of Mifflin's (1982) line search rules (2.9) may be misleading, in fact his algorithm calculates, in addition to $g^j=g_f(y^j)$, all the subgradients $g_f(x^j)$, and uses many of them for search direction finding at the k-th iteration. First, let us note that $g^{k+1}=g_f(x^{k+1})$ at some iterations. This is the case when

$$< g_f(x^k), d^k > \ge m_R \hat{v}^k, \tag{6.17}$$

since then the line search rule (2.9c) yields $y^{k+1}=x^{k+1}=x^k$ and $g^{k+1}= g_f(y^{k+1})=g_f(x^{k+1})$. The test (6.17) is related to the desirable relation (6.9), since if (6.17) holds then the fact that $m_R \in (0,1)$ and $\hat{v}^k < 0$ at line searches yield

$$< g_f(x^k), d^k > \ge m_R \hat{v}^k > \hat{v}^k,$$

so that (6.9) cannot be satisfied. Of course, if $y^{k+1}=x^{k+1}=x^k$ then at the next iteration one has

$$< g_f(x^{k+1}), d^k > = -\alpha(x^{k+1}, y^{k+1}) + < g^{k+1}, d^{k+1} > \le \hat{v}^{k+1},$$

since $(d^{k+1}, \hat{v}^{k+1})$ solves the $(k+1)$-st subproblem of the form (2.34) with $k+1 \in J^{k+1}=\{1,\dots,k+1\}$. Thus relation (6.9) is satisfied for k increased by 1. We may add that the line search rule (2.9c) plays a crucial role in Mifflin's (1982) convergence analysis, who proved that the method has <u>at least one</u> stationary accumulation point if the sequences $\{x^k\}$ and $\{y^j\}$ are bounded (or if $\{x^k\}$ is bounded and the line rule (3.11) is used together with (2.9)). One may remark that under such assumptions <u>every</u> accumulation point of the Mifflin algorithm is stationary. This claim can be easily verified; see the preceding discussion of the method with subgradient accumulation and the following remark.

<u>Remark 6.2</u>. It is worth noting that the definitions of subgradient locality measures used in the methods described so far are arbitrary to a large extent. For instance, in Algorithm 3.1 one may use, instead of (3.2), (3.3) and (3.5), the following definitions

$$\alpha_j^k = \max\{f(x^k)-f_j^k,\gamma(s_j^k)^2\}, \tag{6.18a}$$

$$\alpha_p^k = \max\{f(x^k)-f_p^k,\gamma(s_p^k)^2\}, \tag{6.18b}$$

$$\tilde{\alpha}_p^k = \max\{f(x^k)-\tilde{f}_p^k,\gamma(\tilde{s}_p^k)^2\}, \tag{6.18c}$$

which correspond to the following modified definition of α

$$\alpha(x,y)= \max\{f(x)-\overline{f}(x;y),\ \gamma|x-y|^2\} \quad \text{for all } x,y, \tag{6.18d}$$

which is equivalent to Mifflin's definition (2.3). As before, in (6.18) γ is a positive parameter, which can be set to zero if f is convex. The above modified definitions of subgradient locality measures may be used in all methods described so far. It is straightforward to check that such modifications do not impair the preceding convergence results. One may also use in all of the above-described methods (excepting Algoritm 6.1) the following simplified version of (6.18)

$$\begin{aligned} \alpha_j^k &= \gamma(s_j^k)^2, \\ \alpha_p^k &= \gamma(s_p^k)^2, \\ \tilde{\alpha}_p^k &= \gamma(\tilde{s}_p^k)^2, \\ \alpha(x,y)&=\gamma|x-y|^2 \quad \text{for all } x \text{ and } y, \end{aligned} \tag{6.19}$$

where this time γ is fixed and positive even if f convex. One can easily verify that global convergence results of the form of Theorem 4.17 remain valid for definitions (6.19). However, these results can no longer be strengthened in the convex case to the form of Theorem 4.18, because subgradient locality measures (6.19) neglect linearization errors. For this reason, definitions (6.19) are inferior to the two definitions discussed above.

CHAPTER 4

Methods with Subgradient Deletion Rules for Unconstrained Nonconvex Minimization

1. Introduction

This chapter is a continuation of Chapter 3, in which we began discussing methods for minimizing a locally Lipschitzian function $f:R^N \to R$, which is not necessarily convex or differentiable.

We shall present a class of readily implementable methods of descent, extending the algorithms of Chapter 2 to the nonconvex case in a way different from the one followed in Chapter 3. In effect, the methods generalize Pshenichny's method of linearizations for minimax problems (see Section 2.2). Simplified versions of the methods may be regarded as extensions of the Wolfe (1975) conjugate subgradient method to the nonconvex case, and as modifications of algorithms due to Mifflin (1977b) and Polak, Mayne and Wardi (1983).

The methods for convex minimization described in Chapter 2 use linearizations of f obtained by evaluating f and its subgradient at certain trial points. At each iteration, the pointwise maximum of several past linearizations defines the current polyhedral approximation to f that is used for search direction finding. A fundamental property of the convex case is that the current polyhedral approximation depends on the past trial points only implicity via the corresponding linearizations, so that any two trial points are equivalent if they yield the same linearization, i.e. the same supporting hyperplane to the epigraph of f.

Since in the nonconvex case each past linearization is not necessarily a lower approximation to f, in Chapter 3 we modified the methods of Chapter 2 by using Mifflin's (1982) ideas for ensuring that the search direction generated at any point is based mainly on local past subgradient information, i.e. information collected at trial points close to the current point. In effect, we considered polyhedral approximations defined by past subgradients and their locality measures, which depend on so-called distance measures that majorize distances from the corresponding trial points to the current point. Thus these distance measures directly influence each search direction finding subproblem.

In this chapter we adopt a different approach to extending the methods of Chapter 2 to the nonconvex case. Our approach is motivated by the observation that the polyhedral approximations to f used in Chapter 2 remain useful even if f is nonconvex, provided that they are based on only local past subgradient information. Thus we retain definitions of polyhedral approximations in terms of past subgradients and the corresponding linearization errors. Although in this case, in contrast with the methods of Chapter 3, distance measures are not directly involved in each search direction finding subproblem, they are used for deciding which of the past subgradients should be retained for the next search direction finding. Thus distance measures are employed only in <u>subgradient deletion rules</u> for localizing the past subgradient information by resets. This locality resetting is associated here with estimating the degree of stationarity of the current point, while in Chapter 3 resets were used only for ensuring locally uniform boundedness of the subgradients stored by the methods.

We shall show that each of our methods with subgradient deletion rules is globally convergent in the sense that all its accumulation points are stationary. In the convex each method generates a minimizing sequence of points, which converges to a solution if f attains its infimum. These convergence results, as well as the techniques for deriving them, are similar to those of Chapter 3.

To sum up, the methods with subgradient deletion rules may be regarded as alternatives of the methods with subgradient locality measures. Theoretical results on their global convergence are the same, but they are based on constructions with fundamentally different motivations. More importantly, our preliminary numerical experience indicates that those two classes of methods perform differently in practice. Therefore, one has to address the question of relative advantages and drawbacks of both classes of methods in specific applications. We leave this important question open for future theoretical and experimental investigations.

It turns out that the framework for convergence analysis presented here can easily accomodate methods that neglect linearization errors (see Section 1.3 for a discussion of this class of methods). For this reason, we shall establish global convergence results for such methods, which subsume those of Mifflin (1977b) and Polak, Mayne and Wardi (1983).

In Section 2 we derive the methods. Section 3 contains a detailed description of the method with subgradient aggregation. Its convergence is analyzed in Section 4. In Section 5 we establish convergence of the method with subgradient selection. In Section 6 we dicuss modified subgradient deletion rules and their practical implications. Section 7 is devoted to methods that neglect linearization errors.

2. Derivation of the Methods

In this section we derive two methods for minimizing a locally Lipschitzian function f defined on R^N. Detailed descriptions of the methods will be given in subsequent sections.

In order to implement the methods, we suppose that we have a finite process that can calculate $f(x)$ and a subgradient $g_f(x) \in \partial f(x)$ at each $x \in R^N$.

The algorithms will generate sequences of points $\{x^k\} \subset R^N$, search directions $\{d^k\} \subset R^N$ and nonnegative stepsizes $\{t_L^k\} \subset R_+$, related by

$$x^{k+1}=x^k+t_L^k d^k \quad \text{for} \quad k=1,2,\ldots,$$

where x^1 is a given starting point in R^N. The algorithms are descent methods in the sense that

$$f(x^{k+1}) < f(x^k) \quad \text{if} \quad x^{k+1} \neq x^k, \quad \text{for all } k,$$

and the sequence $\{x^k\}$ is intended to converge to a minimum point of f.

To deal with nondifferentiability of f, the methods will use a two-point line search, similar to Line Search Procedure 3.3.2, for detecting discontinuities in the gradient of f. Thus each algorithm will calculate auxiliary stepsizes $\{t_R^k\}$, $t_R^k \geq t_L^k$ for all k, and trial points

$$y^{k+1}=x^k+t_R^k d^k \quad \text{for} \quad k=1,2,\ldots, \quad y^1=x^1,$$

and evaluate subgradients

$$g^j = g_f(y^j) \quad \text{for} \quad j=1,2,\ldots .$$

With each such subgradient we associate the corresponding linearization of f

$$f_j(x) = f(y^j)+\langle g^j,x-y^j\rangle \quad \text{for all } x. \tag{2.1}$$

The two algorithms to be described differ in the way in which they make use of the past subgradient information. In order to employ only a finite number of the past subgradients for search direction finding at each iteration, the methods will use modified versions of the subgradient selection strategy and the subgradient aggregation strategy introduced in Chapter 2, and extended in Chapter 3. For convenience of the reader, relevant features of these strategies are briefly recalled below.

We start with the method with subgradient selection. Suppose that at the k-th iteration we have a nonempty set $J^k \subset \{1,\ldots,k\}$ and the corresponding past subgradients (g^j,f_j^k), $j \in J^k$, which determine the linearizations f_j, $j \in J^k$, via the relation

$$f_j(x) = f_j^k + \langle g^j, x-x^k \rangle \quad \text{for all } x, \tag{2.2}$$

where $f_j^k = f_j(x^k)$ for all $j \le k$. In the convex case we found it convenient to use the following polyhedral approximation to f

$$\hat{f}_s^k(x) = \max\{f_j(x) : j \in J^k\} \quad \text{for all } x, \tag{2.3}$$

since

$$f(x) \ge \hat{f}_s^k(x) \quad \text{for all } x, \tag{2.4a}$$

$$f(y^j) = \hat{f}_s^k(y^j) \quad \text{for all } j \in J^k \tag{2.4b}$$

if f is convex. These properties followed from the fact that

$$f(x) \ge f_j(x) \quad \text{for all } x \text{ and } j \tag{2.5}$$

in the convex case. In view of (2.4), we had reasons to suppose that $\hat{f}_s(x^k+d)$ is close to $f(x^k+d)$ if $|d|$ is small, so we found the k-th search direction d^k to

$$\text{minimize } \hat{f}_s^k(x^k+d) + \tfrac{1}{2}|d|^2 \quad \text{over all } d \in R^N. \tag{2.6}$$

We also noted in Section 3.2 that in terms of linearization errors

$$\alpha_j^k = f(x^k) - f_j^k \tag{2.7}$$

the k-th selective polyhedral approximation to f could be written as

$$\hat{f}_s^k(x) = \max\{f(x^k) - \alpha_j^k + \langle g^j, x-x^k \rangle : j \in J^k\} \quad \text{for all } x, \tag{2.8}$$

so that d^k could be found by solving the following quadratic programming problem for $(d^k, \hat{v}^k) \in R^N \times R$:

$$\begin{aligned} &\underset{(d,\hat{v}) \in R^{N+1}}{\text{minimize}} \quad \tfrac{1}{2}|d|^2 + \hat{v}, \\ &\text{subject to } -\alpha_j^k + \langle g^j, d \rangle \le \hat{v},\ j \in J^k. \end{aligned} \tag{2.9}$$

The Lagrange multipliers λ_j^k, $j \in J^k$, of (2.9) solve the following dual of (2.9)

$$\begin{aligned} &\underset{\lambda}{\text{minimize}} \ \tfrac{1}{2}\Big|\sum_{j \in J^k} \lambda_j g^j\Big|^2 + \sum_{j \in J^k} \lambda_j \alpha_j^k, \\ &\text{subject to } \lambda_j \ge 0,\ j \in J^k,\ \sum_{j \in J^k} \lambda_j = 1, \end{aligned} \tag{2.10}$$

and yield d^k via the relation

$$-d^k = \sum_{j\in J^k} \lambda_j^k g^j , \tag{2.11}$$

and hence only linearizations f_j with small values of linearization errors $\alpha_j^k \geq 0$ are significantly active in the determination of d^k, i.e. λ_j^k must be small if α_j^k is large in comparison with α_i^k for $i\neq j$, $i\in J^k$. Thus in the convex case the nonnegative linearization errors (2.7) could be used for weighing the past subgradients at each search direction finding. Moreover, each linearization error α_j^k measures the distance of g^j from $\partial f(x^k)$ in the sense that

$$g^j \in \partial_\varepsilon f(x^k) \quad \text{for} \quad \varepsilon=\alpha_j^k . \tag{2.12}$$

In the nonconvex case considered in Chapter 3, we had to modify the above constructions for the following reasons. If f is nonconvex then each linearization f_j need not globally approximate f from below, so that a polyhedral approximation of the form (2.3) does not, in general, satisfy (2.4). Moreover, it is important to observe that if one used linearization errors defined via (2.7) in the k-th search direction finding subproblem (2.9) then such linearization errors would no longer consistently weigh the past subgradients as in the convex case. This follows from the fact that although d^k would be expressed in the form of (2.11) in terms of a solution to (2.10), we could have negative α_j^k. For this reason, in Chapter 3 we used polyhedral approximations of the form (2.8) and the corresponding k-th search direction finding subproblem (2.9) with

$$\alpha_j^k = \max\{|f(x^k)-f_j^k|, \gamma(s_j^k)^2\}, \tag{2.13}$$

where γ was a positive parameter (which could be set to zero in the convex case) and the distance measures

$$s_j^k = |x^j-y^j| + \sum_{i=j}^{k-1} |x^{i+1}-x^i|$$

were used for estimating $|x^k-y^j|$ without storing $\{y^j\}$ (although one could use

$$\alpha_j^k = \max\{|f(x^k)-f_j^k|, \gamma|x^k-y^j|^2\} \tag{2.14}$$

if $\{y^j\}$ were stored).

Observe that if $y^j\neq x^k$ for all $j\in J^k$ then $\hat{f}_s^k$ defined by (2.8) and (2.13) is locally a lower approximation to f at x^k in the sense that

$$f(x^k+d) \geq \hat{f}_s^k(x^k+d) \quad \text{for sufficiently small} \quad |d|, \tag{2.15}$$

since then each α_j^k is positive $(s_j^k \geq |x^k-y^j| > 0)$, and so

$$f(x^k)-\alpha_j^k+\langle g^j,d\rangle \leq f(x^k)-\alpha_j^k+|g^j||d^k|$$

is less than $f(x^k+d)$ for small $|d|$ because of the Lipschitz continuity of f. We also have (2.4a) for $\hat{f}_s^k$ defined by (2.8) and (2.13) if f is convex and $\gamma=0$. Thus (2.15) may be regarded as a local version of (2.4a) in the nonconvex case.

Relation (2.13) defines subgradient locality measures, since in the convex case $(\gamma=0)$ (2.13) reduces to (2.7) and we have (2.12), while in the nonconvex case we may use the following definition of the Goldstein (1977) ε-subdifferential

$$\partial f(x;\varepsilon)=\mathrm{con}\{\partial f(y):|y-x|\leq\varepsilon\} \tag{2.16}$$

to deduce that

$$g^j\in\partial f(x^k;\varepsilon) \qquad \text{for} \qquad \varepsilon=(\alpha_j^k/\gamma)^{1/2}, \tag{2.17}$$

since

$$|y^j-x^k|\leq s_j^k\leq(\alpha_j^k/\gamma)^{1/2}.$$

Moreover, the above inequality implies that if d^k is obtained by solving (2.9) with α_j^k defined via (2.13), then in (2.11) we have small λ_j^k corresponding to large $|x^k-y^j|$, i.e. d^k is based mainly on the local past subgradients. We summarize our observations in

Remark 2.1. The methods considered in Chapter 3 will automatically ensure that each past subgradient g^j becomes progressively less active in successive search direction finding subproblems, i.e. the values of λ_j^k decrease when the algorithm moves away from the point y^j, so that eventually we have $\lambda_j^k=0$ and the subgradient g^j is dropped $(j\notin J^{k+1})$. This mechanism depends on the value of the distance measure parameter γ. If the value of γ is too large then the algorithm will keep deleting even the subgradients collected in relatively small neighborhoods of iterates x^k. In this case the algorithm proceeds in "leaps", i.e. after each long serious step almost all the past subgradients are dropped, so the next long serious step can occur only after the algorithm has accumulated sufficiently many subgradients by executing a series of null or very short serious steps. This should, of course, be avoided. At the same time, too small a value of γ may not ensure proper weighing of the past subgradients at the search direction finding; for instance, (2.15) may hold only for very small $|d|$. Then the algorithm may select for retaining even nonlocal past subgradients in preference to the local ones, which is inefficient.

We conclude from the above remark that in practice it may be difficult to choose a suitable value of the distance measure parameter γ in the methods with subgradient locality measures. For this reason, we shall assume in this chapter that $\gamma=0$ in (2.13), i.e. we shall use the following linearization errors

$$\alpha_j^k = |f(x^k)-f_j^k| \tag{2.18}$$

and polyhedral approximations of the form (2.8) even in the nonconvex case. In effect, the direction d^k found by solving the k-th primal subproblem (2.9) (or, equivalently, the k-th dual subproblem (2.10)) will satisfy (2.11) as in the corresponding method with subgradient selection of Chapter 3. However, since linearization errors will no longer measure locality of the past subgradients, and since $\hat{f}_s^k$ given by (2.8) and (2.18) is a useful local approximation of f at x^k only if y^j is close to x^k for all $j \in J^k$, we shall modify the subgradient selection strategy of Chapter 3 to ensure that the sets J^k are chosen so that we have sufficiently small values of $|y^j-x^k|$ for all $j \in J^k$. To this end we shall use subgradient deletion rules for deciding which of the past subgradients should be retained for the next search direction finding. Namely, we shall first find Lagrange multipliers of (2.9) satisfying

$$\hat{J}^k = \{j \in J^k:\ \lambda_j^k \neq 0\}$$

$$|\hat{J}^k| \leq N+1,$$

and set $J^{k+1}=\hat{J}^k \cup \{k+1\}$ as in Chapter 3. Next, we shall use suitable rules for deciding whether J^{k+1} should be reduced by deleting indices j corresponding to large values of distance measures s_j^k. Since specific deletion rules are applicable also to the method with subgradient aggregation described below, we postpone their discussion till the end of this section.

Let us now pass to the method with subgradient aggregation. As in Chapter 3, for search direction finding at the k-th iteration the aggregate subgradient (p^{k-1},f_p^k), satisfying

$$(p^{k-1},f_p^k) \in \operatorname{conv}\{(g^j,f_j^k):\ j \in \hat{J}_p^{k-1}\},\quad \hat{J}_p^{k-1} \subset \{1,\dots,k-1\}, \tag{2.19}$$

may replace the past subgradients (g^j,f_j^k), $j=1,\dots,k-1$. Since (2.18) corresponds to the following linearization error

$$\alpha(x,y) = |f(x)-\overline{f}(x;y)|, \tag{2.20}$$

we may define the linearization error

$$\alpha_p^k = |f(x^k)-f_p^k| \tag{2.21}$$

associated with the (k-1)-st aggregate linearization

$$\tilde{f}^{k-1}(x) = f_p^k + \langle p^{k-1}, x-x^k \rangle \quad \text{for all } x.$$

(Observe that (2.20) and (2.21), can be obtained by setting $\gamma=0$ in the corresponding definitions of Section3.2.). Then, as in Chapter 3, we may define the k-th aggregate polyhedral approximation to f

$$\hat{f}_a^k(x) = \max\{f(x^k)-\alpha_j^k + \langle g^j, x-x^k \rangle : j \in J^k;$$

$$f(x^k)-\alpha_p^k + \langle p^{k-1}, x-x^k \rangle\} \text{ for all } x \tag{2.22}$$

and find the k-th search direction d^k to

$$\text{minimize} \quad \hat{f}_a^k(x^k+d) + \tfrac{1}{2}|d|^2 \quad \text{over all } d \in R^N. \tag{2.23}$$

This can be done by solving the following quadratic programming problem for $(d^k, \hat{v}^k) \in R^N \times R$:

$$\underset{(d,v) \in R^{N+1}}{\text{minimize}} \quad \tfrac{1}{2}|d|^2 + \hat{v}, \tag{2.24a}$$

$$\text{subject to} \quad -\alpha_j^k + \langle g^j, d \rangle \le \hat{v}, \; j \in J^k, \tag{2.24b}$$

$$-\alpha_p^k + \langle p^{k-1}, d \rangle \le \hat{v}. \tag{2.24c}$$

Moreover, the Lagrange multipliers λ_j^k, $j \in J^k$, and λ_p^k of (2.24), satisfying

$$\lambda_j^k \ge 0, \; j \in J^k, \; \lambda_p^k \ge 0, \; \sum_{j \in J^k} \lambda_j^k + \lambda_p^k = 1, \tag{2.25}$$

determine the current aggregate subgradient

$$(p^k, \tilde{f}_p^k) = \sum_{j \in J^k} \lambda_j^k (g^j, f_j^k) + \lambda_p^k (p^{k-1}, f_p^k) \tag{2.26}$$

such that

$$d^k = -p^k,$$

$$(p^k, \tilde{f}_p^k) \in \text{conv}\{(g^j, f_j^k) : j \in \hat{J}_p^{k-1} \cup J^k\}. \tag{2.27}$$

Therefore, by setting

$$f_p^{k+1} = \tilde{f}_p^k + \langle p^k, x^{k+1} - x^k \rangle$$

one can proceed to the next iteration

One may observe that the method with subgradient aggregation described so far differs from the corresponding method of Chapter 3 only in the choice of $\gamma=0$, i.e. here we use linearization errors instead of the corresponding subgradient locality measures. For this reason, in the nonconvex case the function $\hat{f}_a^k$ defined by (2.22) is a useful local approximation to f around x^k only if it is based on sufficiently local subgradient information, i.e. if the sets J^k and $\hat{J}_p^{k-1}$ (see (2.19)) are such that the values of the following variables

$$\begin{aligned} \hat{a}_J^k &= \max\{|x^k - y^j| : j \in J^k\}, \\ \hat{a}_p^k &= \max\{|x^k - y^j| : j \in \hat{J}_p^{k-1}\}, \\ \hat{a}^k &= \max\{\hat{a}_J^k, \hat{a}_p^k\} \end{aligned} \tag{2.28}$$

are sufficiently small. This can be ensured by a suitable choice of J^k and $\hat{J}_p^{k-1}$. In this context we observe that the value of $\hat{a}_J^k$ can be made as small as desired at the k-th iteration by choosing J^k on the basis of the values $s_j^k \geq |x^k - y^j|$, provided that we have $s_j^k = 0$ for some j. On the other hand, since the set $\hat{J}_p^{k-1}$ is defined by (see (3.4.2))

$$\hat{J}_p^{k-1} = J^{k_p(k-1)} \cup \{j : k_p(k-1) < j \leq k\},$$

$$k_p(k-1) = \max\{j : j \leq k-1 \text{ and } \lambda_j^k = 0\},$$

the only way of reducing $\hat{a}_p^k$, if this is necessary, is to reset $\hat{J}_p^{k-1}$ to an empty set, so that $\hat{a}_p^k$ vanishes. This is equivalent to discarding the (k-1)-st aggregate subgradient in the definition (2.22) of $\hat{f}_a^k$, and to dropping the last constraint of (2.24) at the search direction finding.

We shall now describe a simple strategy for localizing the past subgradient information used for search direction finding. It is based on an idea due to Wolfe (1975) in the convex case. The concept of the Goldstein subdifferential (2.16) is useful in the nonconvex case, because it is defined directly in terms of neighboorhoods of a given point. Moreover, $\partial f(x;0) = \partial f(x)$ and, owing to the definition (1.2.14) of ∂f, $\partial f(x;\varepsilon)$ is a close approximation to $\partial f(x)$ if the value of ε is small. Our aim is to obtain at some iteration $p^k \in \partial f(x^k;\varepsilon)$ for some small values of ε and $|p^k|$; then x^k will be approximately stationary

(stationary points $\bar{x}$ satisfy $0 \in \partial f(\bar{x}) \subset \partial f(\bar{x};\varepsilon)$ for all $\varepsilon \ge 0$). From Chapter 3 we know how to construct locality radii a^k of the aggregate subgradients, by using only distance measures s_j^k, such that (see Lemma 4.1)

$$p^k = \sum_{j \in \hat{J}_p^k} \tilde{\lambda}_j g^j, \tag{2.29a}$$

$$\tilde{\lambda}_j^k \ge 0,\ j \in \hat{J}_p^k,\quad \sum_{j \in \hat{J}_p^k} \tilde{\lambda}_j^k = 1, \tag{2.29b}$$

$$\max\{|y^j - x^k|:\ j \in \hat{J}_p^k\} \le a^k, \tag{2.29c}$$

$$\hat{J}_p^k = J^{k_p(k)} \cup \{j:\ k_p(k) < j \le k\}, \tag{2.29d}$$

$$k_p(k) = \max\{j : j \le k \text{ and } \lambda_p^j = 0\}. \tag{2.29e}$$

Relations (2.29a)-(2.29c) and definition (2.16) lead to the following fundamental property

$$p^k \in \partial f(x^k; a^k). \tag{2.30}$$

In view of (2.30) we want to obtain small values of both $|p^k|$ and a^k at some iteration. This will occur if both

$$|p^k| \le m_a a^k \tag{2.31}$$

and the value of a^k is small, where $m_a > 0$ is a fixed scaling parameter. However, if the variables a^k were generated as in the methods of Chapter 3 with no resettings, i.e. by the formula

$$a^{k+1} = \max\{a^k + |x^{k+1} - x^k|, |y^{k+1} - x^{k+1}|\},$$

then we would always have $a^{k+1} \ge a^k$. Therefore (2.31) would hold only at later iterations for large values of a^k.

Thus a mechanism is needed for decresing the value of a^k if (2.31) occurs for some k. In this case we shall discard the aggregate subgradient by resetting the algorithm. The resetting will involve a repeated calculation of d^k and p^k from the k-th subproblem (2.24a)-(2.24b), i.e. the aggregate constraint (2.24c) is ignored at resets. By setting the corresponding Lagrange multiplier λ_p^k to zero, we again obtain (2.30) from (2.29), but this time for the reduced value of a^k given by

$$a^k = \max\{s_j^k:\ j \in J^k\}. \tag{2.32}$$

Next, if (2.31) holds once again, the resetting procedure should be repeated with a further reduction of the search direction finding subproblem (2.24a)-(2.24b), which consists in deleting some elements j of J^k with large values of s_j^k, so that the value of a^k given by (2.32) decreases. Finally, we should have $|p^k| > m_a a^k$ after a reset. Since $|p^k|=|d^k|$, this may be interpreted as ensuring that the length of the current search direction is comparable to the value of the current locality radius.

Except for possible reductions due to resets, the sets J^k can be chosen as follows. Let $M_g \geq 2$ denote a user-supplied, fixed upper bound on the number of the past subgradients (including the aggregate subgradient) that the algorithm may use for each search direction finding. By choosing the sets J^k recursively so that

$$k+1 \in J^{k+1} \subset J^k \cup \{k+1\},$$

$$J^{k+1} \subset \{j: k-M_g+3 \leq j \leq k+1\}$$

one can control storage and work per iteration.

Having described the subgradient aggregation strategy, we may now return to the method with subgradient selection. In this method the aggregate subgradient $(p^k, \tilde{f}_p^k)$ is defined by (2.26), but with $\lambda_p^k=0$. If the resetting test (2.31) is fulfilled with a^k given by (2.32), then the set J^k is reduced and the search direction finding subproblem (2.10) is solved again. The resetting is repeated until either $|p^k|>m_a a^k$ or the algorithm stops with $|p^k| \leq m_a a^k$ and a small value of a^k. The next index set is of the form $J^{k+1}=\hat{J}^k \cup \{k+1\}$, which is analogous to the one used in Chapter 3, i.e. we have $|\hat{J}^k| \leq N+1$, so that the method can store $M_g=N+2$ past subgradients.

Remark 2.2. In the above-described method with subgradient selection one may use, instead of (2.31), the following test for resetting the method

$$|p^k| \leq m_a \hat{a}^k,$$

where

$$\hat{a}^k = \max\{|y^j-x^k|: j \in \hat{J}^k\}.$$

The subsequent convergence results remain valid for this modification. However, the calculation of $\hat{a}^k$ would require storing N+2 points y^j, $j \in J^k$. For this reason, we perfer to use the easily computable variable a^k.

3. The Algorithm with Subgradient Aggregation

We now state an algorithmic procedure for solving the problem in question. Ways of implementing each step of the method are discussed below.

Algorithm 3.1.

Step 0 (Initialization). Select the starting point $x^1 \in R^N$ and a final accuracy parameter $\varepsilon_s \geq 0$. Choose fixed positive line search parameters $m_L, m_R, \bar{a}, \bar{t}$ and $\bar{\theta}$ with $\bar{t} \leq 1$, $m_L < m_R < 1$ and $\bar{\theta} < 1$. Set M_g equal to the fixed maximum number of subgradients that the algorithm may use for search direction finding; $M_g \geq 2$. Choose a predicted shift in x at the first iteration $s^1 > 0$ and set $\theta^1 = \bar{\theta}$. Select a positive reset tolerance m_a and set the reset indicator $r_a^1 = 1$. Set

$$J^1 = \{1\}, \ y^1 = x^1, \ p^0 = g^1 = g_f(y^1), \ f_p^0 = f_1^1 = f(y^1), \ s_1^1 = a^1 = 0.$$

Set the counters $k=1$, $l=0$ and $k(0)=1$.

Step 1 (Direction finding). Find the solution $(d^k, \hat{v}^k)$ to the following k-th quadratic programmnig problem

$$\begin{aligned} &\underset{(d,\hat{v}) \in R^{N+1}}{\text{minimize}} \quad \tfrac{1}{2}|d|^2 + \hat{v}, \\ &\text{subject to} \quad -\alpha_j^k + \langle g^j, d \rangle \leq \hat{v}, \ j \in J^k, \\ &\qquad\qquad\quad -\alpha_p^k + \langle p^{k-1}, d \rangle \leq \hat{v} \quad \text{if} \quad r_a^k = 0, \end{aligned} \tag{3.1}$$

where

$$\alpha_j^k = |f(x^k) - f_j^k| \quad \text{for} \quad j \in J^k, \tag{3.2a}$$

$$\alpha_p^k = |f(x^k) - f_p^k|. \tag{3.2b}$$

Find Lagrange multipliers λ_j^k, $j \in J^k$, and λ_p^k of (3.1), setting $\lambda_p^k = 0$ if $r_a^k = 1$. Set

$$(p^k, \tilde{f}_p^k) = \sum_{j \in J^k} \lambda_j^k (g^j, f_j^k) + \lambda_p^k (p^{k-1}, f_p^k), \tag{3.3}$$

$$\tilde{\alpha}_p^k = |f(x^k) - \tilde{f}_p^k|, \tag{3.4}$$

$$v^k = -\{|p^k|^2 + \tilde{\alpha}_p^k\}. \tag{3.5}$$

If $\lambda_p^k = 0$ set

$$a^k = \max\{s_j^k : j \in J^k\}. \tag{3.6}$$

Step 2 (Stopping criterion). If $\max\{|p^k|, m_a a^k\} \le \varepsilon_s$ then terminate. Otherwise, go to Step 3.

Step 3 (Resetting test). If $|p^k| \le m_a a^k$ then go to Step 4; otherwise, go to Step 5.

Step 4 (Resetting). (i) If $r_a^k=0$ then set $r_a^k=1$, replace J^k by $J^k \cap \{j : j \ge k-M_g+2\}$ and go to Step 1.

(ii) If $|J^k| > 1$ then delete the smallest number from J^k and go to Step 1.

(iii) Set $y^k=x^k$, $g^k=g_f(y^k)$, $f_k^k=f(y^k)$, $s_k^k=0$, $J^k=\{k\}$ and to Step 1.

Step 5 (Line search). By a line search procedure as discussed below, find two stepsizes t_L^k and t_R^k such that $0 \le t_L^k \le t_R^k$ and such that the two corresponding points defined by

$$x^{k+1}=x^k+t_L^k+d^k \quad \text{and} \quad y^{k+1}=x^{k+t_R^k+d^k}$$

satisfy $t_L^k \le 1$ and

$$f(x^{k+1}) \le f(x^k)+m_L t_L^k v^k, \tag{3.7}$$

$$t_R^k = t_L^k \qquad \text{if} \quad t_L^k \ge \bar{t}, \tag{3.8a}$$

$$-\alpha(x^{k+1},y^{k+1})+ < g_f(y^{k+1}),d^k > \ge m_R v^k \quad \text{if} \quad t_L^k < \bar{t}, \tag{3.8b}$$

$$|y^{k+1}-x^{k+1}| \le \bar{a}, \tag{3.9}$$

$$|y^{k+1}-x^{k+1}| \le \theta^k s^k \qquad \text{if} \quad t_L^k=0, \tag{3.10a}$$

$$|y^{k+1}-x^{k+1}| \le \bar{\theta}|x^{k+1}-x^k| \quad \text{if} \quad t_L^k > 0, \tag{3.10b}$$

where

$$\alpha(x,y) = |f(x)-f(y)- < g_f(y),x-y > |. \tag{3.11}$$

Step 6. If $t_L^k=0$ (null step), set $s^{k+1}=s^k$ and $\theta^{k+1}=\bar{\theta}\theta^k$. Otherwise, i.e. if $t_L^k > 0$ (serious step), set $s^{k+1}=|x^{k+1}-x^k|$, $\theta^{k+1}=\bar{\theta}$, $k(l+1)=k+1$ and increase l by 1.

Step 7 (Subgradient updating). Select a set $\hat{J}^k$ satisfying

$$\hat{J}^k \subset J^k, \tag{3.12a}$$

$$|\hat{J}^k| \le M_g - 2, \tag{3.12b}$$

and set

$$J^{k+1} = \hat{J}^k \cup \{k+1\}. \tag{3.12c}$$

Set $g^{k+1} = g_f(y^{k+1})$ and

$$f_{k+1}^{k+1} = f(y^{k+1}) + \langle g^{k+1}, x^{k+1} - y^{k+1} \rangle, \tag{3.13a}$$

$$f_j^{k+1} = f_j^k + \langle g^j, x^{k+1} - x^k \rangle \quad \text{for} \quad j \in \hat{J}^k, \tag{3.13b}$$

$$f_p^{k+1} = \tilde{f}_p^k + \langle p^k, x^{k+1} - x^k \rangle, \tag{3.13c}$$

$$s_{k+1}^{k+1} = |y^{k+1} - x^{k+1}|, \tag{3.14a}$$

$$s_j^{k+1} = s_j^k + |x^{k+1} - x^k| \quad \text{for} \quad j \in \hat{J}^k. \tag{3.14b}$$

Calculate

$$a^{k+1} = \max\{a^k + |x^{k+1} - x^k|, \, s_{k+1}^{k+1}\} \tag{3.15}$$

and set $r_a^{k+1} = 0$.

<u>Step 8</u>. Increase k by 1 and go to Step 1.

We shall now comment on each step of the algorithm.

By Lemma 2.2.1, the k-th subproblem dual of (3.1) is to find values of the multipliers λ_j^k, $j \in J^k$, and λ_p^k that

$$\begin{aligned} &\underset{\lambda}{\text{minimize}} \ \tfrac{1}{2}\Big|\sum_{j\in J^k} \lambda_j g^j + \lambda_p p^{k-1}\Big|^2 + \sum_{j\in J^k} \lambda_j \alpha_j^k + \lambda_p \alpha_p^k, \\ &\text{subject to} \ \ \lambda_j \ge 0,\ j \in J^k,\ \lambda_p \ge 0,\ \sum_{j\in J^k} \lambda_j + \lambda_p = 1, \\ &\qquad\qquad\quad \lambda_p = 0 \quad \text{if} \quad r_a^k = 1. \end{aligned} \tag{3.16}$$

Any solution of (3.16) is a Lagrange multiplier vector for (3.1) and it yields the unique solution $(d^k, \hat{v}^k)$ of (3.1) as follows

$$d^k = -p^k, \tag{3.17}$$

$$\hat{v}^k = -\{|p^k|^2 + \sum_{j\in J^k} \lambda_j^k \alpha_j^k + \lambda_p^k \alpha_p^k\}, \tag{3.18}$$

where p^k is given by (3.3). Moreover, any Lagrange multipliers of (3.1) also solve (3.16). In particular, we always have

$$\lambda_j^k \geq 0,\ j \in J^k,\ \lambda_p^k \geq 0,\ \sum_{j \in J^k} \lambda_j^k + \lambda_p^k = 1,\ \lambda_p^k = 0 \quad \text{if} \quad r_a^k = 1. \tag{3.19}$$

Thus one may equivalently solve the k-th dual search direction finding subproblem (3.16) in Step 1 of the algorithm.

The algorithm stops at Step 2 when

$$p^k \in \partial f(x^k; \varepsilon_s/m_a) \quad \text{and} \quad |p^k| \leq \varepsilon_s,$$

i.e. when x^k is approximately stationary for f. This follows from the fact that $p^k \in \partial f(x^k; a^k)$ at Step 2, as will be shown below.

The sole purpose of resettings at Step 3 and Step 4 is to ensure that at line searching at Step 5 we have $|d^k| > m_a a^k$. Note that there can be only finitely many returns from Step 4 to Step 1, since Step 4 reduces the number of subgradients that can be used for the next search direction finding. Step 4 (iii) is entered after p^{k-1} has been discarded and J^k has been reduced to $J^k = \{k\}$, while $|g^k| < m_a s_k^k$. Therefore, in this case we have to completely reset the algorithm by redefining y^k to $y^k = x^k$. Then $a^k = s_k^k = 0$ at Step 1, and $\max\{|p^k|, m_a a^k\} = |p^k|$ at Step 2. If the algorithm does not stop at Step 2, i.e. $|p^k| > \varepsilon_s \geq 0$, then $|p^k| > m_a a^k = 0$ in Step 3, hence the algorithm goes to Step 5.

Observe that each reset involves a repetition of search direction finding. To avoid the need for solving too many quadratic programmning problems at resetting, one may delete more than one subgradient in Step 4(i) and Step 4(iii). We shall return to this subject in Section 6.

The line search rules of Step 5 are modifications of the rules (3.3.9)-(3.3.11) of the method with subgradient locality measures. We shall prove in the next section that the line search is always entered with

$$\hat{v}^k = f_a^k(x^k + d^k) - f(x^k) \leq v^k < 0. \tag{3.20}$$

Thus $v^k < 0$ is an estimate of the directional derivative of f at x^k in the direction d^k, and the criteria (3.7)-(3.9) may be interpreted similarly to the line search criteria of Section 3.3. It will be seen that the additional requirement (3.10) ensures that the subgradient $g^{k+1} = g_f(y^{k+1})$ is calculated sufficiently close to the next point x^{k+1}. We show below that the variable θ^k decreases and s^k is constant whenever the algorithm executes a series of null steps at deblocking, i.e. when the algorithm is blocked at x^k due to the nondifferentiability of f. Then (3.10a) ensures that the new subgradient information is increasingly local, which yields better search directions. We may add that the methods with subgradient locality measures of Chapter 3 had no need for such criteria, since the distance $|y^{k+1}-x^{k+1}|$ was accounted for by using a requirement of the form (3.8b), but with $\alpha(x^{k+1}, y^{k+1})$ depending on $|y^{k+1}-x^{k+1}|$.

The following modification of Line Search Procedure 3.3.2 may be used for finding stepsizes $t_L=t_L^k$ and $t_R=t_R^k$ satisfying the requirement of Step 5.

Line Search Procedure 3.2.

(i) Set $t_L=0$ and $t=t_U=\min\{1,\bar{a}/|d^k|\}$.

(ii) If $f(x^k+td^k)\le f(x^k)+m_L t v^k$ set $t_L=t$; otherwise set $t_U=t$.

(iii) If $t_L\ge\bar{t}$ set $t_R=t_L$ and return.

(iv) If $-\alpha(x^k+t_Ld^k,x^k+td^k)+\langle g_f(x^k+td^k),d^k\rangle\ge m_R v$ and either $t_L=0$ and $t|d^k|\le\theta^k s^k$ or $t-t_L\le\bar{\theta}t_L$, then set $t_R=t$ and return.

(v) Set $t=t_L+\bar{\theta}(t_U-t_L)$ and go to (ii).

Convergence of the above procedure can be established similarly to Lemma 3.3.3 under additional semismoothness assumptions.

Lemma 3.3. If f has the semismoothness property (3.3.23) then Line Search Procedure 3.2 terminates with $t_L^k=t_L$ and $t_R^k=t_R$ satisfying (3.7)-(3.10). (Here we use the properties $0<m_L<m_R<1$, $\bar{a}>0$, $\bar{t}>0$, $0<\bar{\theta}<1$, $\theta^k s^k>0$ and $v^k<0$).

Proof. Use the proof Lemma 3.3.3, setting $\zeta=1-\bar{\theta}$ and observing that we must have either $\tilde{t}_L^i=0$ for all i and $t^i|d^k|\le\theta^k s^k$ for large i, or $t^i-\tilde{t}_L^i\le\bar{\theta}\tilde{t}_L^i>0$ for large i. □

We refer the reader to Remark 3.3.4 concerning the semismoothness assumptions.

We may add that in efficient versions of Line Search Procedure 3.2 one may use safeguarded interpolation instead of bisection for choosing trial stepsizes; see Remark 3.3.5.

In Step 6 of the algorithm we update the variables θ^k and s^k

that localize, via the line search requirement (3.10a), the new subgradient information at null steps. Upon completion of Step 6, the current value of l is equal to the number of serious steps taken so far. In general, we have $k(l) < k(l+1)$ and

$$x^k = x^{k(l)} \quad \text{if} \quad k(l) \le k < k(l+1), \tag{3.21a}$$

$$t_L^k = 0 \quad \text{if} \quad k(l) \le k < k(l+1)-1, \tag{3.21b}$$

$$t_L^k > 0 \quad \text{if} \quad k = k(l+1)-1, \tag{3.21c}$$

$$s^k = s^{k(l)} \quad \text{if} \quad k(l) \le k < k(l+1), \tag{3.21d}$$

$$s^k = |x^{k(l+1)} - x^{k(l)}| \quad \text{if} \quad k = k(l+1), \tag{3.21e}$$

where we set $k(l+1)=+\infty$ if the number l of serious steps stays bounded during the run of the algorithm, i.e. if $t_L^k=0$ for some fixed l and all $k \ge k(l)$. By (3.10), (3.21) and the rules of Step 6, we have

$$|y^{k+1}-x^{k+1}| \le (\bar{\theta})^{k+1-k(l)} s^{k(l)} \quad \text{if} \quad k(l) \le k < k(l+1)-1, \tag{3.22a}$$

$$|y^{k+1}-x^{k+1}| \le \bar{\theta} s^{k(l+1)} \quad \text{if} \quad k+1 = k(l+1). \tag{3.22b}$$

Thus the value of $|y^{k+1}-x^{k+1}|$ is always a fraction of the length of the latest serious step.

The rules of Step 7 ensure that the algorithm uses at most M_g subgradients for each search direction finding, so that at most M_g $(N+2)$-vectors of the form (g^j, f_j^k, s_j^k) need to be stored. In fact, it will be seen that, as far as convergence is concerned, the set $\hat{J}^k$ may contain any indices corresponding to the subgradients used since the latest reset, i.e. the requirement (3.12b) is not important. At the same time, the rule of Step 4(i), ensuring that only a finite number of latest subgradients are retained after each reset, is crucial for global convergence. We also note that the latest subgradient is always used for search direction finding, i.e. we have

$$k \in J^k, \; g^k = g_f(y^k) \quad \text{and} \quad |y^k - x^k| \le \bar{a} \quad \text{for all } k. \tag{3.23}$$

This follows from (3.9), (3.12c) and the rules of Step 4.

4. Convergence

In this section we shall establish global convergence of Algorithm 3.1. We suppose that each execution of Line Search Procedure 3.2 is finite, e.g. that f is semismooth in the sense of (3.3.23). For convergence results we assume that the final accuracy tolerance ε_s is set to zero. As usually done, in the absence of convexity we will content ourselves with finding stationary points for f. Our main result states that Algorithm 3.1 either terminates at a stationary point, or generates an infinite sequence $\{x^k\}$ whose accumulation points are stationary for f. If f is convex then $\{x^k\}$ is a minimizing sequence, which converges to a minimum point of f whenever f attains its infimum.

In the following we implicitly assume, unless otherwise stated, that d^k, p^k, v^k, etc. denote the variables at Step 5 of Algorithm 3.1 at the k-th iteration, for any k.

As in Section 3.4, we shall use the following notation for describing resets of the algorithm. Recall that the condition $r_a^k=1$ indicates that a reset occurs at the k-th iteration, while $\lambda_p^k=0$ means that the (k-1)-st aggregate subgradient is dropped. Let

$$k_r(k) = \max\{j : j \le k \quad \text{and} \quad r_a^j=1\}, \tag{4.1a}$$

$$\hat{J}_r^k = J^{k_r(k)} \cup \{j : k_r(k) < j \le k\}, \tag{4.1b}$$

$$k_p(k) = \max\{j : j \le k \quad \text{and} \quad \lambda_p^j=0\}, \tag{4.2a}$$

$$\hat{J}_p^k = J^{k_p(k)} \cup \{j : k_p(k) < j \le k\}, \tag{4.2b}$$

$$\tilde{J}^k = \{j : k_r(k)-M_g \le j \le k\}, \tag{4.3}$$

for all k.

The following lemma shows that the aggregate subgradient is a convex combination of the subgradients that were used for direction finding since the latest reset.

Lemma 4.1. Suppose $k \ge 1$ is such that Algorithm 3.1 did not stop before the k-th iteration. Then

$$(p^k,\tilde{f}_p^k) \in \operatorname{conv}\{(g^j,f_j^k) : j \in \hat{J}_p^k\}, \tag{4.4}$$

$$a^k = \max\{s_j^k : j \in \hat{J}_p^k\}, \tag{4.5}$$

$$|y^j-x^k| \le a^k \quad \text{for all} \quad j \in \hat{J}_p^k, \tag{4.6}$$

$$\hat{J}_p^k \subset \hat{J}_r^k \subset \tilde{J}^k. \tag{4.7}$$

Proof. One can establish (4.4)-(4.6) as in the proof of Lemma 3.4.1. To prove (4.7), observe that $r_a^k=1$ implies $\lambda_p^k=0$, for any k. Therefore (4.1) and (4.2) imply that $k_p(k) \ge k_r(k)$ and $\hat{J}_p^k \subset \hat{J}_r^k$ for all k. Hence it suffices to prove that $\hat{J}_r^k \subset \tilde{J}^k$ for any k. If $r_a^k=1$ for some k, then $k_r(k)=k$ by (4.1a) and $J^k \subset \{j : k-M_g+2 \le j\}$ by the rules of Step 4, hence $\hat{J}_r^k=J^k \subset \tilde{J}^k$ by (4.1) and (4.3). Since $\hat{J}_r^{k+1}=\hat{J}_r^k \cup \{k+1\}$ and $\tilde{J}^{k+1}=\tilde{J}^k \cup \{k+1\}$ if $k_r(k+1)<k+1$, this proves that $\hat{J}_r^k \subset \tilde{J}^k$ for all k. □

It is easy to observe that relations (4.4)-(4.6) imply

$$p^k \in \partial f(x^k;a^k) \tag{4.8}$$

by the definition of the Goldstein subdifferential (see (2.16)). On the other hand, in the convex case we have the following additional result.

Lemma 4.2. Suppose that f is convex and $k \ge 1$ is such that Algorithm 3.1 did not stop before the k-th iteration. Then

$$p^k \in \partial_\varepsilon f(x^k) \quad \text{for} \quad \varepsilon=f(x^k)-\tilde{f}_p^k=\tilde{\alpha}_f^k \ge 0.$$

Proof. As in the proof of Lemma 2.4.2, use (4.4) and the fact that $\tilde{\alpha}_p^k=|f(x^k)-\tilde{f}_p^k|$. □

First, we consider the case when the algorithm terminates.

Lemma 4.3. If Algorithm 3.1 stops at the k-th iteration, then x^k is stationary for f.

Proof. If $\max\{|p^k|,m_a a^k\} \le \varepsilon_s=0$ then (4.8) and the fact that $m_a>0$ yield $0 \in \partial f(x^k)$. □

From now on we suppose that the method constructs an infinite sequence $\{x^k\}$.

We shall now collect a few useful results. In Step 5 of the algorithm we always have

$$|p^k| > m_a a^k \geq 0. \tag{4.9}$$

Since $d^k=-p^k$ by (3.17), $v^k=-\{|p^k|^2+\tilde{\alpha}_p^k\}$ by (3.5), and $\tilde{\alpha}_p^k \geq 0$ by (3.4), we obtain from (4.9) that the line search is always entered with

$$d^k \neq 0, \tag{4.10}$$

$$v^k < 0. \tag{4.11}$$

This establishes (3.20). Moreover, the criterion (3.7) with $m_L > 0$ and $t_L^k \geq 0$ ensures that the sequence $\{f(x^k)\}$ is nonincreasing and

$$f(x^{k+1}) < f(x^k) \quad \text{if} \quad x^{k+1} \neq x^k.$$

Our next result states that the aggregate subgradient can be expressed as a convex combination of $N+2$ (not necessarily different) past subgradients.

Lemma 4.4. At the k-th iteration of Algorithm 3.1 there exist numbers $\hat{\lambda}_i^k$ and vectors $(y^{k,i},f^{k,i}) \in R^N \times R$, $i=1,\ldots,M$, $M=N+2$, satisfying

$$\begin{aligned}
&(p^k,\tilde{f}_p^k) = \sum_{i=1}^{M} \hat{\lambda}_i^k (g_f(y^{k,i}),f^{k,i}),\\
&\hat{\lambda}_i^k \geq 0,\ i=1,\ldots,M, \quad \sum_{i=1}^{M} \hat{\lambda}_i^k=1, \qquad\qquad (4.12)\\
&(y^{k,i},f^{k,i}) \in \{(y^j,f_j^k) : j \in \hat{J}_p^k\},\ i=1,\ldots,M,\\
&\max\{|y^{k,i}-x^k| : i=1,\ldots,M\} \leq a^k.
\end{aligned}$$

Proof. The assertion follows from Lemma 4.1, Caratheodory's theorem (Lemma 1.2.1), and the fact that $g^j=g_f(y^j)$ for all j. □

Comparing Lemma 4.4 with Lemma 3.4.3 we see that the only difference stems from the fact that we are now considering tuples $(p^k,\tilde{f}_p^k)$ instead of triples $(p^k,\tilde{f}_p^k,\tilde{s}_p^k)$.

To deduce stationarity results from the representation (4.12), we shall need the following lemma, which is similar to Lemma 3.4.4.

Lemma 4.5. Let $\bar{x} \in R^N$ be given and suppose that the following hypothesis is fulfilled:

there exist N-vectors $\bar{p}, \bar{y}^i, \bar{g}^i$ for $i=1,\dots,M=N+2$, and numbers $\bar{f}_p, \bar{f}^i, \bar{s}^i$, $i=1,\dots,M$, satisfying

$$(\bar{p}, \bar{f}_p) = \sum_{i=1}^{M} \bar{\lambda}_i (\bar{g}^i, \bar{f}^i), \tag{4.13a}$$

$$\bar{\lambda}_i \geq 0, \quad i=1,\dots,M, \quad \sum_{i=1}^{M} \bar{\lambda}_i = 1, \tag{4.13b}$$

$$\bar{g}^i \in \partial f(\bar{y}^i), \quad i=1,\dots,M, \tag{4.13c}$$

$$\bar{f}^i = f(\bar{y}^i) + \langle \bar{g}^i, \bar{x} - \bar{y}^i \rangle, \quad i=1,\dots,M, \tag{4.13d}$$

$$|\bar{y}^i - \bar{x}| \leq \bar{s}^i, \quad i=1,\dots,M, \tag{4.13e}$$

$$\max\{\bar{s}^i : \bar{\lambda}_i \neq 0\} = 0. \tag{4.13f}$$

Then $\bar{p} \in \partial f(\bar{x})$ and $\bar{f}_p = f(\bar{x})$.

Proof. Since $\bar{s}_p = \sum_{i=1}^{M} \bar{\lambda}_i \bar{s}^i = 0$, we may use part (i) of the proof of Lemma 3.4.4. □

The following lemma states useful asymptotic properties of the aggregate subgradients.

Lemma 4.6. Suppose that there exist a point $\bar{x} \in R^N$, a number $\tilde{a} > 0$ and an infinite set $K \subset \{1,2,\dots\}$ satisfying $x^k \xrightarrow{K} \bar{x}$ and $a^k \leq \tilde{a}$ for all $k \in K$. Then there exist an infinite set $\bar{K} \subset K$ and numbers $\bar{s}^i \leq \liminf_{k \in K} a^k$, $i=1,\dots,N+2$, such that the hypothesis (4.13a)-(4.13b) is fulfilled at $\bar{x}$ and

$$(p^k, \tilde{f}_p^k) \xrightarrow{\bar{K}} (\bar{p}, \bar{f}_p).$$

If additionally $a^k \xrightarrow{K} 0$ then $\bar{p} \in \partial f(\bar{x})$ and $\tilde{\alpha}_p^k \xrightarrow{K} 0$.

Proof. Using Lemma 4.4, and Lemma 4.5 let $s^{k,i}=|y^{k,i}-x^k|$ for $i=1,\ldots,M$ and $k\in K$, and argue as in the proof of Lemma 3.4.6. □

The following result is crucial for establishing convergence of the method. Define the stationarity measure

$$w^k = \tfrac{1}{2}|p^k|^2+\tilde{\alpha}_p^k \tag{4.14}$$

at the k-th iteration (at Step 5) of Algorithm 3.1, for all k.

Lemma 4.7. (i) Suppose that for some point $\bar{x}\in R^N$ we have

$$\liminf_{k\to\infty} \max\{w^k,|\bar{x}-x^k|\}=0, \tag{4.15}$$

or equivalently

there exists an infinite set $K\subset\{1,2,\ldots\}$ such that

$$x^k \xrightarrow{K} \bar{x} \quad \text{and} \quad w^k \xrightarrow{K} 0. \tag{4.16}$$

Then $0\in\partial f(\bar{x})$.

(ii) Relations (4.15) and (4.16) are equivalent to the following

$$\liminf_{k\to\infty} \max\{|p^k|,|\bar{x}-x^k|\}=0. \tag{4.17}$$

Proof. (i) The equivalence of (4.15) and (4.16) follows from the nonnegativity of w^k and $|\bar{x}-x^k|$. If (4.16) holds, then $w^k=\frac{1}{2}|p^k|^2+\tilde{\alpha}_p^k \xrightarrow{K} 0$, so $|p^k|\xrightarrow{K} 0$ by the nonnegativity of $\tilde{\alpha}_p^k$. Since we always have $|p^k|\geq m_a a^k\geq 0$ at Step 5 (see (4.9)) and $m_a>0$ is fixed, we have $|p^k|\xrightarrow{K} 0$, $a^k\xrightarrow{K} 0$ and $x^k\xrightarrow{K}\bar{x}$. Consequently, $0\in\partial f(\bar{x})$ by Lemma 4.6. Also $\max\{|p^k|,|\bar{x}-x^k|\}\xrightarrow{K} 0$, hence we have shown that (4.16) implies (4.17).

(ii) It remains to show that (4.17) implies (4.16). Suppose that (4.17) holds. Then $|p^k|\xrightarrow{K} 0$ and $x^k\xrightarrow{K}\bar{x}$ for some infinite set $K\subset\{1,2,\ldots\}$. Since $0\leq a^k\leq|p^k|/m_a$, we obtain $a^k\xrightarrow{K} 0$. Then Lemma 4.6 yields $\tilde{\alpha}_p^k\xrightarrow{K} 0$, hence $w^k=\frac{1}{2}|p^k|^2+\tilde{\alpha}_p^k\xrightarrow{K} 0$. Thus (4.16) holds, as required. □

The above lemma, which is similar to Lemma 3.4.7, will enable us to reduce further convergence analysis to verifying if the stationarity measures w^k vanish in the neighbourhood of an arbitrary accumulation point $\bar{x}$ of $\{x^k\}$. Therefore, as in the preceding two chapters, it is useful to relate the stationarity measures with the optimal values of the dual search direction finding subproblems.

Let w^k denote the optimal value of the k-th dual search direction finding subproblem (3.16), for all k. Since the Lagrange multipliers of (3.1) solve (3.16) and yield p^k via (3.3), we always have

$$\hat{w}^k = \tfrac{1}{2}|p^k|^2+\hat{\alpha}_p^k, \tag{4.18a}$$

where

$$\hat{\alpha}_p^k = \sum_{j\in J^k}\lambda_j^k\alpha_j^k+\lambda_p^k\alpha_p^k. \tag{4.18b}$$

The following lemma, which can be proved similarly to Lemma 3.4.8, shows that w^k majorizes w^k.

<u>Lemma 4.8</u>. (i) At the k-th iteration of Algorithm 3.1, one has

$$0\le\tilde{\alpha}_p^k\le\hat{\alpha}_p^k, \tag{4.19a}$$

$$0\le w^k\le\hat{w}^k, \tag{4.19b}$$

$$v^k\le -w^k\le 0, \tag{4.19c}$$

$$\hat{v}^k\le v^k. \tag{4.19d}$$

(ii) If f is convex then $\tilde{\alpha}_p^k=\hat{\alpha}_p^k$, $w^k=\hat{w}^k$ and $v^k=\hat{v}^k$, for all k.

Consider the following condition:

there exist a point $\bar{x}\in R^N$ and an infinite set $K\subset\{1,2,\ldots\}$ such that $x^k\xrightarrow{K}\bar{x}$. (4.20)

In view of (3.21a), (4.20) implies that either of the following two disjoint case must arise:

there exists an infinite set $L\subset\{1,2,\ldots\}$ such that

$$x^{k(l)}\longrightarrow\bar{x}\quad\text{as}\quad l\to\infty,\ l\in L; \tag{4.21a}$$

or

there exists a fixed number l such that $x^k = x^{k(l)} = \bar{x}$

for all $k \geq k(l)$. (4.21b)

Conversly, (4.21) implies (4.20), so in fact (4.20) and (4.21) are equivalent.

Our aim is to show that (4.20) implies that $w^k \xrightarrow{K} 0$ for some $\bar{K} \subset K$. To this end we shall first analyze the case of "long" serious steps in the following lemma, which can be established similarly to Lemma 3.4.9 if one uses (3.7), (4.19c) and (3.21a).

Lemma 4.9. (i) If (4.20) holds then

$$f(x^k) \downarrow f(\bar{x}) \quad \text{as} \quad k \to \infty,$$

$$t_L^k v^k \longrightarrow 0 \quad \text{as} \quad k \to \infty.$$

(ii) If (4.21a) is fulfilled and there exist a number $\hat{t} > 0$ and an infinite set $\bar{L} \subset L$ such that $t_L^k \geq \hat{t}$ for all $k = k(l+1)-1$ and $l \in \bar{L}$, then (4.16) holds.

Corollary 4.10. Suppose (4.20) is satisfied, but (4.15) does not hold, i.e.

$$\liminf_{k \to \infty} \max\{w^k, |\bar{x} - x^k|\} \geq \bar{\varepsilon} > 0 \tag{4.22}$$

for some $\bar{\varepsilon}$. Then

$$\max\{t_L^k : k(l) \leq k < k(l+1)\} \longrightarrow 0 \quad \text{as} \quad l \to \infty,\ l \in L. \tag{4.23}$$

Proof. The assertion follows from Lemma 4.9 (ii), (3.21b), and the equivalence of (4.15) and (4.16) on the one hand, and of (4.20) and (4.21) on the other. □

In view of the above result, we need only consider the case of arbitrarily small stepsizes. Therefore, we shall now show that whenever the value of t_L^k is small, i.e. no significant improvement in the objective value occurs at the k-th iteration (cf. (3.7)), then w^{k+1} is a fraction of w^k.

Lemma 4.11. Suppose that $t_L^{k-1} < \bar{t}$ and $r_a^k = 0$ for some $k > 1$. Then

$$w^k \leq \hat{w}^k \leq \phi_C(w^{k-1}) + |\alpha_p^k - \tilde{\alpha}_p^{k-1}|, \tag{4.24}$$

where the function ϕ_C is defined by

$$\phi_C(t)=t-(1-m_R)^2t^2/(8C^2),$$

and C is any number satisfying

$$\max\{|p^{k-1}|,|g^k|,\tilde{\alpha}_p^{k-1},1\}\le C. \tag{4.25}$$

Proof. Use the proof of Lemma 3.4.11. □

To obtain locally uniform bounds of the form (4.24) we shall need the following result, which is a consequence of (3.23). Its proof, which is similar to the proof of Lemma 3.4.13, is left to the reader.

Lemma 4.12. The assertions of Lemma 3.4.13 remain valid for Algorithm 3.1. In particular,

$$\max\{|p^k|,|g^k|,\tilde{\alpha}_p^k,1\}\le C \quad \text{if} \quad |x^k-\bar{x}|\le\bar{a}, \tag{4.26}$$

where $\bar{a}$ is the line search parameter involved in (3.9), and C is the constant defined in Lemma 3.4.13.

Our next result demonstrates that the term $|\alpha_p^k-\tilde{\alpha}_p^{k-1}|$ involved in (4.24) vanishes in the limit.

Lemma 4.13. Suppose that (4.20) holds. Then

$$|\alpha_p^{k+1}-\tilde{\alpha}_p^k|\longrightarrow 0 \quad \text{as} \quad k\to\infty.$$

Proof. Since $\alpha_p^{k+1}=|f(x^{k+1})-f_p^{k+1}|$ and $\tilde{\alpha}_p^k=|f(x^k)-\tilde{f}_p^k|$ for all k, one may use Lemma 4.9(i) to obtain the desired conclusion as in the proof of Lemma 3.4.14. □

The following lemma will imply that the decrease of stationarity measures established in Lemma 4.11 will take place for sufficiently many consecutive iterations, provided that no reset occurs.

Lemma 4.14. Suppose that (4.21a) and (4.22) hold. Then:

(i) For any fixed integer $m\ge 0$ there exists a number l_m such that

$$\max\{|\bar{x}-x^k| : k(l) \le k \le k(l+m)\} \longrightarrow 0 \quad \text{as} \quad l\to\infty,\ l\in L, \tag{4.27a}$$

$$\min\{w^k : k(l) \le k < k(l+m+l)\} \ge \bar{\varepsilon}/2 \quad \text{for all} \quad l \ge \hat{l}_m,\ l\in L, \tag{4.27b}$$

$$\max\{t_L^k : k(l) \le k < k(l+m+l)\} \longrightarrow 0 \quad \text{as} \quad l\to\infty,\ l\in L, \tag{4.27c}$$

$$\max\{s^k : k(l+l) \le k < k(l+m+l) \longrightarrow 0 \quad \text{as} \quad l\to\infty,\ l\in L, \tag{4.28a}$$

$$\max\{|y^{k+1}-x^{k+1}| : k(l+l) \le k < k(l+m+l) \longrightarrow 0 \quad \text{as} \quad l\to\infty,\ l\in L. \tag{4.28b}$$

(ii) For any numbers $\hat{k}$, $\tilde{N}$ and $\varepsilon_\alpha > 0$ there exists a number $\tilde{k} \ge \hat{k}$ such that $\tilde{k} \in K=\{k(l+l)-l : l \in L\}$ and

$$\begin{aligned} & w^k \ge \bar{\varepsilon}/2 \quad \text{for} \quad k=\tilde{k},\dots,\tilde{k}+\tilde{N}, \\ & \max\{|p^{k-1}|, |g^k|, \tilde{\alpha}_p^{k-1}, l\} \le C \quad \text{for} \quad k=\tilde{k},\dots,\tilde{k}+\tilde{N}, \\ & |\alpha_p^k-\tilde{\alpha}_p^{k-1}| \le \varepsilon_\alpha \quad \text{for} \quad k=\tilde{k},\dots,\tilde{k}+\tilde{N}, \\ & t_L^{k-1} < \bar{t} \quad \text{for} \quad k=\tilde{k},\dots,\tilde{k}+\tilde{N}, \end{aligned} \tag{4.29}$$

where C is the constant defined in Lemma 3.4.13.

<u>Proof</u>. (4.27) can be proved by using Corollary 4.10 and Lemma 4.12 as in the proof of Lemma 3.4.12. Then (4.28) follows from (4.27), (3.21), (3.22) and the fact that $\bar{\theta} \in (0,1)$. Part (ii) of the lemma follows from part (i), (4.26) and Lemma 4.13. □

Since the above lemma dealt with the case of an infinite number of serious steps, we now have to analyze the remaining case.

<u>Lemma 4.15</u>. Suppose that (4.21b) holds. Then $t_L^k=0$ for each $k \ge k(l)$ and $y^k \longrightarrow \bar{x}$ as $k\to\infty$. If additionally (4.22) holds then the second assertion of Lemma 4.14 is true.

<u>Proof</u>. Suppose that (4.21b) holds. Then, since we always have $x^{k+1}= x^k+t_L^k d^k$ and $d^k \ne 0$ (see (4.10)), we deduce from the line search rule (4.10) that $t_L^k=0$ and $|y^{k+1}-x^{k+1}|=|y^{k+1}-\bar{x}| \le \theta^k s^k$ for all $k \ge k(l)$. But $t_L^k=0$ always yields $\theta^{k+1}=\bar{\theta}\theta^k$ and $s^{k+1}=s^k$, hence $\theta^k \downarrow 0$ $(0<\bar{\theta}<1)$

and $y^k \longrightarrow \bar{x}$ as $k\to\infty$. Having established that $|x^k-\bar{x}|\to 0$, $t_L^k \to 0$ and $|y^{k+1}-x^{k+1}|\to 0$ as $k\to\infty$, we see that the proof of the second assertion of Lemma 4.14 remains valid in this case. □

Up till now we have concentrated on the relations between the variables at Step 5 of the algorithm and their impact on convergence. We now turn our attention to the properties of variables generated at each search direction finding. Recall that Step 1 of the method may be repeated if a reset occurs at any iteration.

Lemma 4.16. For any $k \geq 1$, let b^k denote the minimum value taken on by $\max\{|p^k|, m_a a^k\}$ for the successive values of p^k and a^k calculated at each execution of Step 1 of Algorithm 3.1 at the k-th iteration. (The variable b^k is well-defined, because there can be only finitely many returns to Step 1 at any iteration.) Suppose that

$$\liminf_{k\to\infty} \max\{b^k, |\bar{x}-x^k|\}=0 \tag{4.30}$$

for some $\bar{x} \in R^N$. Then $0 \in \partial f(\bar{x})$.

Proof. For any $k \geq 1$, let a^k and p^k have the values calculated at the first execution of Step 1 such that $b^k=\max\{|p^k|, m_a a^k\}$. One easily check that Lemma 4.1, Lemma 4.4 and Lemma 4.6 remain valid for the variables generated by this particular execution of Step 1, for all k. Therefore one may use the proof of Lemma 4.7(i) to obtain the desired conclusion. □

Remark 4.17. One may check that, except for Lemma 4.7, all the preceing results of this section hold also for variables generated by each execution of Step 1 at any iteration. Lemma 4.7 assumes that $|p^k|>m_a a^k$ hence it deals only with the variables calculated by the last execution of Step 1 at any iteration.

We are now ready to prove the principal result of this section.

Lemma 4.18. Suppose that (4.20) holds. Then at least one of relations (4.15) and (4.30) is satisfied.

Proof. Suppose that (4.20) holds. For purposes of a proof by contradiction, assume that neither (4.15) nor (4.30) is fulfilled. Thus suppose that there exist positive constants $\bar{\varepsilon}$ and $\bar{\varepsilon}_p$ satisfying (4.22) and

$$\liminf_{k\to\infty} \max\{b^k, |\bar{x}-x^k|\} \geq \bar{\varepsilon}_p. \qquad (4.31)$$

(i) Let $\varepsilon_w=\bar{\varepsilon}/2>0$ and choose $\varepsilon_\alpha=\varepsilon_\alpha(\varepsilon_w,C)$ and $\bar{N}=\bar{N}(\varepsilon_w,C)<+\infty$, $\bar{N}\geq 1$, as specified in Lemma 3.4.12, where C is the constant defined in Lemma 3.4.13.

(ii) Let $\tilde{N}=10(\bar{N}+M_g)$. Combining (4.22), (4.31), Lemma 4.14 and Lemma 4.15 with the fact that $\bar{\varepsilon}_p/(2m_a)>0$, we deduce the existence of $\tilde{k}$ satisfying (4.29) and

$$\max\{s_j^j : j=\tilde{k},\dots,\tilde{k}+\tilde{N}\} + \sum_{j=\tilde{k}}^{\tilde{k}+\tilde{N}} |x^{j+1}-x^j| < \bar{\varepsilon}_p/(2m_a), \qquad (4.32a)$$

$$b^k \geq \bar{\varepsilon}_p/2 \quad \text{for} \quad k=\tilde{k},\dots,\tilde{k}=\tilde{N}, \qquad (4.32b)$$

since $s_j^j=|y^j-x^j|$ for all j, $m_a>0$, $\bar{\varepsilon}_p>0$ and $\tilde{N}<+\infty$.

(iii) Suppose that there exists a number $\hat{k}$ satisfying

$$\tilde{k}\leq\hat{k}\leq\tilde{k}+\tilde{N}-2\bar{N}, \qquad (4.33)$$

$$r_a^k = 0 \quad \text{for all} \quad \hat{k}\in[\hat{k},k+\bar{N}]. \qquad (4.34)$$

Then (4.29), (4.34), Lemma 4.11, Lemma 3.4.12 and our choice of ε_α and $\bar{N}$ imply that $w^k<\varepsilon_w=\bar{\varepsilon}/2$ for some $k\in[\hat{k},\hat{k}+\bar{N}]$, which contradicts (4.33) and (4.29). Consequently, we have shown that for any number $\hat{k}$ satisfying (4.33) we have $r_a^k=1$ for some $k\in[\hat{k},\hat{k}+\bar{N}]$.

(iv) Let $k_1=\tilde{k}+2M_g$. Then part (iii) of the proof implies that $r_a^{k_2-1}=1$ for some $k_2\in[k_1,k_1+\bar{N}]$. Since $k_2-M_g>\tilde{k}$ and $r_a^{k_2-1}=1$, we obtain from (4.1), (4.3) and Lemma 4.1 (see Remark 4.17) that

$$\tilde{J}^k\subset[\tilde{k},\tilde{k}+\tilde{N}] \quad \text{for all} \quad k\in[k_2,k_2+\bar{N}], \qquad (4.35a)$$

$$a^k\leq\max\{s_j^k : j\in\tilde{J}^k\} \quad \text{for all} \quad k\in[k_2,k_2+\bar{N}]. \qquad (4.35b)$$

We want to stress that (4.35) holds for the values of a^k and $\tilde{J}^k$ computed by every execution of Step 1 at the k-th iteration, $k=k_2,\dots,k_2+\bar{N}$. Also in view of (4.32b) and the definition of b^k (see Lemma 4.16) we have

$$\max\{|p^k|, m_a a^k\}\geq\bar{\varepsilon}_p/2 \quad \text{for} \quad k=k_2,\dots,k_2+\bar{N}, \qquad (4.36)$$

i.e. the inequality in (4.36) holds on each entrance to Step 3 at the k-th iteration, for $k=k_2,\ldots,k_2+\bar{N}$. From (4.35), (4.32a) and the fact that we always have $s_j^k=s_j^j+\sum_{i=j}^{k-1}|x^{i+1}-x^i|$, we deduce that at Step 3

$$a^k \le \max\{s_j^j : j\in \tilde{J}^k\}+\sum_{j=k}^{k-1}|x^{j+1}-x^j| < \bar{\varepsilon}_p/(2m_a) \tag{4.37}$$

for $k=k_2,\ldots,k_2+\bar{N}$. Using this in (4.36), we get that $|p^k|>m_a a^k$ on each entrance to Step 3 at the k-th iteration, for $k=k_2,\ldots,k_2+\bar{N}$, so

$$r_a^k = 0 \quad \text{for} \quad k=k_2,\ldots,k_2+\bar{N}. \tag{4.38}$$

(v) Since $\hat{k}=k_2$ satisfies (4.33), from part (iv) of the proof we deduce a contradiction with (4.38). Therefore, either (4.15) or (4.30) must hold, □

Combining Lemma 4.18 with Lemma 4.7 and Lemma 4.16, we obtain

Theorem 4.19. Every accumulation point of the sequence $\{x^k\}$ generated by Algorithm 3.1 is stationary for f.

In the convex case, the above result can be strengthened as follows.

Theorem 4.20. If f is convex then Algorithm 3.1 constructs a minimizing sequence $\{x^k\}$, i.e. $f(x^k)\downarrow \inf\{f(x) : x\in R^N\}$. Moreover, if f attains its infimum then $\{x^k\}$ converges to a minimum point of f.

Proof. Similar to the proof of Theorem 3.4.18. □

The following result validates the stopping test of the method.

Corollary 4.21. If the level set $S=\{x\in R^N : f(x)\le f(x^1)\}$ is bounded and the final accuracy tolerance ε_s is positive, then Algorithm 3.1 terminates in a finite number of iterations.

Proof. If the assertion were false, then the infinite sequence $\{x^k\}\subset S$ would have an accumulation point, say $\bar{x}$. Then Lemma 4.18, Lemma 4.7 and Lemma 4.16 would yield $\liminf_{k\to\infty} \max\{|p^k|,m_a a^k\}=0$, so the algorithm should stop owing to $\max\{|p^k|,m_a a^k\}<\varepsilon_s$ for large k. □

5. The Algorithm with Subgradient Selection

In this section we state and analyze the method with subgradient selection.

Algorithm 5.1.

Step 0 (Initialization). Select $x^1 \in R^N$ and $\varepsilon_s \geq 0$. Choose positive parameters m_a, m_L, m_R, $\bar{a}$, $\bar{t}$, $\bar{\theta}$, M_g and s^1 with $\bar{t} \leq 1$, $m_L < m_R < 1$, $\bar{\theta} < 1$ and $M_g \geq N+2$. Set $\theta^1 = \bar{\theta}$, $r_a^1 = 0$, $J^1 = \{1\}$, $y^1 = x^1$, $g^1 = g_f(y^1)$, $f_1^1 = f(y^1)$ and $s_1^1 = 0$. Set $k=1$, $l=0$ and $k(0)=1$.

Step 1 (Direction finding). Find the solution $(d^k, \hat{v}^k)$ to the following k-th quadratic programmnig problem

$$\begin{aligned} &\underset{(d,\hat{v}) \in R^{N+1}}{\text{minimize}} \quad \tfrac{1}{2}|d|^2 + \hat{v}, \\ &\text{subject to} \quad -\alpha_j^k + \langle g^j, d \rangle \leq \hat{v}, \ j \in J^k, \end{aligned} \tag{5.1}$$

where

$$\alpha_j^k = |f(x^k) - f_j^k| \quad \text{for} \quad j \in J^k. \tag{5.2}$$

Find Lagrange multipliers λ_j^k, $j \in J^k$, of (5.1) and a set $\hat{J}^k$ satisfying

$$\hat{J}^k = \{j \in J^k : \lambda_j^k \neq 0\}, \tag{5.3a}$$

$$|\hat{J}^k| \leq M_g - 1. \tag{5.3b}$$

Set

$$a^k = \max\{s_j^k : j \in \hat{J}^k\}. \tag{5.4}$$

Step 2 (Stopping criterion). If $\max\{|d^k|, m_a a^k\} \leq \varepsilon_s$ then terminate. Otherwise, go to Step 3.

Step 3 (Resetting test). If $|d^k| \leq m_a a^k$ then go to Step 4; otherwise , go to Step 5.

Step 4 (Resetting). (i) Replace J^k by $J^k \cap \{j : j \geq k - M_g + 1\}$ and set $r_a^k = 1$.
(ii) If $|J^k| > 1$ then delete the smallest number from J^k and go to Step 1.
(iii) Set $y^k = x^k$, $g^k = g_f(y^k)$, $f_k^k = f(y^k)$, $s_k^k = 0$, $J^k = \{k\}$ and go to Step 1.

Step 5 (Line search). Find two nonnegative stepsizes $t_L^k \le t_R^k$ and the two corresponding points $x^{k+1}=x^k+t_L^k d^k$ and $y^{k+1}=x^k+t_R^k d^k$ such that

$$f(x^{k+1}) \le f(x^k)+m_L t_L^k \hat{v}^k, \tag{5.5a}$$

$$t_R^k = t_L^k \le 1 \qquad \text{if} \quad t_L^k > \bar{t}, \tag{5.5b}$$

$$-\alpha(x^{k+1},y^{k+1})+ < g_f(y^{k+1}),d^k > \ge m_R \hat{v}^k \quad \text{if} \quad t_L^k \le \bar{t}, \tag{5.5c}$$

$$|y^{k+1}-x^{k+1}| \le \bar{a}, \tag{5.5e}$$

$$|y^{k+1}-x^{k+1}| \le \theta^k s^k \quad \text{if} \quad t_L^k = 0, \tag{5.5f}$$

$$|y^{k+1}-x^{k+1}| \le \bar{\theta}|x^{k+1}-x^k| \quad \text{if} \quad t_L^k > 0. \tag{5.5g}$$

Step 6. If $t_L^k=0$ set $s^{k+1}=s^k$ and $\theta^{k+1}=\bar{\theta}\theta^k$. If $t_L^k>0$ then set $s^{k+1}=|x^{k+1}-x^k|$, $\theta^{k+1}=\bar{\theta}$, $k(l+1)=k+1$ and increase l by 1.

Step 7 (Subgradient updating). Set $J^{k+1}=\hat{J}^k \cup \{k+1\}$, $g^{k+1}=g_f(y^{k+1})$,

$$f_{k+1}^{k+1} = f(y^{k+1})+ < g^{k+1},x^{k+1}-x^k > ,$$

$$f_j^{k+1} = f_j^k+ < g^j,x^{k+1}-x^k > \quad \text{for} \quad j \in \hat{J}^k,$$

$$s_{k+1}^{k+1} = |y^{k+1}-x^{k+1}|,$$

$$s_j^{k+1} = s_j^k+|x^{k+1}-x^k| \quad \text{for} \quad j \in \hat{J}^k.$$

Step 8. Set $r_a^{k+1}=0$, increase k by 1 and go to Step 1.

We shall now compare the above method with Algorithm 3.1. It is easy to observe that Algorithm 5.1 is related with Algorithm 3.1 in the same way as Algorithm 3.5.1 with Algorithm 3.3.1. Therefore, we may shorten our discussion of Algorithm 5.1 by using suitable modifications of the results and remarks of Section 3.5.

We refer the reader to Remark 2.5.2 for a discussion of possible ways of finding the k-th Lagrange multipliers satisfying the requirement (5.3). To this end we observe that such multipliers exist, since M_g satisfies $N+1 \le M_g-1$ by assumption (see also Remark 3.5.2). In Step 1 of the method one may also solve the dual of the k-th search direction finding subproblem (3.1), which is of the form (3.5.11), and then recover $(d^k, \hat{v}^k)$ via (3.5.12)-(3.5.14).

As in Section 3.5, we may derive useful relations betwen variables generated by Algorithm 5.1 by setting

$$\lambda_p^k = 0 \quad \text{for all } k$$

in the corresponding results of Section 3 and Section 4. Thus, defining at the k-th iteration of Algorithm 5.1 the variables

$$
\begin{aligned}
(p^k, \tilde{f}_p^k) &= \sum_{j \in J^k} \lambda_j^k (g^j, f_j^k),\\
\tilde{\alpha}_p^k &= |f(x^k) - \tilde{f}_p^k|,\\
\hat{\alpha}_p^k &= \sum_{j \in J^k} \lambda_j^k \alpha_j^k, \qquad (5.6)\\
w^k &= \tfrac{1}{2}|p^k|^2 + \tilde{\alpha}_p^k,\\
\hat{w}^k &= \tfrac{1}{2}|p^k|^2 + \hat{\alpha}_p^k,\\
v^k &= -\{|p^k|^2 + \tilde{\alpha}_p^k\}
\end{aligned}
$$

for all k, we obtain

$$d^k = -p^k, \tag{5.7a}$$

$$\tilde{\alpha}_p^k \le \tilde{\alpha}_p^k, \tag{5.7b}$$

$$w^k \le \hat{w}^k, \tag{5.7c}$$

$$\hat{v}^k \le v^k, \tag{5.7d}$$

cf. (3.5.12)-(3.5.25).

In view of (5.7a), the stopping criteria and the resetting tests of Algorithm 3.1 and Algorithm 5.1 are equivalent.

Observe that the line search criteria (5.5) can be derived by substituting $\hat{v}^k$ for v^k in the criteria (3.7)-(3.10). Therefore, by replacing v^k with $\hat{v}^k$ in Line Search Procedure 3.2 we obtain a procedure for executing Step 5 of Algorithm 5.1. Since the line search is always entered with $\hat{v}^k < 0$ (cf. (3.20) and (5.7)), Lemma 3.3 remains valid for this modification.

We also note that the subgradient deletion rules of Algorithm 5.1

ensure that at most M_g latest subgradients are retained after each reset, and that the latest subgradient $g^k=g_f(y^k)$ with $|y^k-x^k|\le \bar{a}$ is always used for search direction finding, i.e. (3.23) holds.

We shall now analyze convergence of Algorithm 5.1.

Theorem 5.2. Theorem 4.19, Theorem 4.20 and Corollary 4.21 are true for Algorithm 5.1.

Proof. One can prove the theorem by modifying the results of Section 4 similarly as we modified in Section 3.5 the results of Section 3.4 to establish convergence of Algorithm 3.5.1. For instance, Lemma 4.11 should be replaced by Lemma 3.5.3, and the expression $|\alpha_p^k-\tilde{\alpha}_p^{k-1}|$ in (4.29) by Δ_α^k defined in (3.5.29), with the corresponding part of the proof of Lemma 4.14 being changed to the form of the proof of Lemma 3.5.4. To save space, we leave details to the reader. □

6. Modified Resetting Strategies

The resetting of Algorithm 3.1 and Algorithm 5.1 is crucial for obtaining strong results on their convergence. In this section we shall consider earlier resetting strategies due to Wolfe (1975) and Mifflin (1977b). It turns out that these strategies can be easily analyzed. However, at present it seems impossible to establish for the resulting algorithms, which include those in (Wolfe, 1975; Mifflin, 1977b) as special cases, global convergence results similar to Theorem 4.19, even under additional assumptions. We also propose a new resetting strategy based on aggregating distance measures s_j^k as in Algorithm 3.3.1. This strategy, on the one hand, may be more efficient in practice, and on the other hand, retains all the preceding global convergence results.

To motivate the subsequent theoretical developments, we start with the following practical observation. Algorithm 3.1 has a certain drawback that slows down its convergence in practice. This drawback stems from the definition of the locality radius a^k (estimating the radius of the ball around x^k from which the past subgradient information was accumulated), which resets the algorithm whenever $|p^k|\le m_a a^k$. Namely, the values of a^k are nondecreasing between every two consecutive resets, while in the neighborhood of a solution the values of $|p^k|$ decrease rapidly, thus forcing frequent resets due to $|p^k|\le m_a a^k$. Too frequent reduction of the past subgradient information by discarding the aggregate subgradient hinders convergence, especially when, due to storage limitations, only a small number $M_g (M_g << N)$

of past subgradients are used for search direction finding. This drawback is eliminated to a certain extent in the following modification of Algorithm 3.1.

Before stating the modified method, let us briefly recall the basic tasks of subgradient deletion rules. In this chapter we concentrated on rules for localizing the accumulated subgradient information. Such rules ensure that polyhedral functions of the form (2.8) and (2.22) are close approximations to f in a neighborhood of x^k. On the other hand, in Chapter 3 we used resetting test of the form $a^k \le \bar{a}$ to ensure locally uniform boundedness of the subgradients that were aggregated at any iteration. Observe that in the methods of this chapter there was no need for distance resets through $a^k \le \bar{a}$, since we had $a^k \le |p^k|/m_a$ and $|p^k|$ was locally bounded (cf. Lemma 4.12). However, if we substitute the resetting test $|p^k| \le m_a a^k$ by some other test, then we shall no longer have estimates of the form $a^k \le |p^k|/m_a$. Therefore, we shall additionally use a distance resetting test of the form $a^k \le \bar{a}$. For a sufficiently large value of $\bar{a}$, say $\bar{a}=10^3$, a reduction of the past subgradient information due to a^k being larger than $\bar{a}$ will occur infrequently.

To derive a new resetting strategy, suppose that in Step 1 of Algorithm 3.1 we calculate the aggregate distance measure $\tilde{s}_p^k$ by setting

$$(p^k,\tilde{f}_p^k,\tilde{s}_p^k)=\sum_{j\in J^k}\lambda_j^k(g^j,f_j^k,s_j^k)+\lambda_p^k(p^{k-1},f_p^k,s_p^k) \tag{6.1}$$

and then calculating

$$s_p^{k+1} = \tilde{s}_p^k+|x^{k+1}-x^k| \tag{6.2}$$

in Step 7, for all k. Then according to Lemma 3.4.1, we shall have

$$(p^k,\tilde{f}_p^k,\tilde{s}_p^k)=\sum_{j\in \hat{J}_p^k}\tilde{\lambda}_j^k(g^j,f_j^k,s_j^k), \tag{6.3a}$$

$$\tilde{\lambda}_j^k \ge 0,\ j\in \hat{J}_p^k,\quad \sum_{j\in \hat{J}_p^k}\tilde{\lambda}_j^k=1, \tag{6.3b}$$

$$\max\{|y^j-x^k| : j\in \hat{J}_p^k\} \le \max\{s_j^k : j\in \hat{J}_p^k\}=a^k, \tag{6.3c}$$

where

$$\hat{J}_p^k = J^{k_p(k)} \cup \{j : k_p(k) < j \le k\},$$

$$k_p(k) = \max\{j : j \le k \text{ and } \lambda_p^j=0\}.$$

Thus the aggregate distance measure $\tilde{s}_p^k$, being a convex combination of the distance measures s_j^k, is no greater than the locality radius a^k, which is the maximum of the corresponding s_j^k-s. The value of $\tilde{s}_p^k$, in contrast with that of a^k, can decrease even if no reset occurs. Also the value of $\tilde{s}_p^k$ can be small even if the value of s_j^k is large for some $j \in \hat{J}_p^k$, provided that the value of $\tilde{\lambda}_j^k$ is small. This means that if $\tilde{s}_p^k$ has a small value then only local past subgradients g^j (with relatively small $s_j^k \geq |y^j - x^k|$) contribute significantly to the current direction $d^k = -p^k$ (have large $\tilde{\lambda}_j^k$ in (6.3a)). For this reason we shall reset the method only if

$$|p^k| \leq m_a \tilde{s}_p^k. \tag{6.4}$$

This will decrease the frequency of resettings in comparison with the one that would occur if the test $|p^k| \leq m_a a^k$ were used.

To save space, we shall now describe the modified resetting strategy in detail by using the notation of Algorithm 3.1.

Algorithm 6.1 is obtained from Algorithm 3.1 as follows. In Step 1 we substitute relation (3.3) by (6.1). In Step 2 we replace the stopping criterion $\max\{|p^k|, m_a a^k\} \leq \varepsilon_s$ by the criterion

$$\max\{|p^k|, m_a \tilde{s}_p^k\} \leq \varepsilon_s. \tag{6.5}$$

The resetting test $|p^k| \leq m_a a^k$ of Step 3 is replaced by the test (6.4). In Step 7 we update the aggregate distance measure via (6.2). Step 8 is replaced by the following two steps.

Step 8′ (Distance resetting). If $a^{k+1} \leq \bar{a}$ then go to Step 9. Otherwise, keep deleting from J^{k+1} the smallest indices until the reset value of a^{k+1} satisfies

$$a^{k+1} = \max\{s_j^{k+1} : j \in J^{k+1}\} \leq \bar{a}/2.$$

Set $r_a^{k+1} = 1$.

Step 9′. Increase k by 1 and go to Step 1.

Convergence of the above method is established in the following theorem.

Theorem 6.2. Algorithm 6.1 is globally convergent in the sense of Theorem 4.19, Theorem 4.20 and Corollary 4.21.

Proof. Since a formal proof of the theorem would involve lengthy repetitions of the results of Section 4 and Section 3.4, we give only an outline, hoping that the reader can fill in the necessary details. First, we observe that we always have $|p^k| > m_a s_p^k$ and $a^k \le \bar{a}$ at Step 5 of the method. Using this, one may replace Lemma 4.1 through Lemma 4.6 by Lemma 3.4.1 through Lemma 3.4.6, with the condition "$\tilde{\alpha}_p^k \xrightarrow{K} 0$" in Lemma 3.4.6 being replaced by the condition "$\tilde{s}_p^k \xrightarrow{K} 0$". In the proof of Lemma 4.7, use the fact that $\tilde{s}_p^k \le |p^k|/m_a$ at Step 5. In the formulation and the proof of Lemma 4.16, replace a^k by $\tilde{s}_p^k$. Finally, while proving Lemma 4.18 observe that $\tilde{s}_p^k \le a^k$ from (6.3), replace (4.36) with

$$\max\{|p^k|,\ m_a\tilde{s}_p^k\} \ge \bar{\varepsilon}_p/2 \quad \text{for} \quad k=k_2,\dots,k_2+\bar{N},$$

assume with no loss of generality that $\bar{\varepsilon}_p/(2m_a) < \bar{a}/2$, and deduce that in part (iv) of the proof we have $a^k \le \bar{a}$ and $|p^k| > m_a\tilde{s}_p^k$ for $k=k_2,\dots,k_2+\bar{N}$, so that (4.38) holds, as required. □

For the sake of completeness of our discussion of algorithms for nonsmooth minimization, we shall now discuss a simple resetting strategy due to Wolfe (1975).

Consider the following modification of Algorithm 3.1, in which resets will be regulated by a sequence of positive numbers $\{\delta^k\}$. In Step 0 we set δ^1 equal to any positive number. The resetting test $|p^k| \le m_a a^k$ of Step 3 is replaced by the test

$$|p^k| \le \delta^k. \tag{6.6}$$

Part (i) of Step 4 is replaced by the following

Step 4'(i). If $m_a a^k \le \delta^k$ then replace δ^k by $\delta^k/2$. If $r_a^k=0$ then set $r_a^k=1$, replace J^k by $\{j\in J^k : j \ge k - M_g+2\}$ and go to Step 1.

In Step 5 we replace the line search requirements (3.9)-(3.10) by the following

$$|y^{k+1}-x^{k+1}| \le \delta^k/(2m_a). \tag{6.7}$$

Of course, Line Search Procedure 3.2 is easily modified to handle the above requirement, since $\delta^k/(2m_a) > 0$. The remaining steps of Algorithm 3.1 remain unchanged, except that in Step 8 we set $\delta^{k+1}=\delta^k$.

The aim of the above modification of Algorithm 3.1 is to calculate in a finite number of iterations a point x^k such that

$\max\{|p^k|, m_a a^k\} \le \delta^k$ for a small value of δ^k, so that the method can stop at Step 2 owing to δ^k being smaller than ε_s. Note that δ^k is halved at Step 4'(i) only after we have $|p^k| \le \delta^k$ and $m_a a^k \le \delta^k$, i.e. when $\max\{|p^k|, m_a a^k\} \le \delta^k$, which means that the algorithm has found a significantly better approximation to a stationary point. When δ^k decreases, the line search requirement (6.7) ensures that the algorithm collects progressively more local subgradient information.

We may add that in the original version of the Wolfe (1975) strategy, one would use only part (iii) of Step 4 of Algorithm 3.1, i.e. each reset would involve restarting the method from the current point x^k, and discarding all the past subgradients. As observed by Mifflin (1977b), such a strategy is inefficient, because it leads to many null steps accumulating subgradient information to compensate for total resets,

The following result describes convergence properties of this modification of Algorithm 3.1.

Theorem 6.3. Suppose that the above-described modification of Algorithm 3.1 (with the Wolfe resetting strategy) calculates an infinite sequence of points $\{x^k\}$ with the stopping parameter ε_s set to zero. Let K denote the (possibly empty) set of indices of iterations k at which δ^k is decreased at Step 4'(i), i.e.

$$K = \{k : \delta^{k+1} \le \delta^k/2\}.$$

Then either of the following two case arises:

(i) $f(x^k)\downarrow-\infty$ as $k\to\infty$;

(ii) K is an infinite set, $\delta^k\downarrow 0$ as $k\to\infty$ and every accumulation point of the subsequence $\{x^k\}_{k\in K}$ is stationary for f.

Moreover, if the sequence $\{x^k\}$ is bounded (e.g. the level set $S_f=\{x\in R^N: f(x) \le f(x^1)\}$ is bounded) then case (ii) above occurs and $\{x^k\}$ has at least one stationnary accumulation point.

Proof. If $f(x^k)\downarrow-\infty$ as $k\to\infty$ then there is nothing to prove. So suppose that $f(x^k) \ge c$ for some fixed c and all k.

(i) We shall first prove that K is infinite. For purposes of a proof by contradition, suppose that K is finite, i.e. $\delta^k=\bar{\delta} > 0$ for some fixed k_δ and all $k \ge k_\delta$. Then at Step 5 we have $|p^k| \ge \delta^k=\bar{\delta}$ and

$$t_L^k|p^k|^2 = |p^k|t_L^k|d^k| = |p^k||x^{k+1}-x^k| \geq \bar{\delta}|x^{k+1}-x^k|,$$

i.e.

$$t_L^k|p^k|^2 \geq \bar{\delta}|x^{k+1}-x^k| \quad \text{for all} \quad k \geq k_\delta. \tag{6.8}$$

From the assumption that $f(x^k) \geq c$ for all k we obtain, as in Lemma 2.4.7, that $\sum_{k=1}^{\infty} t_L^k|p^k|^2 < +\infty$, hence (6.8) yields

$$\sum_{k=1}^{\infty} |x^{k+1}-x^k| < +\infty. \tag{6.9}$$

Therefore, by the triangle inequality,

$$|x^{k+n}-x^k| \leq \sum_{i=k}^{k+n-1} |x^{i+1}-x^i| \longrightarrow 0 \quad \text{as} \quad k,n \longrightarrow \infty$$

and $\{x^k\}$, being a Cauchy sequence, has a limit

$$x^k \longrightarrow \bar{x} \quad \text{as} \quad k \to \infty. \tag{6.10}$$

Since $\delta^k = \bar{\delta}$ for all $k \geq k_\delta$, we have

$$m_a a^k > \bar{\delta} \quad \text{at Step 4}' \quad \text{for all} \quad k \geq k_\delta, \tag{6.11}$$

$$b^k = \max\{|p^k|, m_a a^k\} > \bar{\delta} \quad \text{at Step 4}' \quad \text{for all} \quad k \geq k_\delta, \tag{6.12}$$

and, because $|p^k| \geq \delta^k$ and $w^k = \frac{1}{2}|p^k|^2 + \tilde{\alpha}_p^k \geq \frac{1}{2}|p^k|^2$ at Step 5,

$$w^k \geq \bar{\varepsilon} > 0 \quad \text{at Step 5} \quad \text{for all} \quad k \geq k_\delta, \tag{6.13}$$

where $\bar{\varepsilon} = (\bar{\delta})^2/2$. By (6.7), we always have $|y^{k+1}-x^{k+1}| \leq \bar{a}$ for $\bar{a} = \delta^1/(2m_a)$, hence Lemma 4.12 remains valid. Therefore, we deduce from (6.10) and (6.13) that the second assertion of Lemma 4.14 is true, so, in view of (6.10), (6.12) and (6.13), we may use the proof of Lemma 4.18. To this end, observe that, since for large j we have $s_j^j = |y^j - x^j| \leq \delta/(2m_a)$ from (6.7), and $s_j^k = s_j^j + \sum_{i=j}^{k-1} |x^{i+1}-x^i|$, (6.9) yields

$$s_j^k < \bar{\delta}/m_a \quad \text{for all large} \quad j \leq k. \tag{6.14}$$

Then one may set $\bar{\varepsilon}_p=2\bar{\delta}$ and proceed as in the proof of Lemma 4.18, deleting (4.30a) and using (6.14) to replace (4.37) by

$$a^k = \max\{s_j^k : j \in \tilde{J}^k\} < \bar{\delta}/m_a \quad \text{for} \quad k=k_2,\dots,k_2+\bar{N}, \tag{6.15}$$

to get (4.38) from (6.11) and (6.15). Thus we obtain a contradiction, showing that K must be infinite.

(ii) If K is infinite, then for infinitely many $k \in K$ we have $\delta^{k+1}=\delta^k/2$, and $0 \le \delta^{k+1} \le \delta^k$ for all k, so $\delta^k \downarrow 0$ as $k\to\infty$. Suppose that $x^k \xrightarrow{\bar{K}} \bar{x}$ for some point $\bar{x}$ and an infinite set $\bar{K} \subset K$. Then, since

$$b^k = \max\{|p^k|,\ m_a a^k\} \le \delta^k \quad \text{for all} \quad k \in K$$

and $\delta^k \downarrow 0$, we obtain from Lemma 4.16 that $0 \in \partial f(\bar{x})$.

(iii) If $\{x^k\}$ is bounded then $\{f(x^k)\}$ is bounded(f is continuous), so case (ii) cannot occur and the bounded subsequence $\{x^k\}_{k\in K}$ must have at least one accumulation point, which is stationary by the preceding results. □

We conclude from the above theorem that if f is bounded from below and the stopping parameter ε_s is positive, then the above-described modification of Algorithm 3.1 with the Wolfe resetting strategy stops after a finite number of iterations, finding an approximately stationary point x^k with $p^k \in \partial f(x^k;\varepsilon_s/m_a)$ and $|p^k| \le \varepsilon_s$.

Let us now see what happens if the Wolfe resetting strategy is used in Algorithm 6.1. To this end, consider the following modification of Algorithm 6.1. In Step 0 we choose a positive δ^1 satisfying $\delta^1 < m_a\bar{a}$, so that we shall have

$$\delta^k/(2m_a) \le \bar{a}/2 \quad \text{for all } k.$$

In Step 3 we replace the resetting test $|p^k| \le m_a\tilde{s}_p^k$ by the test $|p^k| \le \delta^k$. In Step 4(i) we insert the following additional instruction

$$\text{if} \quad m_a\tilde{s}_p^k \le \delta^k \quad \text{then replace} \quad \delta^k \quad \text{by} \quad \delta^k/2.$$

We also employ the line search requirement (6.7) in place of (3.9)-(3.10) in Step 5 of the method. In Step 9 we set $\delta^{k+1}=\delta^k$. Thus $\delta^{k+1} \le \delta^k/2$ only if

$$\max\{|p^k|, m_a\tilde{s}_p^k\} \le \delta^k,$$

and $\delta^{k+1}=\delta^k$ otherwise.

Using the preceding results, one may check that convergence properties of the above modification of Algorithm 6.1 can be expressed in the form of Theorem 6.3.

We shall now describe a resetting strategy based on the ideas of Mifflin (1977b). Thus consider the following modification of Algorithm 3.1. In Step 0 we choose a positive δ^1. Step 3 and Step 4 are replaced by the following

Step 3"(Resetting test). If $|p^k| \le \bar{\theta}\delta^k$ then go to Step 4"; otherwise, replace δ^k by $\min\{\delta^k, |p^k|\}$ and go to Step 5.

Step 4" (Resetting). Replace δ^k by $\bar{\theta}\delta^k$, and then J^k by $\{j \in J^k : s_j^k \le \delta^k/m_a\}$. If $k \notin J^k$ then set $y^k = x^k$, $g^k = g_f(y^k)$, $f_k^k = f(y^k)$, $s_k^k = 0$ and add k to J^k. Set $r_a^k = 1$ and go to Step 1.

In Step 5 we replace the line search requirements (3.9)-(3.10) by the following

$$|y^{k+1} - x^{k+1}| \le \bar{\theta}\delta^k/m_a. \tag{6.16}$$

Step 8 is substituted by the following two steps.

Step 8" (Distance resetting). Set $\delta^{k+1} = \delta^k$. If $m_a a^{k+1} \le \delta^{k+1}$ then go to Step 9. Otherwise, set $r_a^{k+1} = 1$ and replace J^{k+1} by the set $\{j \in J^{k+1} : s_j^{k+1} \le \bar{\theta}\delta^{k+1}/m_a\}$ so that the reset value of a^{k+1} satisfies

$$a^{k+1} = \max\{s_j^{k+1} : j \in J^{k+1}\} \le \bar{\theta}\delta^{k+1}/m_a. \tag{6.17}$$

Step 9". Increase k by 1 and go to Step 1.

To compare the above modification of Algorithm 3.1 with the original method, we observe that, since $\bar{\theta} \in (0,1)$ by assumption, in the above algorithm we have

$$m_a a^k \le \delta^k \quad \text{for all k.} \tag{6.18}$$

Therefore, if no reset occurs at the k-th iteration at Step 4", i.e. if $|p^k| > \bar{\theta}\delta^k$, then we have the relation $|p^k| > \bar{\theta} m_a a^k$, which is similar to the corresponding relation $|p^k| > m_a a^k$ of Algorithm 3.1. In general, the resetting strategy ensures that

$$|p^k| > m_a' a^k \quad \text{at Step 5} \quad \text{for all } k, \tag{6.19}$$

where $m_a' = \bar{\theta}\, m_a > 0$, and

$$|p^k| \geq \delta^k \quad \text{at Step 5} \quad \text{for all } k. \tag{6.20}$$

At the same time, one can have a reset at Step 8" due to $m_a a^{k+1} \leq \delta^k$ even if the value of $|p^k| \geq \delta^k$ is large. In this case a reset occurs even though $|p^k| > m_a a^k$, in contrast with the rules of Algorithm 3.1. We note that on each entrance to Step 4" one has

$$\max\{|p^k|, m_a a^k\} \leq \delta^k \tag{6.21}$$

for successively smaller δ^k-s, so in this case δ^k measures the stationarity of x^k. Moreover, the line search requirement (6.16) and the rules of Step 8 ensure that the latest subgradient g^{k+1} is never deleted at Step 8. This is similar to the corresponding property of Algorithm 3.3.1.

We shall now establish convergence of the above modification of Algorithm 3.1.

Theorem 6.4. Suppose that the above-described modification of Algorithm 3.1 (with the Mifflin resetting strategy) generates an infinite sequence of points $\{x^k\}$ with the stopping parameter ε_s set to zero. Let K denote the possibly emply set of iterations k at which $|p^k| < \delta^k$ at Step 3", i.e. $K=\{k : \delta^{k+1} < \delta^k\}$. Then either of the following two cases arises:

(i) $f(x^k) \downarrow -\infty$ as $k \to \infty$;

(ii) K is an infinite set, $\delta^k \downarrow 0$ as $k \to \infty$ and every accumulation point of the subsequence $\{x^k\}_{k \in K}$ is stationary for f.

Moreover, if the sequence $\{x^k\}$ is bounded (e.g. the level set $S_f = \{x : f(x) \leq f(x^1)\}$ is bounded), then case (ii) above occurs and $\{x^k\}$ has at least one stationary accumulation point.

Proof. Suppose $\{f(x^k)\}$ is bounded from below. There exists a number $\bar{\delta} \geq 0$ such that $\delta^k \downarrow \bar{\delta}$, because we have $0 \leq \delta^{k+1} \leq \delta^k$ for all k.

(i) We shall first prove that $\bar{\delta}=0$. To obtain a contradiction, suppose that $\bar{\delta} > 0$. Then (6.20) and the boundedness of $\{f(x^k)\}$ yield, as in the proof of Theorem 6.3, that

$$\sum_{k=1}^{\infty} |x^{k+1}-x^k| < +\infty , \tag{6.22}$$

$$x^k \longrightarrow \bar{x} \quad \text{as} \quad k\to\infty . \tag{6.23}$$

Note that Step 4" cannot be entered infinitely often, because, since $\bar{\theta} \in (0,1)$, the change of δ^k at Step 4" would imply $\delta^k \downarrow 0$, a contradiction. Hence there exists a number k_δ such that Step 4" is never entered at iterations $k \geq k_\delta$. Thus we have $|p^k| \geq \delta^k \geq \bar{\delta}$ and $w^k = \frac{1}{2}|p^k|^2 + \tilde{\alpha}_p^k \geq \frac{1}{2}|p^k|^2$, i.e.

$$w^k \geq \bar{\varepsilon} = \tfrac{1}{2}(\bar{\delta})^2 > 0 \quad \text{for all} \quad k \geq k_\delta . \tag{6.24}$$

By (6.16), $|y^{k+1}-x^{k+1}| \leq \bar{\theta}\delta^k/m_a \leq \bar{\theta}\delta^1/m_a$, so Lemma 4.12 remains valid. Therefore, from (6.23) and (6.24) we deduce that the second assertion of Lemma 4.14 is true, and that we may now use the proof of Lemma 4.18 (deleting (4.32a), since we no longer have (4.28b)). To this end, we note that, since $\bar{\theta} \in (0,1)$ and $\delta^k \downarrow \bar{\delta} > 0$, we may choose $\eta \in (\bar{\theta},1)$ and $k_\eta \geq k_\delta$ such that $\bar{\theta}\delta^k < \eta\bar{\delta}$ for all $k \geq k_\eta$. Then the line search requirement (6.16) yields

$$s_{k+1}^{k+1} = |y^{k+1}-x^{k+1}| \leq \bar{\theta}\delta^k/m_a < \eta\bar{\delta}/m_a \quad \text{for all} \quad k \geq k_\delta . \tag{6.25a}$$

In view of (6.22) and the fact that $(1-\eta)\bar{\delta}/m_a > 0$, we may choose $k_s > k_\delta$ such that

$$\sum_{k=k_s}^{\infty} |x^{k+1}-x^k| < (1-\eta)\bar{\delta}/m_a . \tag{6.25b}$$

Now, using parts (i)-(iii) of the proof of Lemma 4.18, we find $k_2 > k_s$ such that $r_a^{k_2}=1$ and

$$\max\{s_j^{k_2} : j \in J^{k_2}\} \leq \bar{\theta}\delta^{k_2}/m_a < \eta\bar{\delta}/m_a \tag{6.25c}$$

from (6.17). Since $s_j^k = s_j^n + \sum_{i=n}^{k-1} |x^{i+1}-x^i|$ for all $k > n \geq j$, from (6.25) we obtain

$$\max\{s_j^k : j \in J^{k_2} \text{ or } j \geq k_2\} < \bar{\delta}/m_a \quad \text{for all} \quad k \geq k_2 . \tag{6.26}$$

But Lemma 4.1 shows that on each entrance to Step 8" we have

$$a^{k+1} = \max\{s_j^{k+1} : j \in \hat{J}_p^k \text{ or } j=k+1\},$$

hence (6.26) and the fact that $\hat{J}_p^k \subset J^{k_2} \cup \{j : j \geq k_2\}$ yield

$$m_a a^{k+1} < \bar{\delta} \leq \delta^{k+1} \quad \text{for all} \quad k \geq k_2, \tag{6.27}$$

showing that no reset due to $m_a a^{k+1} > \delta^{k+1}$ can occur at Step 8" for any $k > k_2$. Thus (4.38) holds and we obtain a contradiction. Therefore, $\bar{\delta}=0$.

(ii) Note that for each $k \in K$ we have $|p^k| < \delta^k$ and $m_a a^k \leq \delta^k$ from (6.18), so

$$b^k = \max\{|p^k|, m_a a^k\} \leq \delta^k \quad \text{for all} \quad k \in K.$$

Using this relation one can complete the proof analogously to parts (ii)-(iii) of the proof of Theorem 6.3. □

We conclude from the preceding results that the resetting strategies of Wolfe (1975) and Mifflin (1977b) have certain theoretical drawbacks. Namely, they lead to algorithms that are convergent in the sense that they have at least one stationary accumulation point, provided that they generate bounded sequences $\{x^k\}$. Such convergence results are weaker than the ones established for Algorithm 3.1 and Algorithm 6.1, which say that every accumulation point of these algorithms is stationary, under no additional assumptions on f.

The difference between these two kinds of convergence results may be explained as follows. The resetting strategies of Wolfe (1975) and Mifflin (1977b) are based on the use of certain monotonically nonincreasing sequences $\{\delta^k\}$, such that δ^k decreases at each reset, so that each reset in influenced by all the preceding resets. On the other hand, resetting tests of the form $|p^k| \leq m_a a^k$ or $|p^k| \leq m_a \tilde{s}_p^k$ are more local in nature in the sense that they do not depend explicitly on any of the preceding resets.

This difference directly influences the required techniques of convergence analysis. Namely, the local character mentioned above is well-suited to proving that an arbitrary accumulation point of $\{x^k\}$ is stationary. On the other hand, contradicting the statement that every accumulation point is nonstationary requires only the global arguments mentioned earlier.

It is worthwhile to point out that, from the theoretical point of view, in the convex case the above-described modifications of Algorithm

3.1 and Algorithm 6.1 with the resetting strategies of Wolfe and Mifflin have the same global convergence properties as the orginal methods. Namely, one may augment Theorem 6.3 and Theorem 6.4 with the statement that if f is convex then $\{x^k\}$ is a minimizing sequence, which converges to a minimum point of f if f attains its infimum. This statement can be proved similarly to Theorem 4.20. In this context we observe, once again, that resetting strategies have no impact on convergence in the convex case, since in fact the convex case needs no resettings; see Chapter 2.

We shall now use the preceding results of this section to design a modified resetting strategy for Algorithm 3.1. This time our motivation is practical and stems from the following observation. Each resetting of Algorithm 3.1 involves solving a quadratic programming subproblem. Hence we may have to solve many search direction finding subproblems at any iteration, especially if few past subgradients are deleted at each reset. For simplicity, at Step 4 of Algorithm 3.1 we discard only one past subgradient, although the preceding convergence results are not impaired if one deletes more subgradients. At the same time, a premature reduction of the past subgradient information can be wasteful, since then the algorithm has to accumulate new subgradients at null steps, when no movement towards a solution occurs. Thus we need a rule for detecting the case when it is more efficient to discard several past subgradients, rather than wait until they are dropped after we have solved several quadratic programming subproblems. Of course, such a rule should not impair the preceding global convergence results. Guided by these observations, we shall now present a subgradient deletion rule based on the ideas of Mifflin (1977b).

Consider the following modification of Algorithm 3.1. In Step 0 we choose a positive δ^1 and a parameter $\bar{\theta} \in (0,1)$. Step 3 and Step 4 are replaced by the following

<u>Step 3"(Resetting test)</u>. Set

$$\tilde{\delta}^k = \max\{|p^k|, m_a a^k\}. \tag{6.28}$$

If $\tilde{\delta}^k < \bar{\theta}\delta^k$ set $\delta^{k+1}=\tilde{\delta}^k$; otherwise set $\delta^{k+1}=\delta^k$. If $|p^k| \le m_a a^k$ then go to Step 4‴; otherwise, go to Step 5.

<u>Step 4‴ (Resetting)</u>. Replace J^k by the set $\{j \in J^k : s_j^k \le \bar{\theta}\delta^{k+1}/m_a\}$. If $k \notin J^k$ then set $y^k=x^k$, $g^k=g_f(y^k)$, $f_k^k=f(y^k)$, $s_k^k=0$ and add k to J^k. Set $r_a^k=1$ and go to Step 1.

The line search requirements (3.9)-(3.10) of Step 5 are replaced by the following

$$|y^{k+1}-x^{k+1}| \leq \bar{\theta}\delta^{k+1}/m_a. \tag{6.29}$$

Finally, Step 8 is substituted by

Step 8" (Distance resetting). If $\delta^{k+1}=\delta^k$ then go to Step 9. Otherwise, replace J^{k+1} by the set $\{j \in J^{k+1} : s_j^{k+1} \leq \bar{\theta}\delta^{k+1}/m_a\}$. If $a^{k+1} > \bar{\theta}\delta^{k+1}/m_a$ then set $r_a^{k+1}=1$ and

$$a^{k+1} = \max\{s_j^{k+1} : j \in J^{k+1}\} \leq \bar{\theta}\delta^{k+1}/m_a. \tag{6.30}$$

Step 9". Increase k by 1 and go to Step 1.

For comparing the above modification of Algorithm 3.1 with the original method we shall use the fact that, as will be proved below, $\delta^k \downarrow 0$ whenever $\{x^k\}$ has at least one accumulation point. Thus the two methods differ in several aspects. First, the above criteria for retaining past subgradients g^j at a reset are formulated directly in terms of subgradient distance measures $s_j^k \geq |y^j-x^k|$, while in Algorithm 3.1 a suitable reduction of the locality radius a^k at a reset was ensured by retaining a limited number of the latest subgradients, so that g^j was retained only if both $j \geq k-M_g+2$ and the value of s_j^k was smaller than $|p^k|/m_a$ after a reset. Secondly, the line search rule (6.29) makes y^{k+1} sufficiently close to x^{k+1} when the algorithm approaches a solution, as indicated by a small value of δ^k, while in Algorithm 3.1 the value of $|y^{k+1}-x^{k+1}|$ is controlled by a combination of the line search criteria (3.10) and the rules of Step 6. Thirdly, the subgradient deletion rules of Step 8" reduce the number of resets at Step 4"', thus saving work requred by quadratic programming subproblems. Namely, at Step 8" we detect the situation when the algorithm approaches a solutions, i.e. $\delta^{k+1} < \bar{\theta}\delta^k$, and then discard a few (seemingly) irrelevant subgradients, trying to forestall a reset at Step 4" at the next iteration.

We shall now show that the preceding global convergence results cover this modification.

Theorem 6.5. Suppose that the above-described modification of Algorithm 3.1 (with Step 3", Step 4", etc.) generates an infinite sequence $\{x^k\}$. Then every accumulation point of $\{x^k\}$ is stationary. Moreover, Theorem 4.20 and Corollary 4.21 are true.

Proof.(i) Suppose that there exist a point $\bar{x} \in R^N$ and an infinite set $K \subset \{1,2,...\}$ such that $x^k \xrightarrow{K} \bar{x}$. We claim that $\delta^k \downarrow 0$. The reader may verify this claim by using the proof of Theorem 6.4.

(ii) We now claim that Lemma 4.18 and its proof remain valid if one replaces in Algorithm 3.1 Step 4 and the line search requirements (3.9)-(3.10) by Step 4''' and (6.29), respectively, provided that $\delta^k \downarrow 0$. To justify this claim, use (6.29) and the assumption that $\delta^k \downarrow 0$ for showing that Lemma 4.12, Lemma 4.14 with (4.28a) deleted, and Lemma 4.15 are true.

(iii) The theorem will be proved if we show how to modify the proof of Lemma 4.18. Thus suppose that (4.20), (4.22) and (4.31) hold. Let $K_\delta = \{k : \delta^{k+1} < \bar{\theta}\delta^k\}$. From part (i) above and the assumption (4.20) we know that K_δ is infinite and $\delta^k \downarrow 0$. Therefore, in view of part (ii) above, in the proof of Lemma 4.18 we need only consider additional resets occuring at Step 8" for $k \in K_\delta$. To this end, suppose that $\tilde{k}$ is so large that $\bar{\theta}\delta^{\tilde{k}} < \bar{\varepsilon}_p/2$, where $\bar{\varepsilon}_p > 0$ is the constant involved in (4.36). Then (4.36) and the rules of Step 3" yield $\delta^{k+1} = \delta^k$ for $k = k_2, ..., k_2 + \tilde{N}$, so that $k \notin K_\delta$ for $k = k_2, ..., k_2 + \tilde{N}$. Thus no resets occur at Step 8" for $k = k_2, ..., k_2 + \tilde{N}$, and part (iv) of the proof of Lemma 4.18 remains valid. Thus Lemma 4.18, and hence also Theorem 4.19, Theorem 4.20 and Corollary 4.21 are true. □

Remark 6.6. We conclude from the above proof that the global convergence results established in Section 4 for Algorithm 3.1 are not impaired if one replaces Step 4 by Step 4''' above, and the line search requirements (3.9)-(3.10) by (6.29), provided that the rules for choosing $\{\delta^k\}$ are such that $\{\delta^k\}$ is bounded and $\delta^k \to 0$ whenever $\{x^k\}$ has at least one accumulation point. This observation may be used in designing rules different from the ones of Algorithm 3.1 and its above-described modification.

Let us now consider another modification of Algorithm 3.1, which is similar to Algorithm 6.1. Thus suppose that we use Step 3''' with

$$\tilde{\delta}^k = \max\{|p^k|, m_a \tilde{s}_p^k\} \tag{6.31}$$

instead of (6.28), and replace the resetting test $|p^k| \le m_a a^k$ by $|p^k| \le m_a \tilde{s}_p^k$, where the aggregate distance measures $\tilde{s}_p^k$ are generated via (6.1)-(6.2). We also use Step 4''' and substitute the line search requirements (3.9)-(3.10) by (6.29), and Step 8 by Step 8''' with the condition "$\delta^{k+1}=\delta^k$" replaced by "$\delta^{k+1}=\delta^k$ and $a^{k+1} \le \bar{a}$".

The resulting method is globally convergent in the sense of Theorem 6.5. To see this, use a combination of the proofs of Theorem 6.2 and Theorem 6.5. At the same time, the use of Step 8''' and of the resetting test $|p^k| \le m_a \tilde{s}_p^k$ decreases the frequency of resets occuring at Step 4".

We may add that each of the modified subgradient deletion rules and line search requirements of this section may be incorporated in Algorithm 5.1. The corresponding results on convergence are the same.

Remark 6.7. As observed in Section 3.6, in many cases it is efficient to calculate subgradients also at points $\{x^k\}$, and use such additional subgradients for each search direction finding. This idea can be readily incorporated in the methods discussed so far in this chapter. Namely, using notation of (3.6.11)-(3.6.16), one may evaluate additional subgradients

$$g^{-j} = g_f(y^{-j}) = g_f(x^j) \quad \text{for} \quad j=1,2,\ldots,$$

and choose sets J^{k+1} subject to the preceding requirements and the following

$$-(k+1) \in J^{k+1} \quad \text{for all } k.$$

Then there is no need for Step 4(iii) in Algorithm 3.1, Algorithm 6.1 and their various modifications discussed so far. Also the preceding global convergence results are not influenced, although in practice the use of additional subgradients may enhance faster convergence.

7. Simplified Versions That Neglect Linearization Errors

In this section we shall consider simplified versions of the previously discussed methods that are obtained by neglecting linearization errors at each search direction finding, i.e. by setting $\alpha_j^k=\alpha_p^k=0$ in the search direction finding subproblems (3.1) and (5.1). The resulting dual search direction finding subproblems (3.16) and (3.5.11) have special structure, which enables one to use efficient quadratic programming subroutines. Particular variants of such methods include the algorithms of Wolfe (1975) and Mifflin (1977b).

We may add that methods that neglect linearization errors, proposed by Wolfe (1975) and extended to the nonconvex case by Mifflin

(1977b), seem to be less efficient in practice than other algorithms (Lemarechal, 1982). However, they are relatively simple to implement and still attract considerable theoretical attention (Polak, Mayne and Wardi, 1983).

Let us, therefore, consider the following modification of Algorithm 3.1. In the primal search direction finding subproblem (3.1) we set

$$\alpha_j^k=0 \quad \text{for} \quad j\in J^k, \text{ and } \quad \alpha_p^k=0, \tag{7.1}$$

i.e. we use (7.1) instead of (3.2). Then the corresponding dual search direction finding subproblem (3.16) is of the form

$$\begin{aligned} &\underset{\lambda,\lambda_p}{\text{minimize}}\ \tfrac{1}{2}\Big|\sum_{j\in J^k}\lambda_j g^j+\lambda_p p^{k-1}\Big|^2, \\ &\text{subject to} \quad \lambda_j\geq 0,\ j\in J^k,\ \lambda_p\geq 0,\ \sum_{j\in J^k}\lambda_j+\lambda_p=1, \\ &\qquad\qquad \lambda_p=0 \quad \text{if} \quad r_a^k=1. \end{aligned} \tag{7.2}$$

Solving (7.2) is equivalent to finding the vector of minimum length in the convex hull of $\{p^{k-1}, g^j : j\in J^k\}$ if $r_a^k=0$ (or $\{g^j : j\in J^k\}$ if $r_a^k=1$). This can be done by using the very efficient and stable Wolfe (1976) algorithm, designed specially for quadratic programming problems of the form (7.2). Another advantage of this simplified version is that linearization values f_j^k and f_p^k are not needed, hence one can save the effort previously required by updating linearization values by (3.13). We shall also neglect linearization errors at line searches by setting

$$v^k = -|p^k|^2, \tag{7.3}$$

$$\alpha(x,y) = 0 \quad \text{for all} \quad x \quad \text{and} \quad y, \tag{7.4}$$

instead of using (3.5) and (3.11). Then $v^k<0$, so that Line Search Procedure 3.2 can be used as before, with Lemma 3.3 remaining valid.

As far as convergence of the above algorithm is concerned, one may reason as if the values of $\tilde{\alpha}_p^k$ were zero for all k, since in this case (7.3) is equivalent to the previously employed relation (3.5). Then it is easy to check that the modification defined by (7.1)-(7.4) does not impair Theorem 4.19 and Corollary 4.21; in fact, the relevant proofs of Section 4 are simplified. We conclude that in the nonconvex case the above-described version of Algorithm 3.1 has the same global convergence properties as the original method.

At the same time, it is not clear whether Theorem 4.20 holds for the above-described method, since the additional results on convergence in the covex case depended strongly on the property that we had

$$\sum_{k=1}^{\infty} t_L^k\{|p^k|^2+\tilde{\alpha}_p^k\} < +\infty \tag{7.5}$$

whenever $\{f(x^k)\}$ was bounded from below; see the proofs of Theorem 2. 4.15 and Theorem 2.4.16. We recall fromm the proof of Lemma 2.4.7 that (7.5) resulted from summing the inequalities due to the line search criterion (3.7)

$$t_L^k(-v^k) \le [f(x^k)-f(x^{k+1})]/m_L \tag{7.6}$$

with $-v^k=|p^k|^2+\tilde{\alpha}_p^k$, for all k. But now we have $-v^k=|p^k|^2$ in the simplified version of the method, so that (7.6) yields

$$\sum_{k=1}^{\infty} t_L^k|p^k|^2 < \infty \tag{7.7}$$

if $\{f(x^k)\}$ is bounded from below, and, unfortunately, (7.5) need not hold.

In view of the difficulties mentioned above, we shall now show that by using a suitable resetting strategy one may neglect linearization errors in Algorithm 3.1 without sacrificing strong global convergence properties in the convex case.

Our new resetting strategy can be motivated as follows. We want to ensure (7.5) if $\{f(x^k)\}$ is bounded from below. In view of (7.7), this will be the case if we always have

$$|p^k|^2 > m_\alpha \alpha_p^k \tag{7.8}$$

at line searches, where m_α is a fixed, positive parameter. To this end, we shall reset the method whenever (7.8) is violated.

Thus consider the version of Algorithm 3.1 that uses (7.1)-(7.4) and the following modification of Step 3.

<u>Step 3' (Resetting test)</u>. If $|p^k| \le m_a a^k$ or $|p^k|^2 \le m_\alpha \tilde{\alpha}_p^k$ then go to Step 4; otherwise, go to Step 5.

Reasoning as in Section 3, one may check that only finitely many resets can occur at any iteration of the modified method.

We shall now show that the modified method retains all the convergence properties of Algorithm 3.1. Since Lemma 4.3 remains true, we may suppose that the method does not terminate.

Theorem 7.1. The above-described modification of Algorithm 3.1 (with (7.1)-(7.4) and Step 3') is globally convergent in the sense of Theorem 4.19, Theorem 4.20 and Corollary 4.21.

Proof. In view of the preceding results, we only need to establish Theorem 4.19, since then, by (7.7)-(7.8), Theorem 4.20 and Corollary 4.21 will follow as before. To this end, we show how to modify the proof of Lemma 4.18. Suppose that in addition to (4.35) we can assert that

$$\tilde{\alpha}_p^k < \bar{\varepsilon}_p^2/(4m_\alpha) \quad \text{for all} \quad k \in [k_2, k_2+\bar{N}]. \tag{7.9}$$

Then (4.36), (4.37) and (7.9) yield $|p^k| > m_a a^k$ and $|p^k|^2 > m_\alpha \tilde{\alpha}_p^k$ for all $k \in [k_2,\ldots,k_2+\bar{N}]$, so (4.38) holds and we may complete the proof of Lemma 4.18. Thus it suffices to show that in part (ii) of the proof of Lemma 4.18 we can choose $\tilde{k}$ such that we have (4.29), (4.32), and then (4.35) and (7.9) in part (iv) of the proof. This is easy if we observe that, by Lemma 4.4,

$$\begin{aligned}
\tilde{\alpha}_p^k &= |f(x^k)-\tilde{f}_p^k| = \\
&= \Big|\sum_{i=1}^{M} \hat{\lambda}_i^k[f(x^k)-f(y^{k,i})-\langle g_f(y^{k,i}),x^k-y^{k,i}\rangle]\Big| \le \\
&\le \max\{|f(x^k)-f(y^{k,i})-\langle g_f(y^{k,i}),x^k-y^{k,i}\rangle| : i=1,\ldots,M\} \le \\
&\le \max_i|f(x^k)-f(y^{k,i})| + \max_i|g_f(y^{k,i})|\,|x^k-y^{k,i}| \le \\
&\le 2L\max\{|x^k-y^{k,i}| : i=1,\ldots,M\} \le 2La^k,
\end{aligned}$$

so, by (4.5) and (4.7),

$$\tilde{\alpha}_p^k \le 2L\max\{s_j^k : j\in \tilde{J}^k\}, \tag{7.10}$$

where L is the Lipschitz constant of f on $\tilde{B}$

$$|f(x)-f(y)| \le L|x-y| \quad \text{for all } x \text{ and } y \text{ in } \tilde{B},$$

and $\tilde{B}$ is any bounded set containing x^k and $y^{k,i}$, $i=1,\ldots,M$. Let

$\tilde{B}=\{y : |\bar{x}-y| \le 2\}$. From (4.22), (4.31), Lemma 4.14, Lemma 4.15 and (4.20) we deduce the existence of $\tilde{k}$ satisfying (4.29), (4.32) and

$$\max\{s_j^k : \tilde{k} \le j \le k < \tilde{k}+\tilde{N}\} < \min\{1, \bar{\varepsilon}_p^2/(8m_\alpha)\}, \tag{7.11a}$$

$$|\bar{x}-x^k| < 1 \quad \text{for} \quad k=\tilde{k},\dots,\tilde{k}+\tilde{N}. \tag{7.11b}$$

Then in part (iv) we have, for $k=k_2,\dots,k_2+\bar{N}$ and $i=1,\dots,M$, that

$$|\bar{x}-y^{k,i}| \le |\bar{x}-x^k|+|x^k-y^{k,i}| < 1+1=2$$

from (7.11), since $\{y^{k,i} : i=1,\dots,M\} \subset \{y^j : j \in \tilde{J}^k\}$ and $|x^k-y^j| \le s_j^k$. Thus $x^k \in \tilde{B}$ and $y^{k,i} \in \tilde{B}$ for $k=k_2,\dots,k_2+\bar{N}$ and $i=1,\dots,M$. Then (4.35a), (7.10) and (3.11a) yield (7.9), as required. □

CHAPTER 5

Feasible Point Methods for Convex Constrained Minimization Problems

1. Introduction

In this chapter we consider the following convex constrained minimization problem

$$\text{minimize} \quad f(x), \quad \text{subject to} \quad F(x) \le 0,$$

where the functions $f : R^N \to R$ and $F : R^N \to R$ are convex, but not necessarily differentiable. We assume that the Slater constraint qualification is fulfilled, i.e. there exists $\tilde{x} \in R^N$ satisfying $F(\tilde{x}) < 0$. This implies that the feasible set

$$S = \{x \in R^N : F(x) \le 0\}$$

has a nonempty interior.

We present a class of readily implementable algorithms, which differ in complexity and efficiency. The algorithms require only the calculation of f or F and one subgradient of f or F at designated points. Each algorithm yields a minimizing sequence of points. Moreover, this sequence converges to a solution of problem (1.1) whenever the solution set of (1.1)

$$\overline{X} = \operatorname*{Arg\,min}_S f = \{\overline{x} \in S : f(\overline{x}) \le f(x) \quad \text{for all} \quad x \in S\}$$

is nonempty. Storage requirements and work per iteration of the algorithms can be controlled by the user.

The algorithms are <u>feasible point descent methods</u>, i.e. they generate a sequence of points $\{x^k\}$ satisfying

$$x^k \in S \quad \text{and} \quad f(x^{k+1}) < f(x^k) \quad \text{if} \quad x^{k+1} \ne x^k, \text{ for } \quad k=1,2,\ldots,$$

where $x^1 \in S$ is the starting point. They may be viewed as extensions of the Pironneau and Polak (1972,1973) method of centers and method of feasible directions to the nondifferentiable case.

The methods generate search directions by using the subgradient selection and aggregation strategies introduced in Chapter 2. The important extension to the constrained case is that we use <u>separate selection and aggregation</u> of subgradients of each of the problem functions. This enables us to construct suitable polyhedral approximations of f and F. One of the algorithms presented (the method of feasible directions from Section 6) may be obtained by employing subgradient aggregation in the algorithm of Mifflin (1982). We may add that another type of aggregation was used by Lemarechal, Strodiot and Bihain (1981) for linearly constrained problems, but with no global convergence results.

The algorithms require a feasible starting point $x^1 \in S$, but require no knowledge of f at infeasible points. Also at feasible points we require no knowledge of F (other than F being nonpositive); this is important in certain applications, see Section 1.3. Each of the algorithms can find a feasible starting point by minimizing F, starting from any point.

We shall also present phase I - phase II methods which can be used when the initial approximation to a solution is infeasible. These methods stem from an algorithm of Polak, Trahan and Mayne (1979) for smooth problems. They differ significantly from the phase I - phase II algorithm of Polak, Mayne and Wardi (1983) for nonsmooth problems, both in their construction and in stronger global convergence properties.

In Section 2 we derive basic versions of the methods and give comparisons with other algorithms. An algorithmic procedure is presented in Section 3, and its global convergence is demonstrated in Section 4. Section 5 is devoted to methods with subgradient selection. Useful modifications of the methods are given in Section 6. Phase I - phase II methods are described in Section 7.

2. Derivation of the Algorithm Class

We start by recalling the necessary and sufficient optimality conditions for problem (1.1); see Section 1.2. For any x and y in R^N, let

$$H(y;x) = \max\{f(y)-f(x), F(y)\}, \tag{2.1}$$

and, for any fixed x, let $\partial H(y;x)$ denote the subdifferential of the convex function $H(\cdot;x)$ at y, i.e.

$$\partial H(y;x) = \begin{cases} \partial f(y) & \text{if } f(y)-f(x) > F(y), \\ \text{conv}\{\partial f(y) \cup \partial F(y)\} & \text{if } f(y)-f(x) = F(y), \\ \partial F(y) & \text{if } f(y)-f(x) < F(y). \end{cases}$$

The Slater constraint qualification implies that the feasible set S is nonempty and that the solution set $\overline{X}$ can be characterized as follows; see Lemma 1.2.16.

Lemma 2.1. The condition $\overline{x} \in \overline{X}$ is equivalent to

$$\min\{H(y;x) : y \in R^N\} = H(\overline{x};\overline{x})=0, \tag{2.2}$$

which in turn is equivalent to

$$0 \in \partial H(\overline{x};\overline{x}).$$

Remark 2.2. There is no loss of generality in requiring that F be scalar-valued. If the original formulation of the problem involves a finite number of convex constraints $\tilde{F}_j(x) \le 0$, $j \in J_F$, then one may set

$$F(x) = \max\{\tilde{F}_j(x) : j \in J_F\}.$$

By Corollary 1.2.6, $\partial F(x)=\mathrm{conv}\{g \in \partial \tilde{F}_i(x) : \tilde{F}_i(x)=F(x)\}$.

Observe that if $H(x;x)=0$, i.e. x is feasible, and $H(y;x) < H(x;x)$, then $f(y)-f(x) < 0$ and $F(y) < 0$, hence y is both feasible and has a strictly lower objective value than x. For this reason the function H is sometimes called an improvement function for problem (1.1).

To calculate a point $\overline{x}$ satsfying (2.2), one may use the following method of centers due to Huard (1967):

Step 0. Set $k=1$ and find a point $x^1 \in S$.

Step 1. Find a direction d^k solving the problem

$$\text{minimize } H(x^k+d;x^k) \quad \text{over all } d \in R^N \tag{2.3}$$

and set $y^{k+1}=x^k+d^k$.

Step 2. If

$$H(y^{k+1};x^k) < H(x^k;x^k)$$

then set $x^{k+1}=y^{k+1}$; otherwise set $x^{k+1}=x^k$.

Step 3. Increase k by 1 and go to Step 1.

By (Polak, 1970; Section 4.2), if a sequence $\{x^k\} \subset S$ generated by the method of centers has an accumulation point $\bar{x}$ then (2.2) holds, i.e. $\bar{x}$ solves problem (1.1).

Note that (2.3) is a nonsmooth unconstrained minimization problem that generally cannot be carried out. Therefore, in implementable versions of the method it is replaced by a simpler subproblem based on a suitable local approximation to the function $H(\cdot;x^k)$.

Since the algorithms to be described are structurally similar to the Pironneau and Polak (1972,1973) method of centers for smooth problems, we shall now review an extension of that method to constrained minimax problems (Kiwiel, 1981a). To this end, suppose momentarily that

$$f(x)=\max\{f_j(x) : j \in J_f\} \quad \text{and} \quad F(x)=\max\{F_j(x) : j \in J_F\}, \tag{2.4}$$

where f_j, $j \in J_f$, and F_j, $j \in J_F$, are convex functions with continuous gradients ∇f_j and ∇F_j, respectively, and the sets J_f and J_F are finite. To calculate a point $\bar{x}$ satisfying (2.2), the method of centers (Kiwiel, 1981a) proceeds as follows. Given the k-th approximation to a solution $x^k \in S$, a search direction d^k is found from the solution $(d^k,v^k) \in R^N \times R$ to the following problem

$$\begin{aligned} &\text{minimize } \tfrac{1}{2}|d|^2+v, \\ &\text{subject to } f_j(x^k)-f(x^k)+ < \nabla f_j(x^k),d > \le v, \quad j \in J_f, \\ &\qquad\qquad F_j(x^k)+ < \nabla F_j(x^k),d > \le v, \quad j \in J_F. \end{aligned} \tag{2.5}$$

The above problem may be interpreted as a local first order approximation to the problem of minimizing $H(x^k+d;x^k)$ with respect to $d \in R^N$. To see this, define the following polyhedral approximations to f, F and $H(\cdot;x^k)$

$$\begin{aligned} \hat{f}_c^k(x) &= \max\{f_j(x^k)+ < \nabla f_j(x^k),x-x^k > : j \in J_f\}, \\ \hat{F}_c^k(x) &= \max\{F_j(x^k)+ < \nabla F_j(x^k),x-x^k > : j \in J_F\}, \\ \hat{H}_c^k(x) &= \max\{\hat{f}_c^k(x)-f(x^k),\ \hat{F}_c^k(x)\}, \end{aligned} \tag{2.6}$$

respectively. Then subproblem (2.5) is equivalent to the following

$$\text{minimize } \hat{H}_c^k(x^k+d)+ \tfrac{1}{2}|d|^2 \quad \text{over all } d, \tag{2.7}$$

and we have

$$v^k = \hat{H}_c^k(x^k+d^k). \tag{2.8}$$

We note that subproblems (2.5) and (2.7) are extensions of the search direction finding subproblems of the method of linearizations described in Section 2.2, see (2.2.2) and (2.2.4). Also if $|J_f|=|J_F|=1$ then subproblem (2.5) reduces to subproblems of the method of centers and method of feasible directions due to Pironneau and Polak (1972,1973). The regularizing term $\frac{1}{2}|d|^2$ in (2.7) serves to keep x^k+d^k in the region where $\hat{H}^k(\cdot)$ is a close approximation to $H(\cdot;x^k)$.

Reasoning as in Section (2.2), one may prove that v^k majorizes the directional derivative of $H(\cdot;x^k)$ at x^k in the direction d^k. Therefore the method of centers can choose a stepsize $t^k > 0$ as the largest number in $\{1, \frac{1}{2}, \frac{1}{4}, \ldots\}$ satisfying

$$H(x^k+t^kd^k;x^k) \le H(x^k;x^k)+m\, t^kv^k, \tag{2.9}$$

where $m \in (0,1)$ is a fixed line search parameter. This is possible when $v^k < 0$, which is the case if $x^k \notin \overline{X}$. The method, of course, stops if $x^k \in \overline{X}$. Otherwise (2.9) yields the next point x^{k+1} satisfying

$$f(x^{k+1}) < f(x^k) \quad \text{and} \quad x^{k+1} \in S.$$

This follows from the nonnegativity of mt^kv^k and the fact that $H(x^k;x^k)=0$ owing to $x^k \in S$.

It is known (Kiwiel, 1981a) that the above method of centers is globally convergent (to stationary points of (1.1) if the problem function are nonconvex), and that the rate of convergence is at least linear under standard second order sufficiency conditions of optimality. This justifies our efforts to extend the method to more general nondifferentiable problems.

Although the methods given below will not require the special form (2.4) of the problem functions, they are based on similar representations

$$f(x) = \max\{f(y)+\langle g_{f,y}, x-y\rangle : g_{f,y} \in \partial f(y),\ y \in R^N\},$$

$$F(x) = \max\{F(y)+\langle g_{F,y}, x-y\rangle : g_{F,y} \in \partial F(y),\ y \in R^N\},$$

which are due to convexity. Since such implicit representations cannot be computed, the methods will use their approximate versions constructed as follows.

We suppose that we have subroutines that can evaluate subgradient functions $g_f(x)\in\partial f(x)$ at each $x\in S$, and $g_F(x)\in\partial F(x)$ at each $x\in R^N\setminus S$. For simplicity, we shall temporarily assume that $g_f(x)\in\partial f(x)$ and $g_F(x)\in\partial F(x)$ for each $x\in R^N$.

Suppose that at the k-th iteration we have the current point $x^k\in S$ and some auxiliary points y^j, $j\in J_f^k\cup J_F^k$, and subgradients $g_f^j=g_f(y^j)$, $j\in J_f^k$, and $g_F^j=g_F(y^j)$, $j\in J_F^k$, where J_f^k and J_F^k are some subsets of $\{1,...,k\}$. Define the linearizations

$$
\begin{aligned}
f_j(x) &= f(y^j)+\langle g_f^j,x-y^j\rangle, \ j\in J_f^k\\
F_j(x) &= F(y^j)+\langle g_F^j,x-y^j\rangle, \ j\in J_F^k,
\end{aligned}
\tag{2.10}
$$

and the current polyhedral approximations $\hat{f}_s^k$ and $\hat{F}_s^k$ to f and F, respectively,

$$
\begin{aligned}
\hat{f}_s^k(x) &= \max\{f_j(x): j\in J_f^k\},\\
\hat{F}_s^k(x) &= \max\{F_j(x): j\in J_F^k\}.
\end{aligned}
\tag{2.11}
$$

Noting the similarities in (2.4) and (2.10)-(2.11), we see that applying one iteration of the method of centers to $\hat{f}_s^k$ and $\hat{F}_s^k$ at x^k would lead to the following search direction finding subproblem

$$
\begin{aligned}
&\text{minimize} \quad \tfrac{1}{2}|d|^2+v,\\
&\text{subject to} \quad f_j^k-f(x^k)+\langle g_f^j,d\rangle \le v, \ j\in J_f^k,\\
&\qquad\qquad\quad F_j^k+\langle g_F^j,d\rangle \le v, \ j\in J_F^k,
\end{aligned}
\tag{2.12}
$$

where $f_j^k=f_j(x^k)$, $j\in J_f^k$, and $F_j^k=F_j(x^k)$, $j\in J_F^k$. Defining the following polyhedral approximation to $H(\cdot;x^k)$

$$\hat{H}_s^k(x) = \max\{\hat{f}_s^k(x)-f(x^k),\hat{F}_s^k(x)\}, \tag{2.13}$$

we deduce that (2.12) is a quadratic programming formulation of the subproblem

$$\text{minimize} \quad \hat{H}_s^k(x^k+d)+\tfrac{1}{2}|d|^2 \quad \text{over all } d, \tag{2.14}$$

with the solution (d^k,v^k) of (2.12) satisfying

$$\hat{H}_s^k(x^k+d^k) = v^k, \tag{2.15}$$

cf. (2.6)-(2.8). Thus subproblem (2.12) is a local approximation to the problem of minimizing $H(x^k+d;x^k)$ over all d.

In the next section we prove that $v^k \le 0$ and that $v^k=0$ only if $x^k \in \bar{X}$. Therefore we may now suppose that v^k is negative. The line search rule of the above-described method of centers must be modified here, because v^k need no longer be an upper estimate of the directional derivative of $H(\cdot,x^k)$ at x^k in the direction d^k. This is due to the fact that $\hat{H}_s^k(\cdot)$ may poorly approximate $H(\cdot;x^k)$. However, we still have

$$\hat{H}_s^k(x^k+td^k) \le H(x^k;x^k)+tv^k \quad \text{for all} \quad t \in [0,1], \tag{2.16}$$

owing to (2.15), the convexity of $\hat{H}^k$ and the fact that $\hat{H}^k(x^k) \le H(x^k;x^k)$, which is zero by assumption ($x^k \in S$). Therefore the variable

$$v^k = \hat{H}_s^k(x^k+d^k)-H(x^k;x^k) \tag{2.17}$$

may be thought of as an <u>approximate directional derivative</u> of $H(\cdot,x^k)$ at x^k in the direction d^k. Consequently, if we use the rules of Section 2.2 for searching from x^k along d^k for a stepsize that gives a reduction in $H(\cdot;x^k)$, we obtain the following line search rules.

Let $m \in (0,1)$ and $\bar{t} \in (0,1]$ be fixed line search parameters. We shall start by searching for the largest number $t_L^k \ge \bar{t}$ in $\{1, \frac{1}{2}, \frac{1}{4}, \ldots\}$ that satisfies

$$H(x^k+t_L^k d^k;x^k) \le H(x^k;x^k)+mt_L^k v^k. \tag{2.18}$$

This requires a finite number of the problem function evaluations. For instance, if $\bar{t}=1$ then (2.18) reduces to the following test

$$H(x^k+d^k)-H(x^k;x^k) \le m[\hat{H}_s^k(x^k+d^k)-H(x^k;x^k)]. \tag{2.19}$$

If a stepsize $t_L^k \ge \bar{t}$ satisfying (2.18) is found, then the method can execute a serious step by setting $x^{k+1}=x^k+t_L^k d^k$ and $y^{k+1}=x^{k+1}$. Otherwise a null step is taken by setting $x^{k+1}=x^k$. In this case an auxiliary stepsize $t_R^k \ge \bar{t}$ satisfying

$$H(x^k+t_R^k d^k;x^k) > H(x^k;x^k)+mt_R^k v^k$$

is known from the search for t_L^k. Then the trial point $y^{k+1}=x^k+t_R^k d^k$ and the subgradients $g_f^{k+1}=g_f(y^{k+1})$ and $g_F^{k+1}=g_F(y^{k+1})$ will define the corresponding linearizations f_{k+1} and F_{k+1} by (2.4) that will significantly modify the next polyhedral approximations $\hat{f}^{k+1}$ and $\hat{F}^{k+1}$, provided that $k+1 \in J_f^{k+1} \cup J_F^{k+1}$, see Section 6. Thus after a null step the method will improve its model of the problem functions, increasing the chance of generating a descent direction for $H(\cdot;\, x^{k+1})$.

We shall now discuss how to select the next subgradient index sets J_f^{k+1} and J_F^{k+1}. In order for the algorithm to use the latest subgradient information, we should have $k+1 \in J_f^{k+1} \cap J_F^{k+1}$, which is satisfied if

$$J_f^{k+1} = \hat{J}_f^k \cup \{k+1\} \quad \text{and} \quad J_F^{k+1} = \hat{J}_F^k \cup \{k+1\} \tag{2.20}$$

for some sets $\hat{J}_f^k \subset J_f^k$ and $\hat{J}_F^k \subset J_F^k$. From Chapter 1 and Chapter 2 we know that at least three approaches to the selection of $\hat{J}_f^k$ and $\hat{J}_F^k$ are possible. First, one may use subgradient accumulation by choosing $\hat{J}_f^k = J_f^k$ and $\hat{J}_F^k = J_F^k$, which results in

$$J_f^k = \{1,\ldots,k\} \quad \text{and} \quad J_F^k = \{1,\ldots,k\} \quad \text{for all } k.$$

This strategy, which is employed in the algorithm of Mifflin (1982) (see Section 1.3), encounters serious difficulties in storage and computation after a large number of iterations. For this reason, we shall now present extensions of the other two approaches from Chapter 2: the subgradient selection strategy and the subgradient aggregation strategy. Both strategies are based on analyzing Lagrange multipliers of the search direction finding subproblems. Therefore we shall need the following generalization of Lemma 2.2.1.

Lemma 2.3.(i) The unique solution (d^k, v^k) of subproblem (2.12) always exists.
(ii) (d^k, v^k) solves (2.12) if and only if there exist Lagrange multipliers λ_j^k, $j \in J_f^k$, and μ_j^k, $j \in J_F^k$, and a vector $p^k \in R^N$ satisfying

$$\lambda_j^k \geq 0,\ j \in J_f^k,\ \mu_j^k \geq 0,\ j \in J_F^k,\ \sum_{j \in J_f^k} \lambda_j^k + \sum_{j \in J_F^k} \mu_j^k = 1, \tag{2.21a}$$

$$[f_j^k - f(x^k) + \langle g_f^j, d^k \rangle - v^k]\,\lambda_j^k = 0,\ j \in J_f^k, \tag{2.21b}$$

$$[F_j^k + \langle g_F^j, d^k \rangle - v^k]\,\mu_j^k = 0,\ j \in J_F^k, \tag{2.21c}$$

$$p^k = \sum_{j \in J_f^k} \lambda_j^k g_f^j + \sum_{j \in J_F^k} \mu_j^k g_F^j, \tag{2.21d}$$

$$d^k = -p^k, \tag{2.21e}$$

$$v^k = -\{|p^k|^2 + \sum_{j \in J_f^k} \lambda_j^k [f(x^k) - f_j^k] - \sum_{j \in J_F^k} \mu_j^k F_j^k\}, \tag{2.21f}$$

$$f_j^k - f(x^k) + \langle g_f^j, d^k \rangle \le v^k, \; j \in J_f^k, \tag{2.21g}$$

$$F_j^k + \langle g_F^j, d^k \rangle \le v^k, \; j \in J_F^k. \tag{2.21h}$$

(iii) Multipliers λ_j^k, $j \in J_f^k$, μ_j^k, $j \in J_F^k$, satisfy (2.21) if and only they solve the following dual of (2.12)

$$\underset{\lambda,\mu}{\text{minimize}} \; \tfrac{1}{2}\left|\sum_{j \in J_f^k} \lambda_j g_f^j + \sum_{j \in J_F^k} \mu_j g_F^j\right|^2 + \sum_{j \in J_f^k} \lambda_j [f(x^k) - f_j^k] - \sum_{j \in J_F^k} \mu_j F_j^k,$$

$$\text{subject to} \quad \lambda_j \ge 0, \; j \in J_f^k, \; \mu_j \ge 0, j \in J_F^k, \; \sum_{j \in J_f^k} \lambda_j + \sum_{j \in J_F^k} \mu_j = 1. \tag{2.22}$$

(iv) There exists a solution λ_j^k, $j \in J_f^k$, μ_j^k, $j \in J_F^k$, of subproblem (2.22) such that the sets

$$\hat{J}_f^k = \{j \in J_f^k : \lambda_j^k > 0\} \quad \text{and} \quad \hat{J}_F^k = \{j \in J_F^k : \mu_j^k > 0\} \tag{2.23a}$$

satisfy

$$|\hat{J}_f^k| + |\hat{J}_F^k| \le N+1. \tag{2.23b}$$

Such a solution can be obtained by solving the following linear programming problem by the simplex method:

$$\begin{aligned} &\underset{\lambda,\mu}{\text{minimize}} && \sum_{j \in J_f^k} \lambda_j [f(x^k) - f_j^k] - \sum_{j \in J_F^k} \mu_j F_j^k, \\ &\text{subject to} && \sum_{j \in J_f^k} \lambda_j + \sum_{j \in J_F^k} \mu_j = 1, \\ &&& \sum_{j \in J_f^k} \lambda_j g_f^j + \sum_{j \in J_F^k} \mu_j g_F^j = p^k, \\ &&& \lambda_j \ge 0, \; j \in J_f^k, \; \mu_j \ge 0, \; j \in J_F^k, \end{aligned} \tag{2.24}$$

where $p^k = -d^k$. Moreover, (d^k, v^k) solves the following reduced subproblem

$$\underset{(d,v)\in R^{N+1}}{\text{minimize}} \quad \tfrac{1}{2}|d|^2+v,$$

$$\text{subject to} \quad f_j^k-f(x^k)+\langle g_f^j,d\rangle \le v, \; j\in \hat{J}_f^k, \tag{2.25}$$

$$F_j^k+\langle g_F^j,d\rangle \le v, \; j\in \hat{J}_F^k.$$

Proof. It suffices to observe that subproblem (2.12) is structurally similar to subproblem (2.2.11), and that the above lemma is an obvious reformulation of Lemma 2.2.1. □

Lemma 2.3(iv) and the generalized cutting plane idea from Section 2.2 lead us to the following subgradient selection strategy. Subproblem (2.25) is a reduced, equivalent version of subproblem (2.12). Therefore the choice of J_f^{k+1} and J_F^{k+1} specified by (2.20) and (2.23) conforms with the generalized cutting plane concept, because it consists in appending to a reduced subproblem linear constraints generated by the latest subgradients. Thus only those past subgradients that contribute to the current search direction are retained, see (2.21a), (2.21d), (2.21e) and (2.23). Subgradient selection results in implementable algorithms that require storage of at most N+1 past subgradients.

In the subgradient aggregation strategy we shall construct an auxiliary reduced subproblem by forming surrogate contraints with the help of Lagrange multipliers of (2.12). As expounded in Chapter 2, subgradient aggregation consists in forming convex combinations of the past subgradients of a given function on the basis of the corresponding Lagrange multipliers. Here a slight complication arises from the fact that the Lagrange multipliers associated with each of the problem functions (λ^k with f, and μ^k with F) do not form separate convex combinations, see (2.21a). Yet the subgradients of f should be aggregated separately from those of F, since otherwise the mixing of subgradients would spoil crucial properties of subgradient aggregation. For separate subgradient aggregation we shall use scaled versions of Lagrange multipliers of (2.12). A suitable scaling procedure, which yields separate convex combinations, is given below.

Let (λ^k,μ^k) denote any vectors of Lagrange multipliers of (2.12), which do not necessarily satisfy (2.23), and let the numbers ν_f^k, $\tilde{\lambda}_j^k$, $j\in J_f^k$, ν_F^k, $\tilde{\mu}_j^k$, $j\in J_F^k$, satisfy

$$\nu_f^k=\sum_{j\in J_f^k}\lambda_j^k, \quad \lambda_j^k=\nu_f^k\tilde{\lambda}_j^k, \; j\in J_f^k, \tag{2.26a}$$

$$\nu_F^k = \sum_{j \in J_f^k} \mu_j^k, \quad \mu_j^k = \nu_F^k \tilde{\mu}_j^k, \; j \in J_F^k, \tag{2.26b}$$

$$\tilde{\lambda}_j^k \geq 0, \; j \in J_f^k, \quad \sum_{j \in J_f^k} \tilde{\lambda}_j^k = 1, \tag{2.26c}$$

$$\tilde{\mu}_j^k \geq 0, \; j \in J_F^k, \quad \sum_{j \in J_F^k} \tilde{\mu}_j^k = 1. \tag{2.26d}$$

Such numbers exist and can be easily computed as follows. By (2.21a), (2.26a) and (2.26b), we have

$$\nu_f^k \geq 0, \; \nu_F^k \geq 0, \; \nu_f^k + \nu_F^k = 1. \tag{2.27}$$

If $\nu_f^k \neq 0$ then the scaled multipliers

$$\tilde{\lambda}_j^k = \lambda_j^k / \nu_f^k, \; j \in J_f^k,$$

satisfy (2.26a) and (2.26d) in view of (2.21a). If $\nu_f^k = 0$ then $\lambda_j^k = 0$ for all $j \in J_f^k$ by (2.21a), hence (2.26a) is trivially fulfilled by any numbers $\tilde{\lambda}_j^k$ satisfying (2.26c). Similarly one may choose $\tilde{\mu}_j^k$.

The above scaled Lagrange multipliers $(\tilde{\lambda}^k, \tilde{\mu}^k)$ will be used for subgradient aggregation as follows.

<u>Lemma 2.4</u>. Define the <u>aggregate subgradients</u>

$$(p_f^k, \tilde{f}_p^k) = \sum_{j \in J_f^k} \tilde{\lambda}_j^k (g_f^j, f_j^k) \quad \text{and} \quad (p_F^k, \tilde{F}_p^k) = \sum_{j \in J_F^k} \tilde{\mu}_j^k (g_F^j, F_j^k). \tag{2.28}$$

Then

$$p^k = \nu_f^k p_f^k + \nu_F^k p_F^k, \tag{2.29}$$

$$v^k = -\{|p^k|^2 + \nu_f^k [f(x^k) - \tilde{f}_p^k] - \nu_F^k \tilde{F}_p^k\}. \tag{2.30}$$

Moreover, subproblem (2.12) is equivalent to the following reduced problem

$$\underset{(d,v) \in R^{N+1}}{\text{minimize}} \quad \tfrac{1}{2}|d|^2 + v,$$

$$\text{subject to} \quad \tilde{f}_p^k - f(x^k) + \langle p_f^k, d \rangle \leq v, \tag{2.31}$$

$$\tilde{F}_p^k + \langle p_F^k, d \rangle \leq v.$$

Proof. (2.29) and (2.30) follow easily from (2.21) and (2.26). The equivalence of (2.31) and (2.12) can be shown as in the proof of Lemma 2.2.2. □

The constraints of the reduced subproblem (2.31) are generated by the following aggregate linearizations

$$\tilde{f}^k(x) = \tilde{f}_p^k + \langle p_f^k, x-x^k \rangle \quad \text{and} \quad \tilde{F}^k(x) = \tilde{F}_p^k + \langle p_F^k, x-x^k \rangle . \tag{2.32}$$

They are convex combinations of the linearizations f_j and F_j, respectively, because the aggregating scaled multipliers form convex combinations, cf. (2.26c,d) and Lemma 2.2.3.

The rules for updating the linearizations can be taken from Chapter 2:

$$f_j^{k+1} = f_j^k + \langle g_f^j, x^{k+1}-x^k \rangle \quad \text{and} \quad F_j^{k+1} = F_j^k + \langle g_F^j, x^{k+1}-x^k \rangle , \tag{2.33}$$

because for each $x \in R^N$ and $j=1,...,k$ we have

$$f_j(x) = f_j^k + \langle g_f^j, x-x^k \rangle \quad \text{and} \quad F_j(x) = F_j^k + \langle g_F^j, x-x^k \rangle , \tag{2.34}$$

see (2.2.30) and (2.2.32). Denoting $f_p^{k+1}=\tilde{f}^k(x^{k+1})$ and $F_p^{k+1}=\tilde{F}^k(x^{k+1})$ we obtain from (2.32) similar rules for updating the aggregate linearizations:

$$f_p^{k+1} = \tilde{f}_p^k + \langle p_f^k, x^{k+1}-x^k \rangle \quad \text{and} \quad F_p^{k+1} = \tilde{F}_p^k + \langle p_F^k, x^{k+1}-x^k \rangle . \tag{2.35}$$

Also by convexity the linearizations satisfy

$$f(x) \ge f_j(x) \quad \text{and} \quad F(x) \ge F_j(x) \quad \text{for all } x \text{ and } j, \tag{2.36}$$

hence the aggregate linearizations, being their convex combinations, also satisfy

$$f(x) \ge \tilde{f}^k(x) \quad \text{and} \quad F(x) \ge \tilde{F}^k(x) \quad \text{for all } x. \tag{2.37}$$

We also note that at the (k+1)-st iteration the aggregate linearizations

$$\tilde{f}^k(x) = f_p^{k+1} + \langle p_f^k, x-x^{k+1} \rangle \quad \text{and} \quad \tilde{F}^k(x) = F_p^{k+1} + \langle p_F^k, x-x^{k+1} \rangle \tag{2.38}$$

are generated by the updated aggregate subgradients (p_f^k, f_p^{k+1}) and (p_F^k, F_p^{k+1}).

In terms of aggregate linearizations, Lemma 2.4 states that an equivalent formulation of subproblem (2.14) is

$$\text{minimize} \quad \tilde{H}^k(x^k+d)+\tfrac{1}{2}|d|^2 \quad \text{over all } d, \tag{2.39}$$

where

$$\tilde{H}^k(x) = \max\{\tilde{f}^k(x)-f(x^k),\tilde{F}^k(x)\} \quad \text{for all } x. \tag{2.40}$$

Thus the use of separate aggregation enables one to construct aggregate versions of polyhedral approximations to the improvement function $H(\circ;x^k)$.

Following the generalized cutting plane idea one may obtain the next search direction finding subproblem by updating the reduced subproblem (2.31) according to (2.35), and appending constraints generated by the latest subgradients g_f^{k+1} and g_F^{k+1}. For efficiency one may also retain a limited number of linear constraints generated by the past subgradients.

In this way we arrive at the following description of consecutive subproblems of the method with subgradient aggregation. Let (d^k,v^k) denote the solution to the following k-th search direction finding subproblem

$$\begin{aligned}
&\underset{(d,v)\in R^{N+1}}{\text{minimize}} && \tfrac{1}{2}|d|^2+v,\\
&\text{subject to} && f_j^k-f(x^k)+\langle g_f^j,d\rangle \le v,\ j\in J_f^k,\\
& && f_p^k-f(x^k)+\langle p_f^{k-1},d\rangle \le v,\\
& && F_j^k+\langle g_F^j,d\rangle \le v,\ j\in J_F^k,\\
& && F_p^k+\langle p_F^{k-1},d\rangle \le v.
\end{aligned} \tag{2.41}$$

For $k=1$ we shall initialize the method by choosing the starting point $x^1\in S$ and setting $y^1=x^1$ and

$$\begin{aligned}
p_f^0 &= g_f^1 = g_f(x^1),\ f_p^1 = f_1^1 = f(x^1),\ J_f^1 = \{1\},\\
p_F^0 &= g_F^1 = g_F(x^1),\ F_p^1 = F_1^1 = F(x^1),\ J_F^1 = \{1\}.
\end{aligned} \tag{2.42}$$

Subproblem (2.41) is of the form (2.12), hence we may rephrase the preceding results as follows. Let λ_j^k, $j\in J_f^k$, λ_p^k, μ_j^k, $j\in J_F^k$, and μ_p^k denote any Lagrange multipliers of (2.41). Then Lemma 2.4 yields

$$\lambda_j^k\ge 0,\ j\in J_f^k,\ \lambda_p^k\ge 0,\ \mu_j^k\ge 0,\ j\in J_F^k,\ \mu_p^k\ge 0, \tag{2.43a}$$

$$\sum_{j\in J_f^k}\lambda_j^k+\lambda_p^k+\sum_{j\in J_F^k}\mu_j^k+\mu_p^k = 1, \tag{2.43b}$$

$$p^k = \sum_{j\in J_f^k}\lambda_j^k g_f^j+\lambda_p^k p_f^{k-1}+\sum_{j\in J_F^k}\mu_j^k g_F^j+\mu_p^k p_F^{k-1}, \tag{2.43c}$$

$$d^k = -p^k, \tag{2.43d}$$

hence we may calculate scaled multipliers satisfying

$$\nu_f^k = \sum_{j\in J_f^k}\lambda_j^k+\lambda_p^k,\ \lambda_j^k = \nu_f^k\tilde{\lambda}_j^k,\ j\in J_f^k,\ \lambda_p^k = \nu_f^k\tilde{\lambda}_p^k, \tag{2.44a}$$

$$\nu_F^k = \sum_{j\in J_F^k}\mu_j^k+\mu_p^k,\ \mu_j^k = \nu_F^k\tilde{\mu}_j^k,\ j\in J_F^k,\ \mu_p^k = \nu_F^k\tilde{\mu}_p^k, \tag{2.44b}$$

$$\tilde{\lambda}_j^k \geq 0,\ j\in J_f^k,\ \tilde{\lambda}_p^k \geq 0,\ \sum_{j\in J_f^k}\tilde{\lambda}_j^k + \tilde{\lambda}_p^k = 1, \tag{2.44c}$$

$$\tilde{\mu}_j^k \geq 0,\ j\in J_F^k,\ \tilde{\mu}_p^k \geq 0,\ \sum_{j\in J_F^k}\tilde{\mu}_j^k + \tilde{\mu}_p^k = 1, \tag{2.44d}$$

$$\nu_f^k \geq 0,\ \nu_F^k \geq 0,\ \nu_f^k + \nu_F^k = 1, \tag{2.44e}$$

and use them for computing the aggregate subgradients

$$(p_f^k,\tilde{f}_p^k) = \sum_{j\in J_f^k}\tilde{\lambda}_j^k(g_f^j,f_j^k) + \tilde{\lambda}_p^k(p_f^{k-1},f_p^k), \tag{2.45}$$

$$(p_F^k,\tilde{F}_p^k) = \sum_{j\in J_F^k}\tilde{\mu}_j^k(g_F^j,F_j^k) + \tilde{\mu}_p^k(p_F^{k-1},F_p^k).$$

Moreover, relations (2.29) and (2.30) also hold for the method with subgradient aggregation based on (2.43)-(2.45).

One may observe that for the method with subgradient aggregation the last assertion of Lemma 2.4 can be reformulated as follows: subproblem (2.31) is equivalent to subproblem (2.41). Both subproblems are equivalent to the following

$$\text{minimize}\quad \hat{H}_a^k(x^k+d)+\tfrac{1}{2}|d|^2 \quad \text{over all } d, \tag{2.46}$$

where the aggregate polyhedral approximation to the improvement function

$$\hat{H}_a^k(x) = \max\{\hat{f}_a^k(x)-f(x^k),\ \hat{F}_a^k(x)\} \tag{2.47}$$

is defined by the following aggregate polyhedral approximations to f and F

$$\hat{f}_a^k(x) = \max\{\tilde{f}^{k-1}(x), f_j^k(x) : j \in J_f^k\},$$
$$\hat{F}_a^k(x) = \max\{\hat{F}^{k-1}(x), F_j^k(x) : j \in J_F^k\}. \tag{2.48}$$

Remark 2.5. Suppose the problem functions are of the form

$$f(x) = \max\{\tilde{f}_j(x) : j \in J_f\} \quad \text{and} \quad F(x) = \max\{\tilde{F}_j(x) : j \in J_F\}$$

and one can compute subgradients $g_f^{k,j} \in \partial\tilde{f}_j(x^k)$, $j \in J_f$, and $g_F^{k,j} \in \partial\tilde{F}_j(x^k)$, $j \in J_F$. Then one may append the constraints

$$\tilde{f}_j(x^k) - f(x^k) + \langle g_f^{j,k}, d\rangle \le v, \ j \in J_f,$$
$$\tilde{F}_j(x^k) + \langle g_F^{j,k}, d\rangle \le v, \ j \in J_F, \tag{2.49}$$

to the search direction subproblems (2.12) and (2.41), for all k. This enhances faster convergence, but at the cost of more work per iteration. One may also replace the sets J_f and J_F in (2.49) with the sets

$$J_f(\varepsilon) = \{j \in J_f : \tilde{f}_j(x^k) \ge f(x^k) - \varepsilon\}, \ J_F(\varepsilon) = \{j \in J_F : \tilde{F}_j(x^k) \ge F(x^k) - \varepsilon\}$$

for some $\varepsilon \ge 0$. Such augmentations should be used especially if there are many constraints in the original formulation of problem (1.1), cf. Remark 2.2. It is straighforward to extend the subgradient selection and aggregation rules to the augmented subproblems. The subsequent results hold also for such modifications.

We shall now remark on relations of the above methods with other algorithms. The methods generalize the method of centers for inequality constrained minimax problems (Kiwiel, 1981a), which in turn extends the Pironneau and Polak (1972) method of centers for smooth nonlinear programming problems. On the other hand, subproblems (2.12) are reduced versions of the Mifflin (1982) subproblems, cf. (1.3.31).

Remark 2.6. It is worthwile to compare subproblems (2.14) and (2.46) with the "conceptual" search direction finding subproblem (1.2.82). The latter problem always yields a descent direction, provided that the point at which it is defined is nonstationary. However, this problem requires full subdifferentials of the problem functions for defining the corresponding approximations, whereas subproblems (2.14) and (2.41) are based on the polyhedral approximations (2.11), (2.13), (2.47) and (2.48) (see also Remark 2.2.6).

3. The Algorithm with Subgradient Aggregation

The results of the preceding section are formalized in the following procedure for solving problem (1.1).

Algorithm 3.1

Step 0 (Initialization). Select a starting point $x^1 \in S$, a final accuracy tolerance $\varepsilon_s \geq 0$ and a line search parameter $m \in (0,1)$. Set $y^1=x^1$ and initialize the algorithm according to (2.42). Set the counters $k=1$, $l=0$ and $k(0)=1$.

Step 1 (Direction finding). Compute multipliers λ_j^k, $j \in J_f^k$, λ_p^k, μ_j^k, $j \in J_F^k$, and μ_p^k that solve the k-th dual search direction finding subproblem

$$\underset{\lambda,\mu}{\text{minimize}}\ \tfrac{1}{2}\Big|\sum_{j\in J_f^k}\lambda_j g_f^j+\lambda_p p_f^{k-1}+\sum_{j\in J_F^k}\mu_j g_F^j+\mu_p p_F^{k-1}\Big|^2+$$

$$+\sum_{j\in J_f^k}\lambda_j[f(x^k)-f_j^k]+\lambda_p[f(x^k)-f_p^k]-\sum_{j\in J_F^k}\mu_j F_j^k-\mu_p F_p^k, \tag{3.1}$$

$$\text{subject to } \lambda_j \geq 0,\ j\in J_f^k,\ \lambda_p\geq 0,\ \mu_j\geq 0,\ j\in J_F^k,\ \mu_p\geq 0,$$

$$\sum_{j\in J_f^k}\lambda_j+\lambda_p+\sum_{j\in J_F^k}\mu_j+\mu_p=1.$$

Calculate multipliers ν_f^k, $\tilde\lambda_j^k$, $j \in J_f^k$, $\tilde\lambda_p^k$, ν_F^k, $j\in J_F^k$, and $\tilde\mu_p^k$ satisfying (2.44). Compute $(p_f^k,\tilde f_p^k)$ and $(p_F^k,\tilde F_p^k)$ by (2.45), and use (2.29) and (2.30) for calculating p^k and v^k. Set $d^k=-p^k$.

Step 2 (Stopping criterion). Set

$$w^k=\tfrac{1}{2}|p^k|^2+\nu_f^k[f(x^k)-\tilde f_p^k]-\nu_F^k F_p^k. \tag{3.2}$$

If $w^k \leq \varepsilon_s$ terminate; otherwise, go to Step 3.

Step 3 (Line search). Set $y^{k+1}=x^k+d^k$. If

$$H(x^k+d^k;x^k)\leq H(x^k;x^k)+mv^k \tag{3.3}$$

then set $t_L^k=1$ (a serious step), set $k(l+1)=k+1$ and increase l by 1; otherwise, i.e. if (3.3) is violated, set $t_L^k=0$ (a null step).

Step 4 (Linearization updating). Set $x^{k+1}=x^k+t_L^k d^k$. Choose some sets $\hat{J}_f^k \subset J_f^k$ and $\hat{J}_F^k \subset J_F^k$, and calculate the linearization values f_j^{k+1}, $j \in \hat{J}_f^k$, F_j^{k+1}, $j \in \hat{J}_F^k$, f_p^{k+1} and F_p^{k+1} by (2.33) and (2.35). Set $g_f^{k+1}=g_f(y^{k+1})$, $g_F^{k+1}=g_F(y^{k+1})$ and

$$f_{k+1}^{k+1} = f(y^{k+1})+\langle g_f^{k+1}, x^{k+1}-y^{k+1}\rangle \quad \text{and} \quad F_{k+1}^{k+1} = F(y^{k+1})+\langle g_F^{k+1}, x^{k+1}-y^{k+1}\rangle .$$

Set $J_f^{k+1}=\hat{J}_f^k \cup \{k+1\}$ and $J_F^{k+1}=\hat{J}_F^k \cup \{k+1\}$. Increase k by 1 and go to Step 1.

Remark 3.2. It follows from Lemma 2.3 that in Algorithm 3.1 (d^k,v^k) solves subproblem (2.41), and that λ_j^k, $j \in J_f^k$, λ_p^k, μ_j^k, $j \in J_F^k$, μ_p^k are the associated Lagrange multipliers. Thus one may equivalently solve subproblem (2.41) in Step 1 of the above algorithm.

Remark 3.3. As noted in the previous section, the line search guarantees that $x^{k+1} \in S$ if $x^k \in S$. Since $x^1 \in S$ by assumption, it follows that Algorithm 3.1 is a feasible point method. In particular, $H(x^k,x^k)=0$ for all k.

Remark 3.4. For convenience, the above version of the algorithm requires the subgradient mappings g_f and g_F to be defined everywhere. In fact, g_f need not be defined at $x \notin S$. In this case, in Step 4 set $g_f^{k+1}=g_f(x^{k+1})$ and $f_{k+1}^{k+1}=f(x^{k+1})$ if $y^{k+1} \notin S$. If g_F is not defined at feasible points, then the following modifications are necessary. Set $J_F^1=\emptyset$ in Step 0 and impose the additional constraint $\mu_p^k=0$ in subproblem (3.1) (or delete the last constraint of (2.41)) for all k such that $y^j \in S$ for $j=1,...,k-1$. Also let $J_F^{k+1}=\hat{J}_F^k$ in Step 4 if $y^{k+1} \in S$. This amounts to not using the constraint subgradients until the first infeasible trial point is found. One can check that all the subsequent proofs need only minor changes to cover this modification of Step 1 and Step 3. We will not treat this case explicitly, since this would further complicate the notation.

4. Convergence

In this section we show that the sequence $\{x^k\}$ generated by Algorithm 3.1 is a minimizing sequence, i.e. $\{x^k\} \subset S$ and $f(x^k) \downarrow \inf \{f(x) : x \in S\}$ as $k \to \infty$, and that $\{x^k\}$ converges to a solution of problem (1.1), provided that the solution set $\bar{X}$ is nonempty. Naturally, convergence results assume that the final accuracy tolerance ε_s is set to zero.

Since Algorithm 3.1 is an extension of Algorithm 2.3.1 to constrained problems, the analysis given below dwells on the results of Section 2.4. Therefore we shall concentrate on modifications only, omitting certain details that can be found in Chapter 2.

We start by stating that the aggregate subgradients are convex combinations of the past subgradients.

Lemma 4.1. Suppose that Algorithm 3.1 did not stop before the k-th iteration. Then

$$(p_f^k, \tilde{f}_p^k) \in \text{conv}\{(g_f^j, f_j^k) : j=1,\dots,k\},$$
$$(p_F^k, \tilde{F}_p^k) \in \text{conv}\{(g_F^j, F_j^k) : j=1,\dots,k\}. \tag{4.1}$$

If additionally $k > 1$, then

$$(p_f^{k-1}, f_p^k) \in \text{conv}\{(g_f^j, f_j^k) : j=1,\dots,k-1\},$$
$$(p_F^{k-1}, F_p^k) \in \text{conv}\{(g_F^j, F_j^k) : j=1,\dots,k-1\}. \tag{4.2}$$

Proof. Since Algorithm 3.1 uses aggregation rules analogous to those of Algorithm 2.3.1, the proof is similar to the proof of Lemma 2.4.1. □

It follows from the above convex representation of the aggregate subgradients that they can be interpreted as ε-subgradients of the functions associated with problem (1.1), viz. the objective function f, the constraint violation function F_+ defined by

$$F_+(x) = F(x)_+ = \max\{F(x), 0\} \quad \text{for all } x,$$

and the improvement function H. Note that the function F_+ is convex and we always have $F(x^k)_+ = 0$, because $\{x^k\} \subset S$. In the following, suppose that Algorithm 3.1 did not terminate before the k-th iteration. Define the linearization errors

$$\alpha_{f,j}^k=f(x^k)-f_j^k\ ,\quad \alpha_{F,j}^k=-F_j^k\ ,\quad j=1,\ldots,k, \tag{4.3a}$$

$$\alpha_{f,p}^k=f(x^k)-f_p^k\ ,\quad \alpha_{F,j}^k=-F_p^k\ , \tag{4.3b}$$

$$\tilde{\alpha}_{f,p}^k=f(x^k)-\tilde{f}_p^k\ ,\quad \tilde{\alpha}_{F,p}^k=-\tilde{F}_p^k\ , \tag{4.3c}$$

$$\tilde{\alpha}_p^k = \nu_f^k[f(x^k)-\tilde{f}_p^k]-\nu_F^k\tilde{F}_p^k\ . \tag{4.3d}$$

Observe that $F(x^k)_+=0$ implies

$$\alpha_{F,j}^k = F(x^k)_+-F_j^k\ ,\quad \alpha_{F,p}^k = F(x^k)_+-F_p^k\ ,\quad \tilde{\alpha}_{F,p}^k = F(x^k)_+-\tilde{F}_p^k,$$

which justifies calling those variables the linearization errors of the constraint violation function. The linearization errors characterize the subgradients calculated by the algorithm as follows.

Lemma 4.2. At the k-th iteration of Algorithm 3.1, one has

$$g_f^j \in \partial_\varepsilon f(x^k) \quad \text{for} \quad \varepsilon=\alpha_{f,j}^k\ ,\ j=1,\ldots,k, \tag{4.4a}$$

$$g_F^j \in \partial_\varepsilon F(x^k)_+ \quad \text{for} \quad \varepsilon=\alpha_{F,j}^k\ ,\ j=1,\ldots,k, \tag{4.4b}$$

$$p_f^{k-1} \in \partial_\varepsilon f(x^k) \quad \text{for} \quad \varepsilon=\alpha_{f,p}^k\ , \tag{4.4c}$$

$$p_F^{k-1} \in \partial_\varepsilon F(x^k)_+ \quad \text{for} \quad \varepsilon=\alpha_{F,p}^k\ , \tag{4.4d}$$

$$p_f^k \in \partial_\varepsilon f(x^k) \quad \text{for} \quad \varepsilon=\tilde{\alpha}_{f,p}^k\ , \tag{4.4e}$$

$$p_F^k \in \partial_\varepsilon F(x^k)_+ \quad \text{for} \quad \varepsilon=\tilde{\alpha}_{F,p}^k\ , \tag{4.4f}$$

$$p^k \in \partial_\varepsilon H(x^k;x^k) \quad \text{for} \quad \varepsilon=\tilde{\alpha}_p^k \geq 0. \tag{4.4g}$$

Proof. Using (2.34), (2.36), Lemma 4.1 and the fact that we always have $F(x)_+ \geq F(x)$ and $F(x^k)_+=0$, one obtains (4.4a)-(4.4f) as in the proof of Lemma 2.4.2. In particular, similarly to (2.4.5c) we get

$$f(x) \geq f(x^k)+\langle p_f^k,x-x^k\rangle -[f(x^k)-\tilde{f}_p^k],$$

$$F(x) \geq \langle p_F^k,x-x^k\rangle +\tilde{F}_p^k$$

for each $x \in R^N$. The above inequalities, (2.44e), (2.29) and (4.3d) yield

$$\nu_f^k[f(x)-f(x^k)]+\nu_F^k F(x) \geq \langle \nu_f^k p_f^k+\nu_F^k p_F^k, x-x^k \rangle -\nu_f^k[f(x^k)-\tilde{f}_p^k] +$$
$$+ \nu_F^k \tilde{F}_p^k \geq \langle p^k, x-x^k \rangle -\tilde{\alpha}_p^k$$

for each $x \in R^N$. Since $\nu_f^k \geq 0$ and $\nu_F^k \geq 0$ satisfy $\nu_f^k+\nu_F^k=1$, we have

$$H(x;x^k)=\max\{f(x)-f(x^k),F(x)\} \geq \nu_f^k[f(x)-f(x^k)]+\nu_F^k F(x),$$

hence

$$H(x;x^k) \geq H(x^k;x^k)+\langle p^k, x-x^k \rangle -\tilde{\alpha}_p^k \quad \text{for all } x,$$

because $H(x^k;x^k)=0$. Setting $x=x^k$, we complete the proof of (4.4g). □

Remark 4.3. In view of (4.4), the linearization errors (4.3) may also be called subgradient locality measures, because they indicate the distance from subgradients to the corresponding subdifferentials at the current point x^k. For instance, the value of $\tilde{\alpha}_p^k \geq 0$ indicates how much p^k differs from being a member of $\partial H(x^k;x^k)$; if $\tilde{\alpha}_p^k=0$ then $p^k \in \partial H(x^k;x^k)$.

The following result is useful for justifying the stopping criterion of the algorithm.

Lemma 4.4. At the k-th iteration of Algorithm 3.1, one has

$$w^k = \tfrac{1}{2}|p^k|^2+\tilde{\alpha}_p^k, \tag{4.5}$$

$$v^k = -\{|p^k|^2+\tilde{\alpha}_p^k\}, \tag{4.6}$$

$$v^k \leq -w^k \leq 0. \tag{4.7}$$

Proof. This follows easily from (3.2), (2.30), (4.3d) and the nonnegativity of $\tilde{\alpha}_p^k$. □

From relations (4.4g) and (4.5) we deduce easily that

$$p^k \in \partial_\varepsilon H(x^k;x^k) \quad \text{and} \quad |p^k| \leq (2\varepsilon)^{1/2} \quad \text{for} \quad \varepsilon = w^k. \tag{4.8}$$

Thus w^k may be called a stationarity measure of the current point x^k, because $\frac{1}{2}|p^k|^2$ indicates how much p^k differs from the null vector and $\tilde{\alpha}_p^k$ measures the distance from p^k to $\partial H(x^k;x^k)$, and stationary points $\overline{x} \in \overline{X}$ satisfy

$$\overline{p} \in \partial H(\overline{x};\overline{x}) \quad \text{and} \quad \overline{p}=0$$

by Lemma 2.1. The estimate (4.8) shows that x^k is approximately optimal when the value of w^k is small.

In what follows we assume that the final accuracy tolerance ε_s is set to zero. Since the algorithm stops if and only if $0 \le w^k \le \varepsilon_s=0$, (4.8) and Lemma 2.1 yield

Lemma 4.5. If Algorithm 3.1 terminates at the k-th iteration, then $x^k \in \overline{X}$.

From now on we suppose that the algorithm does not terminate, i.e. $w^k > 0$ for all k. Since the line search rules imply that we always have

$$f(x^{k+1})-f(x^k) \le mt_L^k v^k \tag{4.9}$$

with $m > 0$ and $t_L^k \ge 0$, the fact that $v^k \le -w^k < 0$ (see (4.7)) yields that the sequence $\{f(x^k)\}$ is nonincreasing.

We shall need the following properties of the improvement functions H.

Lemma 4.6. The mapping $\partial_{\cdot}H(\cdot;\cdot)$ is locally bounded, i.e. if (y,x,ε) remains bounded in $R^N \times R^N \times R$ then $\partial_\varepsilon H(y;x)$ remains bounded in R^N. Moreover, $\partial_{\cdot}H(\cdot;\cdot)$ is upper semicontinuous, i.e. if the sequences $\{\tilde{\tau}^i\}$, $\{\tau^i\}$, $\{\varepsilon^i\}$, $\{g^i\}$, $g^i \in \partial_{\varepsilon^i}H(\tilde{\tau}^i;\tau^i)$ for all i, tend to $\tilde{\tau},\tau,\varepsilon$ and g, respectively, then $g \in \partial_\varepsilon H(\tilde{\tau};\tau)$.

Proof. Consider a bounded set $B \subset R^N \times R^N \times R$ and let $(y,x,\varepsilon) \in B, g_H \in \partial_\varepsilon H(y,x)$ and $\tau=y+g_H/|g_H|$. Then $H(\tau;x) \ge H(y;x)+\langle g_H,\tau-y\rangle-\varepsilon$ yields

$$\langle g_H,\tau-y\rangle = |g_H| \le H(\tau;x)-H(y,y)+\varepsilon.$$

But τ is bounded, and so are $H(\tau;x)$ and $H(y,x)$ (H is continuous as a convex function), thus proving the local boundedness of $\partial_{\cdot}H(\cdot;\cdot)$. Next, $g^i \in \partial_{\varepsilon^i}H(\tilde{\tau}^i;\tau^i)$ implies

$$H(\tau;\tau^i) \ge H(\tilde{\tau}^i;\tau^i)+\langle g^i,\tau-\tilde{\tau}^i\rangle-\varepsilon^i \quad \text{for all } \tau.$$

Passing to the limit, we obtain the desired conclusion. □

The following property of the stationarity measures is crucial for convergence.

Lemma 4.7. Suppose that there exist an infinite set $K \subset \{1,2,\ldots\}$ and a point $\bar{x} \in R^N$ satisfying $x^k \xrightarrow{K} \bar{x}$ and $w^k \xrightarrow{K} 0$. Then $\bar{x} \in \bar{X}$.

Proof. By (4.5), (4.8) and Lemma 4.6, we have $p^k \xrightarrow{K} 0 \in \partial H(\bar{x};\bar{x})$. Thus $\bar{x} \in \bar{X}$ by Lemma 2.1. □

The following result can be obtained similarly to Lemma 2.4.7.

Lemma 4.8. Suppose that the sequence $\{f(x^k)\}$ is bounded from below. Then

$$\sum_{k=1}^{\infty} \{t_L^k |p^k|^2 + t_L^k \tilde{\alpha}_p^k\} < +\infty. \tag{4.10}$$

As in Section 2.4 (see (2.4.10)), we have

$$x^k = x^{k(l)} \quad \text{for} \quad k=k(l),\ k(l)+1,\ldots,k(l+1)-1, \tag{4.11}$$

where $k(l+1)=+\infty$ if the number l of serious steps stays bounded, i.e. if $x^k = x^{k(l)}$ for some fixed l and all $k \geq k(l)$.

First we deal with the case of infinitely many serious steps.

The following result can be proved similarly to Lemma 2.4.8, since it is an immediate consequence of (4.5), Lemma 4.8 and the fact that $t_L^{k(l)-1}=1$ for all l.

Lemma 4.9. Suppose that there exist an infinite set $L \subset \{1,2,\ldots\}$ and a point $\bar{x} \in R^N$ such that $x^{k(l)} \to \bar{x}$ as $l \to \infty$, $l \in L$. Then $w^{k(l+1)-1} \to 0$ as $l \to \infty$, $l \in L$.

In the case of a finite number of serious steps, we have to show that the stationarity measures $\{w^k\}$ tend to zero. To this end we shall analyze the dual search direction finding subproblems.

Lemma 4.10. At the k-th iteration of Algorithm 3.1, w^k is the optimal value of the following problem

$$\underset{\lambda,\mu}{\text{minimize}}\ \tfrac{1}{2}\Big|\sum_{j \in J_f^k} \lambda_j g_f^j + \lambda_p p_f^{k-1} + \sum_{j \in J_F^k} \mu_j g_F^j + \mu_p p_F^{k-1}\Big|^2 +$$

$$+\sum_{j\in J_f^k}\lambda_j\alpha_{f,j}^k+\lambda_p\alpha_{f,p}^k+\sum_{j\in J_F^k}\mu_j\alpha_{F,j}^k+\mu_p\alpha_{F,p}^k,\tag{4.12}$$

subject to $\lambda_j \geq 0,\ j\in J_f^k,\ \lambda_p \geq 0,\ \mu_j \geq 0,\ j\in J_F^k,\ \mu_p \geq 0,$

$$\sum_{j\in J_f^k}\lambda_j+\lambda_p+\sum_{j\in J_F^k}\mu_j+\mu_p=1,$$

which is equivalent to subproblem (3.1).

Proof. As in the proof of Lemma 2.4.9, the assertion follows from (4.3), (4.5), (2.45), (2.43c) and the fact that the k-th Lagrange multipliers solve (3.1). □

Let us now define the variables

$$g^k = g_F^k \quad\text{and}\quad \alpha^k = \alpha_{F,k}^k \quad\text{if}\quad f(y^k)-f(x^{k-1}) < F(y^k),\tag{4.13a}$$

$$g^k = g_f^k \quad\text{and}\quad \alpha^k = \alpha_{f,k}^k \quad\text{if}\quad f(y^k)-f(x^{k-1}) \geq F(y^k),\tag{4.13b}$$

for all $k > 1$. They will be used in the following extension of Lemma 2. 4.11.

Lemma 4.11. Suppose that $t_L^{k-1}=0$ for some $k > 1$. Then

$$-\alpha^k + \langle g^k, d^{k-1}\rangle \geq mv^{k-1},\tag{4.14}$$

$$w^k \leq \phi_C(w^{k-1}),\tag{4.15}$$

where ϕ_C is defined by (2.4.16) and C is any number satisfying

$$C \geq \max\{|p^{k-1}|, |g^k|, \tilde{\alpha}_p^{k-1}, 1\}.$$

Proof.(i) If $t_L^{k-1}=0$ then the line search rules yield $y^k=x^{k-1}+d^{k-1}$ and $x^k=x^{k-1}$, i.e. $y^k=x^k+d^{k-1}$, and

$$\max\{f(y^k)-f(x^k), F(y^k)\} > mv^{k-1}.\tag{4.16}$$

First, suppose that $F(y^k) \geq mv^{k-1}$. Then (4.13a), (4.3a), the rules of Step 4 and the fact that $y^k=x^k+d^{k-1}$ yield

$$-\alpha^k + \langle g^k, d^{k-1}\rangle = F_k^k + \langle g_F^k, y^k-x^k\rangle = F(y^k) \geq mv^k.\tag{4.17a}$$

Next, suppose that $F(y^k) < mv^{k-1}$. Then (4.16) implies $f(y^k)-f(x^k) > mv^{k-1}$. Hence (4.13b), (4.3a) and the rules of Step 4 yield

$$-\alpha^k + \langle g^k, d^{k-1} \rangle = -[f(x^k)-f(y^k) - \langle g_f^k, x^k-y^k \rangle] + \langle g_f^k, y^k-x^k \rangle$$

$$= f(y^k)-f(x^k) > mv^{k-1}. \tag{4.17b}$$

This completes the proof of (4.14).

(ii) If (4.17a) holds, let $\nu \in [0,1]$ and define the multipliers

$$\begin{aligned} &\lambda_j(\nu) = 0,\ j \in J_f^k,\ \lambda_p(\nu) = (1-\nu)\nu_f^{k-1}, \\ &\mu_k(\nu) = \nu,\ \mu_j(\nu) = 0,\ j \in J_F^k \setminus \{k\},\ \mu_p(\nu) = (1-\nu)\nu_F^{k-1}. \end{aligned} \tag{4.18a}$$

If (4.17b) is satisfied, let

$$\begin{aligned} &\lambda_k(\nu) = \nu,\ \lambda_j(\nu) = 0,\ j \in J_f^k \setminus \{k\},\ \lambda_p(\nu) = (1-\nu)\nu_f^{k-1}, \\ &\mu_j(\nu) = 0,\ j \in J_F^k,\ \mu_p(\nu) = (1-\nu)\nu_F^{k-1}. \end{aligned} \tag{4.18b}$$

Observe that the multipliers (4.18) are feasible for subproblem (4.12) for each $\nu \in [0,1]$, because $k \in J_f^k \cap J_F^k$ and (2.44e) is satisfied. Moreover, for each ν

$$\sum_{j \in J_f^k} \lambda_j(\nu) g_f^j + \lambda_p(\nu) p_f^{k-1} + \sum_{j \in J_F^k} \mu_j(\nu) g_F^j + \mu_p(\nu) p_F^{k-1} = (1-\nu)p^{k-1} + \nu g^k. \tag{4.19a}$$

This follows from (4.18), (2.29) and (4.13). Next, $x^k = x^{k-1}$ implies $\tilde{f}_p^k = f_p^{k-1}$ and $\tilde{F}_p^k = F_p^{k-1}$ by (2.35), hence (4.18), (4.13) and (4.3) yield

$$\begin{aligned} &\sum_{j \in J_f^k} \lambda_j(\nu)\alpha_{f,j}^k + \lambda_p(\nu)\alpha_{f,p}^k + \sum_{j \in J_f^k} \mu_j(\nu)\alpha_{F,j}^k + \mu_p(\nu)\alpha_{F,p}^k = \\ &\qquad = (1-\nu)\tilde{\alpha}_p^{k-1} + \nu\alpha^k, \end{aligned} \tag{4.19b}$$

for each $\nu \in [0,1]$. (4.19) and Lemma 4.10 imply that w^k is not larger than the optimal value of the problem

$$\begin{aligned} &\text{minimize } \tfrac{1}{2}|(1-\nu)p^{k-1} + \nu g^k|^2 + (1-\nu)\tilde{\alpha}_p^{k-1} + \nu\alpha^k, \\ &\text{subject to } \nu \in [0,1]. \end{aligned} \tag{4.20}$$

Since we also have (4.14), one may complete the proof by using Lemma 2.4.10 for bounding the optimal value of (4.20), as in the proof of Lemma 2.4.11. □

We may now complete the analysis of the case of a finite number of serious steps of the algorithm. The following result can be established similarly to Lemma 2.4.12, if one uses Lemma 4.11 and the local boundedness of $\partial H(\circ;\cdot)$ (see Lemma 4.6) together with the definition (4.13).

Lemma 4.12. Suppose that the number l of serious steps of Algorithm 3.1 stays bounded, i.e. $x^k = x^{k(l)}$ for some fixed l and all $k \geq k(l)$. Then $w^k \longrightarrow 0$ as $k\to\infty$.

Combining Lemma 4.7 with Lemma 4.9 and Lemma 4.12, and using (4.11), we obtain

Theorem 4.13. Every accumulation point of the sequence $\{x^k\}$ generated by Algorithm 3.1 is a solution to problem (1.1).

A sufficient condition for the sequence $\{x^k\}$ to have accumulation points is given below.

Lemma 4.14. Suppose that a point $\hat{x} \in S$ satisfies $f(\hat{x}) \leq f(x^k)$ for all k. Then the sequence $\{x^k\}$ is bounded and

$$|\hat{x}-x^k|^2 \leq |\hat{x}-x^n|^2 + \sum_{i=n}^{k} \{|x^{i+1}-x^i|^2 + 2t_L^i \tilde{\alpha}_p^i\} \quad \text{for} \quad k > n \geq 1,$$

$$\sum_{i=1}^{\infty} \{|x^{i+1}-x^i|^2 + 2t_L^i \tilde{\alpha}_p^i\} \longrightarrow 0 \quad \text{as} \quad n\to\infty.$$

Proof. Observe that $f(\hat{x}) \leq f(x^k)$ and $F(\hat{x}) \leq 0$ imply $H(\hat{x};x^k) \leq 0 = H(x^k;x^k)$ $(x^k \in S)$, hence (4.4g) yields

$$\langle p^k, \hat{x}-x^k \rangle \leq \tilde{\alpha}_p^k \quad \text{for all } k.$$

Since the above inequality is of the form (2.4.28), one may complete the proof by using Lemma 4.8 similarly to the proof of Lemma 4.14. □

We may now state the principal result.

Theorem 4.15. If problem (1.1) admits of a solution, then Algorithm 3.1

calculates a sequence $\{x^k\}$ converging to a solution of problem (1.1).

Proof. Let $\hat{x} \in \bar{X}$. Then $\hat{x} \in S$ and $f(\hat{x}) \le f(x^k)$ for all k, hence Lemma 4.14 shows that $\{x^k\}$ is bounded. By Theorem 4.15, $\{x^k\}$ has an accumulation point $\bar{x} \in \bar{X}$. For showing that $x^k \longrightarrow \bar{x}$, use Lemma 4.14 and the proof of Theorem 2.4.15. □

The following results can be proved similarly to Theorem 2.4.16, Lemma 2.4.17 and Corollary 2.4.18.

Theorem 4.16. Each sequence $\{x^k\}$ constructed by Algorithm 3.1 is a minimizing sequence: $\{x^k\} \subset S$ and $f(x^k) \downarrow \inf\{f(x) : x \in S\}$.

Lemma 4.17. Suppose that the sequence $\{f(x^k)\}$ is bounded from below. Then $w^k \to 0$.

Corollary 4.18. Suppose that $\inf \{f(x) : x \in S\} > -\infty$. Then Algorithm 3.1 terminates if its final accuracy tolerance ε_s is positive.

5. The Method with Subgradient Selection

In this section we analyze the method with subgradient selection from Section 2.

Algorithm 5.1 is obtained from Algorithm 3.1 by replacing Step 1 with

Step 1′ (Direction finding). Find multipliers λ_j^k, $j \in J_f^k$, and $\mu_j^k, j \in J_F^k$, that solve the k-th dual subproblem (2.22), and sets $\hat{J}_f^k$ and $\hat{J}_F^k$ satisfying (2.23). Calculate scaled Lagrange multipliers satisfying (2.26). Compute $(p_f^k, \tilde{f}_p^k)$ and $(p_F^k, \tilde{F}_p^k)$ by (2.28), and use (2.29) and (2.30) for calculating p^k and v^k. Set $d^k = -p^k$.

Clearly, Algorithm 5.1 is an extension of Algorithm 2.5.1. Therefore we refer the reader to Remark 2.5.2 on the computation of Lagrange multipliers in Step 1′, and to Remark 2.5.3 on methods that use more than N+3 past subgradients for search direction finding. In particular, the analysis given below applies also to the method of Mifflin (1982) (see Section 1.3), which uses all the past subgradients for search direction finding.

By the results of Section 2, in Algorithm 5.1 (d^k, v^k) solves the k-th primal subproblem (2.12), for any k. Therefore one may equivalently use (2.12) for direction finding.

Convergence of Algorithm 5.1 can be proved by modifying the results of Section 4, One may proceed as in Section 2.5, where the properties of the method with subgradient selection were derived from the results on the convergence of the method with subgradient aggregation from Section 2.4. Therefore we shall outline significant modifications only.

We substitute Lemma 4.10 with the following result.

Lemma 5.2. At the k-th iteration of Algorithm 5.1, w^k is the optimal value of the following problem, which is equivalent to subproblem (2.22):

$$\underset{\lambda,\mu}{\text{minimize}} \quad \tfrac{1}{2}\Big|\sum_{j\in J_f^k}\lambda_j g_f^j + \sum_{j\in J_F^k}\mu_j g_F^j\Big|^2 + \sum_{j\in J_f^k}\lambda_j\alpha_{f,j}^k + \sum_{j\in J_F^k}\mu_j\alpha_{F,j}^k, \tag{5.1}$$

$$\text{subject to} \quad \lambda_j \ge 0,\ j\in J_f^k,\ \mu_j\ge 0,\ j\in J_F^k,\ \sum_{j\in J_f^k}\lambda_j + \sum_{j\in J_F^k}\mu_j = 1.$$

In the proof of Lemma 4.11, the definition (4.18a) should be substituted by

$$\begin{aligned} &\lambda_j(\nu) = (1-\nu)\lambda_j^{k-1}, \quad j\in \hat{J}_f^{k-1}, \quad \lambda_k(\nu) = 0,\\ &\mu_j(\nu) = (1-\nu)\mu_j^{k-1}, \quad j\in \hat{J}_F^{k-1}, \quad \mu_k(\nu) = \nu, \end{aligned} \tag{5.2a}$$

and (4.18b) by

$$\begin{aligned} &\lambda_j(\nu) = (1-\nu)\lambda_j^{k-1}, \quad j\in \hat{J}_f^{k-1}, \quad \lambda_k(\nu) = \nu,\\ &\mu_j(\nu) = (1-\nu)\mu_j^{k-1}, \quad j\in \hat{J}_F^{k-1}, \quad \mu_k(\nu) = 0. \end{aligned} \tag{5.2b}$$

By (2.26) and (2.28), we have

$$\begin{aligned} &\lambda_j^{k-1} \ge 0,\ j\in\hat{J}_f^{k-1},\ \mu_j^{k-1}\ge 0,\ j\in \hat{J}_F^{k-1},\ \sum_{j\in\hat{J}_f^{k-1}}\lambda_j^{k-1} + \sum_{j\in\hat{J}_F^{k-1}}\mu_j^{k-1} = 1,\\ &\nu_f^{k-1}(p_f^{k-1},\tilde{f}_p^{k-1}) = \sum_{j\in\hat{J}_f^{k-1}}\lambda_j^{k-1}(g_f^j, f_j^k),\\ &\nu_F^{k-1}(p_F^{k-1},\tilde{F}_p^{k-1}) = \sum_{j\in\hat{J}_F^{k-1}}\mu_j^{k-1}(g_F^j, F_j^k). \end{aligned} \tag{5.3}$$

Using (5.2) and (5.3), one may replace (4.19) by

$$\sum_{j\in J_f^k}\lambda_j(\nu)g_f^j+\sum_{j\in J_F^k}\mu_j(\nu)g_F^j=(1-\nu)p^{k-1}+\nu g^k,$$

$$\sum_{j\in J_f^k}\lambda_j(\nu)\alpha_{f,j}^k+\sum_{j\in J_F^k}\mu_j(\nu)\alpha_{F,j}^k=(1-\nu)\tilde{\alpha}_p^{k-1}+\nu\alpha^k, \tag{5.4}$$

$$\lambda_j(\nu)\geq 0,\ j\in J_f^k,\ \mu_j(\nu)\geq 0,\ j\in J_F^k,\ \sum_{j\in J_f^k}\lambda_j(\nu)+\sum_{j\in J_F^k}\mu_j(\nu)=1,$$

for all $\nu\in[0,1]$, if $t_L^{k-1}=0$. In view of Lemma 5.2, (5.4) suffices for completing the proof of Lemma 4.11 for Algorithm 5.1. The remaining proofs need not be modified.

We conclude that all the convergence results of Section 4 hold also for Algorithm 5.1.

6. Line Search Modifications

In this section we discuss general line search rules that may be used in efficient procedures for stepsize selection. We also derive a new class of methods of feasible directions from the methods discussed so far.

The practical singnificance of rules that allow much freedom in stepsize selection was discussed in Section 2.6 in the unconstrained case. Most of that discussion applies to the constrained case, too.

As noted in Section 2.6, the requirement $t_L^k=1$ for a serious step may result in too many null steps. For this reason, a lower threshold $\bar{t}\in(0,1]$ for a serious stepsize may be preferable. This leads to replacing Step 3 in Algorithm 3.1 and Algorithm 5.1 by the following more general

Step 3′ (Line search). Select an auxiliary stepsize $t_R^k\in[\bar{t},1]$ and set $y^{k+1}=x^k+t_R^kd^k$. If

$$H(y^{k+1};x^k)\leq H(x^k;x^k)+mt_R^kv^k \tag{6.1}$$

then set $t_L^k=t_R^k$ (a serious step); otherwise set $t_L^k=0$ (a null step).

The search for a suitable value of $t_R^k\in[\bar{t},1]$ may use geometrical contraction, as described in Section 2. Of course, many other procedures can be constructed; see Remark 3.3.5.

One may check that the above line search modification does not impair the preceding convergence results. Only two proofs need changes. In proving Lemma 4.9, take account of the remark in Section 2.6 on the proof of Lemma 2.4.8. To prove (4.14), use Lemma 2.6.1 and the following generalization of a result of Mifflin (1982).

Lemma 6.1. Suppose that a point $y=x^k+td^k$ satisfies $F(y)>mtv^k$ for some $t\in(0,1]$, where $F(x^k)\le 0$. Let $g=g_F(y)\in\partial F(y)$ and $\alpha=-[F(y)+\langle g,x^k-y\rangle]$. Then $-\alpha+\langle g,d^k\rangle>mv^k$.

Proof. By assumption,

$$-\alpha+\langle g,d^k\rangle=F(y)-t\langle g,d^k\rangle+\langle g,d^k\rangle>tmv^k+(1-t)\langle g,d^k\rangle.$$

By the convexity of F, $0\ge F(x^k)\ge F(y)-t\langle g,d^k\rangle$, hence

$$t\langle g,d^k\rangle\ge F(y)>mtv^k,$$

and, since $t>0$, we have $\langle g,d^k\rangle>mv^k$. It follows that

$$-\alpha+\langle g,d^k\rangle>tmv^k+(1-t)mv^k=mv^k,$$

since $t\in(0,1]$. □

We shall now show how a modification of line search rules turns the previously discussed algorithms into new methods of feasible directions that extend the Pironneau and Polak (1973) method to the nondifferentiable case. Again, let $\bar{t}\in(0,1]$ be fixed and replace Step 3 in Algorithm 3.1 and Algorithm 5.1 by the following

Step 3'' (Line search). Select an auxiliary stepsize $t_R^k\in[\bar{t},1]$ and set $y^{k+1}=x^k+t_R^kd^k$. If

$$f(x^k+t_R^kd^k)-f(x^k)\le mt_R^kv^k \quad \text{and} \quad F(x^k+t_R^kd^k)\le 0, \tag{6.2}$$

then set $t_L^k=t_R^k$ (a serious step); otherwise, i.e. if at least one of inequalities (6.2) is violated, set $t_L^k=0$ (a null step).

Step 3'' quarantees that each x^k is feasible. One may implement Step 3'', for instance, as in (Mifflin, 1982) (see Section 6.3).

All the preceding convergence results remain valid for Algorithm 3.1 and Algorithm 5.1 with Step 3''. This follows essentially from the fact that with respect to the objective value the criteria (6.1) and (6.2) are equivalent, whereas if $F(y^{k+1}) > 0$ then necessarily $F(y^{k+1}) > mv^k$, because $m > 0$ and $v^k < 0$ at line searches, i.e. (6.1) is stronger than (6.2).

Our computational experience suggests that the methods of feasible directions (with Step 3″) converge faster than the methods of centers (with Step 3′). The rule $F(x^{k+1}) \le mt_L^k v^k < 0$ for a serious step of the methods of centers hinders progress of $\{x^k\}$ towards the boundary of the feasible set.

We may add that the results on convergence in Section 4 and Section 5 hold also for line search rules that allow for arbitrarily short serious stepsizes (as if $\bar{t}=0$ in Step 3″). Such rules were introduced by Mifflin (1982). The relevant analysis is presented in the next chapter. However, we belive that the rules of Step 3″ are general enough to allow for constructing efficient line search procedures in the convex case.

7. Phase I - phase II methods

The algorithms described in the preceding sections require a feasible starting point. Of course, by minimizing F each of the algorithms can find a feasible point in a finite number of iterations, since they generate minimizing sequence, while $\inf\{F(x) : x \in R^N\} < 0$ by the Slater constraint qualification. However, in certain cases one knows a point, say $\tilde{x} \in R^N$, which is close to a solution, but infeasible. Then it is reasonable to search for a feasible point by moving from $\tilde{x}$ towards the constraint boundary in a way that ensures as small an increase in the objective as possible. This is the aim of phase I - phase II methods (Polak, Trahan and Mayne, 1979; Polak, Mayne and Wardi, 1983). At each iteration of phase I such methods try to decrease the constraint violation while not completely ignoring the objective function. Once a feasible point is found, at phase II the methods proceed as feasible direction algorithms.

In this section we show that it is easy to turn the previously discussed algorithms into phase I - phase II methods. In fact, this requires only minor line search modifications and a slightly more involved convergence analysis. The resulting algorithms may be considered as more advanced versions of the method of Polak, Mayne and Wardi (1983).

Throughout this section we suppose that the objective subgradient mapping g_f is defined on the whole of R^N, i.e. $g_f(x) \in \partial f(x)$ for all

x. Also for simplicity we assume that F and g_F can be evaluated everywhere; see Remark 3.4 for a discussion of how to relax this assumption.

We shall first describe the modified algorithm with subgradient aggregation. Consider the following modification of the k-th primal subproblem (2.41): find $(d^k, v^k) \in R^N \times R$ to

$$\begin{aligned}
&\text{minimize } \tfrac{1}{2}|d|^2+v,\\
&\text{subject to } f_j^k-f(x^k)-F(x^k)_+ + \langle g_f^j, d\rangle \le v,\ j\in J_f^k,\\
&\qquad f_p^k-f(x^k)-F(x^k)_+ + \langle p_f^{k-1}, d\rangle \le v,\\
&\qquad F_j^k-F(x^k)_+ + \langle g_F^j, d\rangle \le v,\ j\in J_F^k,\\
&\qquad F_p^k-F(x^k)_+ + \langle p_F^{k-1}, d\rangle \le v,
\end{aligned} \tag{7.1}$$

and its dual

$$\begin{aligned}
\underset{\lambda,\mu}{\text{minimize}}\ &\tfrac{1}{2}\Big|\sum_{j\in J_F^k}\lambda_j g_f^j+\lambda_p p_f^{k-1}+\sum_{j\in J_F^k}\mu_j g_F^j+\mu_p p_F^{k-1}\Big|^2 +\\
&+\sum_{j\in J_f^k}\lambda_j[f(x^k)-f_j^k+F(x^k)_+]+\lambda_p[f(x^k)-f_p^k+F(x^k)_+] +\\
&+\sum_{j\in J_F^k}\mu_j[F(x^k)_+-F_j^k]+\mu_p[F(x^k)_+-F_p^k],
\end{aligned} \tag{7.2}$$

$$\begin{aligned}
\text{subject to } &\lambda_j\ge 0,\ j\in J_f^k,\ \lambda_p\ge 0,\ \mu_j\ge 0,\ j\in J_F^k,\ \mu_p\ge 0,\\
&\sum_{j\in J_f^k}\lambda_j+\lambda_p+\sum_{j\in J_F^k}\mu_j+\mu_p=1
\end{aligned}$$

with a solution denoted by λ_j^k, $j\in J_f^k$, λ_p^k, μ_j^k, $j\in J_F^k$, μ_p^k. If we denote by (d^k,u^k) the solution of subproblem (2.41) then

$$v^k = u^k-F(x^k)_+, \tag{7.3}$$

so the results of Section 2 imply

$$v^k = \hat{H}_a^k(x^k+d^k)-H(x^k;x^k)\le 0, \tag{7.4}$$

cf. (2.17). Thus v^k may be interpreted as an approximate derivative of $H(\cdot;x^k)$ at x^k in the direction d^k. This is important for line searches.

It is easy to observe that if $F(x^k)\le 0$ then $v^k=u^k$ and subproblems (7.1) and (7.2) reduce to subproblems (2.41) and (3.1), respectively. In fact, even for $F(x^k)>0$ subproblems (7.1) and (2.41) are essentially equivalent in view of (7.3), and can be regarded as quadratic programming formulations of the following

$$\text{minimize} \quad \hat{H}_a^k(x^k+d)+\tfrac{1}{2}|d|^2 \quad \text{over all } d,$$

which in turn is a local approximation to the problem of minimizing $H(\cdot;x^k)$. If $H(x^k;x^k)=F(x^k)_+>0$ then it is reasonable to search for a direction d^k that forms obtuse angles with the constraint subgradients, since then d^k points from x^k towards the feasible region. To see that this is the case, note that $d^k=-p^k$, where

$$p^k = \sum_{j\in J_f^k}\lambda_j^k g_f^j + \lambda_p^k p_f^{k-1} + \sum_{j\in J_F^k}\mu_j^k g_F^j + \mu_p^k p_F^{k-1},$$

$$\lambda_j^k\ge 0,\ j\in J_f^k,\ \lambda_p^k\ge 0,\ \mu_j^k\ge 0,\ j\in J_F^k,\ \mu_p^k\ge 0,$$

$$\sum_{j\in J_f^k}\lambda_j^k + \lambda_p^k + \sum_{j\in J_F^k}\mu_j^k + \mu_p^k = 1,$$

cf. (2.43), and use the <u>linearization errors</u>

$$\begin{aligned}
\alpha_{f,j}^k &= f(x^k)-f_j^k, \quad j=1,\dots,k,\\
\alpha_{f,p}^k &= f(x^k)-f_p^k,\\
\alpha_{F,j}^k &= F(x^k)_+-F_j^k, \quad j=1,\dots,k,\\
\alpha_{F,p}^k &= F(x^k)_+-F_p^k,
\end{aligned} \tag{7.5}$$

for rewriting subproblem (7.2) in the following form

$$\begin{aligned}
\underset{\lambda,\mu}{\text{minimize}}\ & \tfrac{1}{2}\Big|\sum_{j\in J_f^k}\lambda_j g_f^j+\lambda_p p_f^{k-1}+\sum_{j\in J_F^k}\mu_j g_F^j+\mu_p p_F^{k-1}\Big|^2 + \\
&+\sum_{j\in J_f^k}\lambda_j[\alpha_{f,j}^k+F(x^k)_+]+\lambda_p[\alpha_{f,j}^k+F(x^k)_+] + \\
&+\sum_{j\in J_F^k}\mu_j\alpha_{F,j}^k + \mu_p\alpha_{F,p}^k,
\end{aligned} \tag{7.6}$$

subject to $\lambda_j\ge 0,\ j\in J_f^k,\ \lambda_p\ge 0,\ \mu_j\ge 0,\ j\in J_F^k,\ \mu_p\ge 0,$

$$\sum_{j\in J_f^k}\lambda_j + \lambda_p + \sum_{j\in J_F^k}\mu_j + \mu_p = 1.$$

We conclude that if the linearization errors have comparable values then a positive term $F(x^k)_+$ in (7.6) tends to make the constraint subgradients influence d^k more actively than do the objective subgradients. On the other hand, if the constraint violation is not too large then the multipliers λ_j^k, $j\in J_f^k$, and λ_p^k are positive and so the objective subgradients contribute significantly to d^k, deflecting it from directions of ascent of the objective function at x^k.

We may now state the first version of our phase I - phase II method in detail. To save space we shall use the notation of Algorithm 3.1

Algorithm 7.1 is obtained from Algorithm 3.1 by replacing Step 1 and Step 2 with the following steps.

Step 1'' (Direction finding). Find a solution λ_j^k, $j\in J_f^k$, λ_p^k, μ_j^k, $j\in J_F^k$, and μ_p^k to the k-th dual subproblem (7.6). Calculate multipliers ν_f^k, $\tilde{\lambda}_j^k$, $j\in J_f^k$, $\tilde{\lambda}_p^k$, ν_F^k, $\tilde{\mu}_j^k$, $j\in J_F^k$, and $\tilde{\mu}_p^k$ satisfying (2.44). Compute $(p_f^k, \tilde{f}_p^k)$ and (p_F^k, F_p^k) by (2.45) and use (2.29) for calculating p^k. Set $d^k = -p^k$ and

$$v^k = -\{|p^k|^2 + \sum_{j\in J_f^k}\lambda_j^k[f(x^k)-f_j^k + F(x^k)_+] + \lambda_p^k[f(x^k)-f_p^k+F(x^k)_+] + \\ + \sum_{j\in J_F^k}\mu_j^k[F(x^k)_+ - F_j^k] + \mu_p^k[F(x^k)_+ - F_p^k]. \tag{7.7a}$$

Step 2' (Stopping criterion). Set

$$w^k = \tfrac{1}{2}|p^k|^2 + \nu_f^k[f(x^k)-\tilde{f}_p^k+F(x^k)_+] + \nu_F^k[F(x^k)_+ - \tilde{F}_p^k]. \tag{7.7b}$$

If $w^k \le \varepsilon_s$ terminate; otherwise, go to Step 3.

In Step 1'' of Algorithm 7.1 one may equivalently solve for (d^k, v^k) the k-th primal search direction finding subproblem (7.1), which has Lagrange multipliers λ_j^k, $j\in J_f^k$, λ_p^k, μ_j^k, $j\in J_F^k$, and μ_p^k. This observation, together with the representation (7.7) of v^k, follow from Lemma 2.3.

To establish relations between Algorithm 3.1 and Algorithm 7.1, suppose that for some $k\ge 1$ one has $F(x^k)\le 0$. Then the k-th iterations of both algorithms are identical, so $F(x^{k+1}) \le 0$ by the properties of Algorithm 3.1. It follows by induction that also at subsequent iterations Algorithm 7.1 generates feasible points and reduces to Algorithm 3.1. We say that Algorithm 7.1 enters phase II at the k-th iteration if $F(x^k) \le 0$ and $F(x^{k-1}) > 0$. In this case iterations 1,...,k-1 form phase I. Thus we see that at phase II Algorithm 7.1 becomes equivalent to Algorithm 3.1. Hence the convergence results of Section 4 are valid for phase II of Algorithm 7.1. Moreover, to analyze convergence of Algorithm 7.1 we need to consider only the case when phase I is infinitely long, i.e. $F(x^k) > 0$ for all k.

We shall now establish global convergence of Algorithm 7.1. To this end we have to extend certain results of Section 4. First, we observe that Lemma 4.1 is valid also for Algorithm 7.1, since the aggregation rules do not depend on the feasibility of $\{x^k\}$. In the following extension of Lemma 4.2 we use the linearization errors defined by (7.5) and the relations

$$\tilde{\alpha}_{f,p}^k = f(x^k) - \tilde{f}_p^k,$$

$$\tilde{\alpha}_{F,p}^k = F(x^k)_+ - \tilde{F}_p^k, \tag{7.8}$$

$$\tilde{\alpha}_p^k = \nu_f^k[f(x^k) - \tilde{f}_p^k + F(x^k)_+] + \nu_F^k[F(x^k)_+ - \tilde{F}_p^k].$$

Lemma 7.2. At the k-th iteration of Algorithm 7.1, one has (4.4) and

$$g_f^j \in \partial_\varepsilon H(x^k;x^k) \quad \text{for} \quad \varepsilon = \alpha_{f,j}^k + F(x^k)_+,\ j=1,\ldots,k, \tag{7.9a}$$

$$g_F^j \in \partial_\varepsilon H(x^k;x^k) \quad \text{for} \quad \varepsilon = \alpha_{F,j}^k,\ j=1,\ldots,k, \tag{7.9b}$$

$$p_f^{k-1} \in \partial_\varepsilon H(x^k;x^k) \quad \text{for} \quad \varepsilon = \alpha_{f,p}^k + F(x^k)_+, \tag{7.9c}$$

$$p_F^{k-1} \in \partial_\varepsilon H(x^k;x^k) \quad \text{for} \quad \varepsilon = \alpha_{F,p}^k, \tag{7.9d}$$

$$p_f^k \in \partial_\varepsilon H(x^k;x^k) \quad \text{for} \quad \varepsilon = \tilde{\alpha}_{f,p}^k + F(x^k)_+, \tag{7.9e}$$

$$p_F^k \in \partial_\varepsilon H(x^k;x^k) \quad \text{for} \quad \varepsilon = \tilde{\alpha}_{F,p}^k, \tag{7.9f}$$

$$p^k \in \partial_\varepsilon H(x^k;x^k) \quad \text{for} \quad \varepsilon = \tilde{\alpha}_p^k \ge 0. \tag{7.9g}$$

Proof. By (2.36), for any x and $j \le k$ we have

$$f(x) \geq \langle g_f^j, x-x^k \rangle + f_j^k ,$$

hence

$$H(x;x^k) = \max\{f(x)-f(x^k), F(x)\} \geq f(x)-f(x^k) \geq$$

$$\geq \langle g_f^j, x-x^k \rangle - [f(x^k)-f_j^k] \geq$$

$$\geq H(x^k;x^k) + \langle g_f^j, x-x^k \rangle - [\alpha_{f,j}^k + F(x^k)_+] . \tag{7.10a}$$

This yields (7.9a) by the definition of ε-subdifferential. Similarly, from (2.36) we obtain

$$F(x) \geq \langle g_F^j, x-x^k \rangle + F_j^k$$

and

$$H(x;x^k) \geq F(x) \geq \langle g_F^j, x-x^k \rangle + F_j^k \geq$$

$$\geq H(x^k;x^k) + \langle g_F^j, x-x^k \rangle - \alpha_{F,j}^k , \tag{7.10b}$$

which implies (7.9b). In view of Lemma 4.1, one may take convex combinations of (7.10) to obtain (7.9c)-(7.9f). In particular, we have

$$H(x;x^k) \geq H(x^k;x^k) + \langle p_f^k, x-x^k \rangle - [f(x^k) - \tilde{f}_p^k + F(x^k)_+] , \tag{7.11a}$$

$$H(x;x^k) \geq H(x^k;x^k) + \langle p_F^k, x-x^k \rangle - [F(x^k)_+ - \tilde{F}_p^k] \tag{7.11b}$$

for any x. Multiplying (7.11a) by $\nu_f^k \geq 0$ and (7.11b) by $\nu_F^k \geq 0$, adding the results and using the fact that $\nu_f^k + \nu_F^k = 1$ by (2.44e), we obtain

$$H(x;x^k) \geq H(x^k;x^k) + \langle \nu_f^k p_f^k + \nu_F^k p_F^k, x-x^k \rangle +$$

$$-\nu_f^k [f(x^k) - \tilde{f}_p^k + F(x^k)_+] - \nu_F^k [F(x^k)_+ - F_p^k] \geq$$

$$\geq H(x^k;x^k) + \langle p^k, x-x^k \rangle - \tilde{\alpha}_p^k$$

from (2.29). Setting $x=x^k$, we get $\tilde{\alpha}_p^k \geq 0$. This completes the proof of (7.9g). (4.4a)-(4.4f) can be established as in the proof of Lemma 4.2. □

From (7.7) and (7.8) we deduce that Lemma 4.4 holds for Algorithm 7.1. Then relation (4.8) follows from (7.9g) and (4.5), so Lemma 4.5 remains valid for Algorithm 7.1.

As observed above, we may assume that $F(x^k)>0$ for all k. Since the line search rules imply that we always have

$$F(x^{k+1}) \le H(x^{k+1};x^k) \le H(x^k;x^k)+mt_L^k v^k = F(x^k)_+ +mt_L^k v^k,$$

we obtain

$$F(x^{k+1}) \le F(x^k)+mt_L^k v^k \quad \text{for all k.} \tag{7.12}$$

The above relation, which substitutes (4.9), shows that $\{F(x^k)\}$ is nonincreasing. Since $F(x^k)>0$ for all k by assumption, one can obtain (4.10) from (7.12) as in the proof of Lemma 2.4.7.

Lemma 4.7, which is based on (4.4g), remains valid for Algorithm 7.1. Lemma 4.9 can be proved by using (7.12) and the arguments in the proof of Lemma 2.4.8, together with Lemma 4.7.

The following extension of Lemma 4.10 can be easily derived.

Lemma 7.3. At the k-th iteration of Algorithm 7.1, w^k is the optimal value of subproblem (7.6).

To prove Lemma 4.11 for Algorithm 7.1, start by substituting (4.13) by

$$g^k = g_F^k \quad \text{and} \quad \alpha^k = \alpha_{F,k}^k \qquad \text{if} \quad f(y^k)-f(x^{k-1}) < F(y^k), \tag{7.13a}$$

$$g^k = g_f^k \quad \text{and} \quad \alpha^k = \alpha_{f,k}^k + F(x^k)_+ \qquad \text{if} \quad f(y^k)-f(x^{k-1}) \ge F(y^k), \tag{7.13b}$$

and (4.16) by

$$\max\{f(y^k)-f(x^k),F(y^k)\} > F(x^k)_+ +mv^{k-1}. \tag{7.14}$$

Instead of (4.17a), we obtain

$$-\alpha^k + \langle g^k,d^{k-1}\rangle = -[F(x^k)_+ - F(y^k) - \langle g_F^k, x^k-y^k\rangle] + \langle g_F^k,d^k\rangle =$$

$$= F(y^k)-F(x^k)_+ \ge mv^{k-1}, \tag{7.15a}$$

while (4.17b) is replaced by

$$-\alpha^k + \langle g^k,d^k\rangle = -[f(x^k)-f(y^k) - \langle g_f^k, x^k-y^k\rangle + F(x^k)_+] + \langle g_f^k,d^k\rangle =$$

$$= f(y^k)-f(x^k)-F(x^k)_+ > mv^{k-1}. \tag{7.15b}$$

Next, substitute (4.19b) by the following relation

$$\sum_{j\in J_f^k}\lambda_j(\nu)[\alpha_{f,j}^k+F(x^k)_+]+\lambda_p(\nu)[\alpha_{f,p}^k+F(x^k)_+]+$$

$$+\sum_{j\in J_F^k}\mu_j(\nu)\alpha_{F,j}^k+\mu_p(\nu)\alpha_{F,j}^k = (1-\nu)\alpha_p^{k-1}+\nu\alpha^k \tag{7.16}$$

and use it together with (4.19a) to deduce from Lemma 7.3 that w^k is majorized by the optimal value of (4.20), as before.

Since Lemma 4.12 is valid for Algorithm 7.1, we obtain the following result.

Theorem 7.4. Suppose Algorithm 7.1 generates an infinite sequence $\{x^k\}$. Then:

(i) If $F(x^k)>0$ for all k, i.e. the algorithm stays at phase I, then every accumulation point of $\{x^k\}$ is a solution to problem (1.1).

(ii) If $F(x^{\tilde{k}})\leq 0$ for some $\tilde{k}\geq 1$, then $F(x^k)\leq 0$ for all $k\geq\tilde{k}$ and $f(x^k)\downarrow\inf\{f(x):F(x)\leq 0\}$, i.e. $\{x^k\}$ is a minimizing sequence for problem (1.1). If additionally problem (1.1) admits of a solution, then $\{x^k\}$ converges to a solution of problem (1.1).

We shall now discuss modifications of line search rules for Algorithm 7.1. First, we note that Algorithm 7.1 can use, instead of Step 3, Step 3′ described in Section 6, cf. (6.1). This will allow for implementing more efficient line search procedures without impairing the convergence results, since one can easily derive suitable extensions of (7.15) by using Lemma 2.6.1 and Lemma 6.1. Secondly, one may use the following modification of Step 3′, in which $\bar{t}\in(0,1]$ is a fixed parameter of the algorithm.

Step 3‴ (Line search). Select an auxiliary stepsize $t_R^k\in[\bar{t},1]$ and set $y^{k+1}=x^k+t_R^kd^k$. If either of the following two conditions is satisfied

$$F(x^k)>0 \quad\text{and}\quad F(y^{k+1})\leq F(x^k)+mt_R^kv^k \tag{7.17a}$$

or

$$F(x^k)\leq 0 \quad\text{and}\quad H(y^{k+1};x^k)\leq H(x^k;x^k)+mt_R^kv^k \tag{7.17b}$$

then set $t_L^k=t_R^k$ (a serious step); otherwise set $t_L^k=0$ (a null step).

If Step 3‴ is used then at phase I the algorithm will ignore the objective function values at line searches until a feasible point is found. A similar strategy can be employed in the context of feasible directions methods based on Step 3" described in Section 6, cf. (6.2). To this end consider the following extension of Step 3′′ .

Step 3′′′′ (Line search). Select an auxiliary stepsize $t_R^k \in [\bar{t},1]$ and set $y^{k+1}=x^k+t_R^k d^k$. If either of the following two conditions is satisfied

$$F(x^k)>0 \quad \text{and} \quad F(y^{k+1}) \le F(x^k)+mt_R^k v^k \tag{7.18a}$$

or

$$f(y^{k+1}) \le f(x^k)+mt_R^k v^k \quad \text{and} \quad F(y^{k+1}) \le 0 \tag{7.18b}$$

then set $t_L^k=t_R^k$; otherwise set $t_L^k=0$.

If the current point x^k is infeasible then in both Step 3′′′ and Step 3′′′′ we should search for a point that descreases the constraint violation. On the other hand, both of these steps maintain feasibility at phase II. In particular, at phase II the algorithm with Step 3′′′′ reduces to the feasible directions method discussed in Section 6.

We may add that Theorem 7.4 remains valid if Step 3′′′ or Step 3′′′′ is used. To see this, note that phase II is covered by the results of Section 6, while at phase I each null step yields

$$F(y^k) > F(x^k)+mt_R^k v^{k-1},$$

so that we have

$$-\alpha_{F,k}^k + \langle g_F^k, d^{k-1} \rangle > mv^{k-1} \tag{7.19}$$

from Lemma 2.6.1 and the fact that $\alpha_{F,k}^k=F(x^k)_+-F_k^k=F(x^k)-F_k^k$ if $F(x^k)>0$. Thus one can use (7.19) instead of (7.15) in the proof of Lemma 4.11.

We now pass to the phase I - phase II method with subgradient selection, which extends Algorithm 5.1 to the case of infeasible starting points.

Algorithm 7.5 is obtained from Algorithm 7.1 by replacing Step 1′′ with

Step 1′′′ (Direction finding). Find multipliers λ_j^k, $j \in J_f^k$, and μ_j^k, $j \in J_F^k$, that solve the following k-th dual subproblem

$$\text{minimize } \frac{1}{2}\Big|\sum_{j\in J_f^k}\lambda_j g_f^j + \sum_{j\in J_F^k}\mu_j g_F^j\Big|^2 +$$

$$+\sum_{j\in J_f^k}\lambda_j[\alpha_{f,j}^k + F(x^k)_+] + \sum_{j\in J_F^k}\mu_j\alpha_{F,j}^k, \tag{7.20}$$

$$\text{subject to } \lambda_j \ge 0,\ j\in J_f^k,\ \mu_j\ge 0,\ j\in J_F^k,\ \sum_{j\in J_f^k}\lambda_j + \sum_{j\in J_F^k}\mu_j = 1$$

and the corresponding sets $\hat{J}_f^k$ and $\hat{J}_F^k$ that satisfy (2.23). Calculate scaled multipliers satisfying (2.26), compute $(p_f^k,\tilde{f}_p^k)$ and $(p_F^k,\tilde{F}_p^k)$ by (2.28), and use (2.29) for calculating p^k. Set $d^k = -p^k$ and

$$v^k = -\{|p^k|^2 + \nu_f^k[\tilde{\alpha}_{f,p}^k + F(x^k)_+] + \nu_F^k\alpha_{F,p}^k\}. \tag{7.21}$$

Of course, in (7.20) and (7.21) we use the linearization errors defined by (7.5) and (7.8). Also it is readily seen that (7.20) is the dual of the following k-th (primal) search direction finding subproblem:

$$\underset{(d,v)\in R^{N+1}}{\text{minimize}} \quad \frac{1}{2}|d|^2 + v,$$

$$\text{subject to } \quad f_j^k - f(x^k) - F(x^k)_+ + \langle g_f^j, d\rangle \le v,\ j\in J_f^k, \tag{7.22}$$

$$F_j^k - F(x^k)_+ + \langle g_F^j, d\rangle \le v,\ j\in J_F^k,$$

with the solution (d^k,v^k) and the Lagrange multipliers λ_j^k, $j\in J_f^k$, and μ_j^k, $j\in J_F^k$. Therefore at phase II Algorithm 7.5 reduces to Algorithm 5.1.

We may add that one can use the modified line search rules discussed in this section also in Algorithm 7.5. Global convergence of the resulting methods can be expressed in the form of Theorem 7.4. To this end one may combinate the preceding results of this section with the techniques of Section 5.

CHAPTER 6

Methods of Feasible Directions for Nonconvex Constrained Problems

1. Introduction

In this chapter we consider the following constrained minimization problem

$$\text{minimize} \quad f(x), \text{ subject to } \quad F(x) \le 0, \tag{1.1}$$

where the functions $f: R^N \to R$ and $F: R^N \to R$ are locally Lipschitzian but not necessarily convex or differentiable. We assume that the feasible set

$$S = \{x \in R^N : F(x) \le 0\}$$

is nonempty.

We present several readily implementable algorithms for solving problem (1.1), which differ in complexity, storage and speed of convergence. The methods require only the evaluation of f or F and one subgradient of f or F at designated points. Storage requirements and work per iteration of the algorithms can be controlled by the user.

The algorithms are obtained by incorporating in the feasible point methods of Chapter 5 the techniques for dealing with nonconvexity that were developed in Chapter 3 and Chapter 4. Thus the algorithms generate search directions by using separate polyhedral approximations to f and F. To construct such approximations we use the rules for selecting and aggregating separately subgradients of f and F that were introduced in Chapter 5. The polyhedral approximations take nonconvexity into account by using either the subgradient locality measures of Chapter 3, or the subgradient deletion rules of Chapter 4. In the latter case we employ resetting strategies for localizing the past subgradient information on the basis of estimating the degree of stationarity of the current approximation to a solution.

The algorithms are feasible point methods of descent, i.e. they generate sequences of points $\{x^k\}$ satisfying

$$x^k \in S \quad \text{and} \quad f(x^{k+1}) < f(x^k) \quad \text{if} \quad x^{k+1} \ne x^k, \text{ for all } k,$$

where $x^1 \in S$ is the starting point. Under mild assumptions on F, such as nonemptiness of the interior of S, each of the algorithms can find a feasible starting point by minimizing F.

We shall also present phase I - phase II methods that can be employed when the user has a good, but infeasible, initial approximation to a solution. Starting from this point, phase I of such methods tries to find a feasible point without unduly increasing the objective value. At phase II the methods reduce to feasible point algorithms.

The algorithms of this chapter may be viewed as extensions of the Pironneau and Polak (1972; 1973) method of centers and method of feasible directions to the nondifferentiable case. One of the algorithms can be derived by applying our subgradient selection and aggregation rules to the Mifflin (1982) method . Also our extensions of the Polak, Mayne and Trahan (1979) phase I - phase II algorithm differ from those of Polak, Mayne and Wardi (1983).

We shall prove that each of our feasible point methods is globally convergent in the sense that it generates an infinite sequence of points $\{x^k\}$ such that every accumulation point of $\{x^k\}$ is stationary for f on S. If problem (1.1) is convex and satisfies the Slater constraint qualification (i.e. $F(x) < 0$ for some x in R^N), then x^k is a minimizing sequence for f on S, which converges to a solution of problem (1.1) whenever f attains its infimum on S. Similar convergence results hold for our phase I - phase II methods.

In Section 2 we derive the methods. The algorithm with subgradient aggregation is described in detail in Section 3, and its convergence is established in Section 4. Section 5 is devoted to the algorithm with subgradient selection. In Section 6 we study various modifications of the methods with subgradient locality measures. Several versions of methods with subgradient deletion rules are analyzed in Section 7. In Section 8 we discuss methods that neglect linearization errors. Phase I - phase II methods are described in Section 9.

2. Derivation of the Methods

We start by recalling the necessary conditions of optimality for problem (1.1), see Section 1.2.

For any fixed $x \in R^N$, define the improvement function

$$H(y;x) = \max\{f(y)-f(x),\ F(y)\} \quad \text{for all} \quad y \in R^N. \tag{2.1}$$

If $\bar{x} \in S$ is a local solution of (1.1) then $H(\cdot;\bar{x})$ attains a local minimum at $\bar{x}$, so $0 \in \partial H(\bar{x};\bar{x})$, where $\partial H(\bar{x};\bar{x})$ denotes the subdifferential of $H(\cdot;\bar{x})$ at $\bar{x}$. Since $\partial H(x;x) \subset \hat{M}(x)$ for

$$\hat{M}(x) = \begin{cases} \partial f(x) & \text{if } F(x) < 0, \\ \text{conv}\{\partial f(x) \cup \partial F(x)\} & \text{if } F(x) = 0, \\ \partial F(x) & \text{if } F(x) > 0, \end{cases} \tag{2.2}$$

the necessary condition of optimality is $0 \in \hat{M}(x)$. For this reason, a point $\bar{x} \in S$ such that $0 \in \hat{M}(x)$ is called stationary for f on S.

Remark 2.1. There is no loss of generality in requiring that F be scalar-valued. If the original formulation of the problem involves a finite number of constraints $\tilde{F}_j(x) \leq 0$, $j \in J$, with locally Lipschitzian functions $\tilde{F}_j$, then one can let

$$F(x) = \max\{\tilde{F}_j(x) : j \in J\} \quad \text{for all } x. \tag{2.3a}$$

Defining

$$\tilde{\partial}F(x) = \text{conv}\{\partial \tilde{F}_j(x) : j \in J \text{ and } \tilde{F}_j(x) = F(x)\} \quad \text{for all } x, \tag{2.3b}$$

we have (see (1.2.60))

$$\partial F(x) \subset \tilde{\partial}F(x) \quad \text{for all } x. \tag{2.4}$$

Let

$$M(x) = \begin{cases} \partial f(x) & \text{if } F(x) < 0, \\ \text{conv}\{\partial f(x) \cup \tilde{\partial}F(x)\} & \text{if } F(x) = 0, \\ \tilde{\partial}F(x) & \text{if } F(x) > 0, \end{cases} \tag{2.5}$$

for all x. By (2.2) and (2.5), $\hat{M}(\cdot) \subset \tilde{M}(\cdot)$, so, although we may have $\hat{M}(\bar{x}) \neq \tilde{M}(\bar{x})$, if $\bar{x}$ solves(1.1) locally then $0 \in \tilde{M}(\bar{x})$. Therefore, we shall also say that a point $\bar{x} \in S$ is stationary for f on S if $0 \in \tilde{M}(\bar{x})$.

In view of the above results, testing if a point $x \in S$ is stationary for f on S is in a sense equivalent to testing if there exists a direction of descent for $H(\cdot;x)$ at x. At the same time, if we find a point y such that $H(y;x) < H(x;x) = 0$ then $f(y) < f(x)$ and $F(y) < 0$, so y is better than x. Therefore, in theory, one could solve problem (1.1) by the Huard (1968) method of centers described in Section 5.2, which in the present case has stationary accumulation points, if any.

One can, in theory, find a descent direction for $\partial H(\cdot;x)$ at x by finding the subgradient of minimum norm in $\partial H(\bar{x};\bar{x})$ (see Lemma 1.2.18).

This would require the knowledge of full subdifferentials $\partial f(x)$ and $\partial F(x)$. However, we assume only that we have a finite process for calculating $f(x)$ and a certain subgradient $g_f(x) \in \partial f(x)$ at each $x \in S$, and $F(x)$ and an arbitrary subgradient $g_F(x) \in \partial F(x)$ at each $x \notin S$. This assumption is realistic in many applications (Mifflin, 1982). Therefore, we shall compensate for the lack of $\partial H(x;x)$ by using $g_f(y)$ and $g_F(y)$ evaluated at several points y close to x. For simplicity of exposition, we shall temporarily assume that g_f and g_F are defined on the whole of R^N.

Remark 2.2. In the case considered in Remark 2.1 it suffices to assume that $g_F(x) \in \tilde{\partial} F(x)$ at each $x \notin S$. Then for each infeasible x one has to find an index $j \in J$ satisfying $\tilde{F}_j(x)=F(x)$ and an arbitrary subgradient $g_{F_j}(x) \in \partial \tilde{F}_j(x)$, cf. (2.3b). This requirement, formulated directly in terms of subdifferentials of the constraint functions $\tilde{F}_j$, is frequently more practical than the one in terms of ∂F, since $\partial F(x)$ may not be available (because $\partial F(x)$ is, in general, different from $\tilde{\partial} F(x)$).

We shall now derive the first generalization of the feasible direction method of Chapter 5. Our extension of that method to the nonconvex case will use polyhedral approximations based on subgradient locality measures introduced in Chapter 3.

The algorithm will generate sequences of points $\{x^k\} \subset S$, search directions $\{d^k\} \subset R^N$ and stepsizes $\{t_L^k\} \subset R_+$ related by

$$x^{k+1} = x^k + t_L^k d^k \quad \text{for} \quad k=1,2,\ldots,$$

where $x^1 \in S$ is a given starting point. At the k-th iteration d^k is intended to be a direction of descent for $H(\cdot;x^k)$ at x^k, and $H(x^k;x^k)=0$ because $x^k \in S$. Therefore, we shall use a two-point line search for finding two stepsizes t_L^k and t_R^k, $0 \le t_L^k \le t_R^k$, the next point $x^{k+1} \in S$ satisfying

$$f(x^{k+1}) < f(x^k) \quad \text{if} \quad x^{k+1} \ne x^k \ (t_L^k > 0),$$

and the trial point

$$y^{k+1} = x^k + t_R^k d^k$$

such that the subgradients $g_f(y^{k+1})$ and $g_F(y^{k+1})$ modify significantly the next polyhedral approximations to f and F that will be used for finding the next search direction.

Thus the algorithm calculates subgradients

$$g_f^j = g_f(y^j) \quad \text{and} \quad g_F^j = g_F(y^j) \quad \text{for} \quad j=1,2,\ldots,$$

where $y^1=x^1$. Each point y^j defines the linearizations

$$f_j(x) = f(y^j)+\langle g_f^j, x-y^j\rangle \quad \text{for all } x,$$
$$F_j(x) = F(y^j)+\langle g_F^j, x-y^j\rangle \quad \text{for all } x, \tag{2.6}$$

of f and F, respectively. At the k-th iteration the subgradient information collected at the j-th iteration $(j \le k)$ is characterized by the linearization values

$$f_j^k = f_j(x^k),$$
$$F_j^k = F_j(x^k),$$

and the distance measure

$$s_j^k = |y^j-x^j|+\sum_{i=j}^{k-1}|x^{i+1}-x^i|.$$

The linearization values determine the current expression of the linearizations

$$f_j(x) = f_j^k+\langle g_f^j, x-x^k\rangle \quad \text{for all } x,$$
$$F_j(x) = F_j^k+\langle g_F^j, x-x^k\rangle \quad \text{for all } x, \tag{2.7}$$

while the distance measure estimates $|y^j-x^k|$:

$$|y^j-x^k| \le s_j^k.$$

These easily updated quantities enable us not to store the points y^j.

At the k-th iteration we want to find a descent direction for $H(\cdot;x^k)$. Therefore, we need some measures, say $\alpha_{f,j}^k \ge 0$ and $\alpha_{F,j}^k \ge 0$, that indicate how much the subgradients $g_f^j=g_f(y^j)$ and $g_F^j=g_F(y^j)$ differ from being elements of $\partial H(x^k;x^k)$. To this end, we shall use the following subgradient locality measures

$$\alpha_{f,j}^k = \max\{|f(x^k)-f_j^k|, \gamma_f(s_j^k)^2\}, \tag{2.8a}$$

$$\alpha_{F,j}^k = \max\{|F_j^k|,\ \gamma_F(s_j^k)^2\}, \tag{2.8b}$$

where γ_f and γ_F are positive parameters. We shall also set $\gamma_f=0$ if f is convex, and $\gamma_F=0$ if F is convex. This construction can be motivated as follows. In the convex case we have

$$\begin{aligned} g_f^j &\in \partial_\varepsilon H(x^k;x^k) \quad \text{for} \quad \varepsilon=\alpha_{f,j}^k=f(x^k)-f_j^k \geq 0, \\ g_F^j &\in \partial_\varepsilon H(x^k;x^k) \quad \text{for} \quad \varepsilon=\alpha_{F,j}^k=-F_j^k \geq 0, \end{aligned} \tag{2.9}$$

see Lemma 5.7.2. Next, suppose that F is nonconvex and the value of $\alpha_{F,j}^k$ is small. Then $|F_j^k| \approx 0$ and $s_j^k \approx 0$ $(\gamma_F > 0)$, so $|y^j-x^k| \approx 0$ and, by (2.6)-(2.7),

$$F(y^j) = F_j^k - \langle g_F^j, y^j-x^k \rangle \approx F_j^k \approx 0,$$

so the subgradient $g_F^j \in \partial F(y^j)$ is close to $\hat{M}(x^k)$ (see (2.2)), which approximates $\partial H(x^k;x^k)$. Similarly, if the value of $\alpha_{f,j}^k$ is small then the subgradient $g_f^j \in \partial f(y^j)$ is close to $\partial f(x^k)$, and to $\hat{M}(x^k)$ (see (2.2) and note that $F(x^k) \leq 0$).

Suppose that at the k-th iteration we have the subgradients (g_f^j, f_j^k, s_j^k) for $j \in J_f^k$, and (g_F^j, F_j^k, s_j^k) for $j \in J_F^k$, where J_f^k and J_F^k are some nonempty subsets of $\{1,\ldots,k\}$. Let

$$H^k(x) = \max\{f(x) - f(x^k), F(x)\} \quad \text{for all } x. \tag{2.10}$$

In the convex case, the methods of Chapter 5 would use the following search direction finding subproblem

$$\text{minimize} \quad \hat{H}_s^k(x^k+d) + \tfrac{1}{2}|d|^2 \quad \text{over all} \quad d \in R^N, \tag{2.11}$$

where

$$\begin{aligned} \hat{H}_s^k(x) &= \max\{\hat{f}_s^k(x)-f(x^k), \hat{F}_s^k(x)\}, \\ \hat{f}_s^k(x) &= \max\{f_j(x) : j \in J_f^k\}, \\ \hat{F}_s^k(x) &= \max\{F_j(x) : j \in J_F^k\} \end{aligned} \tag{2.12}$$

are polyhedral approximations to H^k, f and F, repectively. If f and F are convex then $\gamma_f=\gamma_F=0$, (2.8) becomes

$$\alpha_{f,j}^k = f(x^k)-f_j^k,$$

$$\alpha_{F,j}^k = -F_j^k,$$

and from (2.7)

$$f_j(x)-f(x^k)=f_j^k+\langle g_f^j,x-x^k\rangle -f(x^k)=-\alpha_{f,j}^k+\langle g_f^j,x-x^k\rangle ,$$

$$F_j(x)=F_j^k+\langle g_F^j,x-x^k\rangle = -\alpha_{F,j}^k+\langle g_F^j,x-x^k\rangle ,$$

so

$$\hat{H}_s^k(x) = \max\big[\max\{-\alpha_{f,j}^k+\langle g_f^j,x-x^k\rangle : j\in J_f^k\},$$
$$\max\{-\alpha_{F,j}^k+\langle g_F^j,x-x^k\rangle : j\in J_F^k\}\big] \tag{2.13}$$

and a quadratic programming formulation of subproblem (2.11) is to find $(d^k,\hat{v}^k)$ to

$$\begin{aligned} &\underset{(d,\hat{v})\in R^{N+1}}{\text{minimize}} && \tfrac{1}{2}|d|^2+\hat{v},\\ &\text{subject to} && -\alpha_{f,j}^k+\langle g_f^j,d\rangle \le \hat{v},\ j\in J_f^k,\\ &&& -\alpha_{F,j}^k+\langle g_F^j,d\rangle \le \hat{v},\ j\in J_F^k. \end{aligned} \tag{2.14}$$

Also

$$\hat{v}^k = \hat{H}_s^k(x^k+d^k)=\hat{H}_s^k(x^k+d^k)-H^k(x^k)$$

may be regarded as an approximate directional derivative of H^k at x^k in the direction d^k. Therefore, a natural extension of the methods of Chapter 5 to the nonconvex case consists in finding d^k by solving (2.14), with $\alpha_{f,j}^k$ and $\alpha_{F,j}^k$ defined by (2.8).

We shall now present another reason for using the search direction finding subproblem (2.14) in the nonconvex case. To this end we recall from Chapter 5 that $(d^k,\hat{v}^k)$ can be found by solving the following dual of (2.14)

$$\begin{aligned} &\underset{\lambda,\mu}{\text{minimize}} && \tfrac{1}{2}\Big|\sum_{j\in J_f^k}\lambda_j g_f^j+\sum_{j\in J_F^k}\mu_j g_F^j\Big|^2+\sum_{j\in J_f^k}\lambda_j\alpha_{f,j}^k+\sum_{j\in J_F^k}\mu_j\alpha_{F,j}^k,\\ &\text{subject to} && \lambda_j\ge 0,\ j\in J_f^k,\ \mu_j\ge 0,\ j\in J_F^k,\ \sum_{j\in J_f^k}\lambda_j+\sum_{j\in J_F^k}\mu_j=1, \end{aligned} \tag{2.15}$$

since if λ_j^k, $j \in J_f^k$, and μ_j^k, $j \in J_F^k$ denote any solution of (2.15) then

$$-d^k = \sum_{j \in J_f^k} \lambda_j^k g_f^j + \sum_{j \in J_F^k} \mu_j^k g_F^j, \tag{2.16a}$$

$$\hat{v}^k = -\{|d^k|^2 + \sum_{j \in J_f^k} \lambda_j^k \alpha_{f,j}^k + \sum_{j \in J_F^k} \mu_j^k \alpha_{F,j}^k\}, \tag{2.16b}$$

and

$$\lambda_j^k \geq 0,\ j \in J_f^k, \mu_j^k \geq 0, j \in J_F^k,\ \sum_{j \in J_f^k} \lambda_j^k + \sum_{j \in J_F^k} \mu_j^k = 1.$$

Thus the past subgradients g_f^j and g_F^j may contribute significantly to d^k (have relatively large values of $\lambda_j^k > 0$ and $\mu_j^k > 0$) only if the values of $\alpha_{f,j}^k$ and $\alpha_{F,j}^k$ are relatively small, i.e. g_f^j and g_F^j are approximate subgradients of H^k at x^k.

Up till now we have not specified how to choose the sets J_f^k and J_F^k involved in (2.13) and (2.15). Since subproblem (2.15) is of the form studied in Chapter 5 (see Lemma 5.2.3), we may use the subgradient selection rules developed in that chapter for choosing J_f^k and J_F^k recursively so that at most $N+3$ past subgradients are used for each direction finding. Thus at the k-th iteration one can find Lagrange multipliers λ_j^k and μ_j^k of (2.15) and sets $\hat{J}_f^k \subset J_f^k$ and $\hat{J}_F^k \subset J_F^k$ such that

$$\hat{J}_f^k = \{j \in J_f^k : \lambda_j^k \neq 0\} \quad \text{and} \quad \hat{J}_F^k = \{j \in J_F^k : \mu_j^k \neq 0\},$$

$$|\hat{J}_f^k \cup \hat{J}_F^k| \leq N+1.$$

Then, since the subgradients (g_f^j, f_j^k, s_j^k) for $j \in \hat{J}_f^k$ and (g_F^j, F_j^k, s_j^k) for $j \in \hat{J}_F^k$ embody, in the sense of Lemma 5.2.3(iv), all the past subgradient information that determined $(d^k, \hat{v}^k)$, one may discard the subgradients indexed by $j \notin \hat{J}_f^k \cup \hat{J}_F^k$ that were inactive at the k-th search direction finding (had null Lagrange multipliers). At the same time, the algorithm should use the latest subgradients. This leads to the choice

$$J_f^{k+1} = \hat{J}_f^k \cup \{k+1\} \quad \text{and} \quad J_F^{k+1} = \hat{J}_F^k \cup \{k+1\}.$$

As in Chapter 3, we shall also use suitable rules for reducing J_f^{k+1} and J_F^{k+1} at some iterations. Such resetting strategies are employed only to ensure locally uniform boundedness of the subgradients stored by the algorithm; see Section 3.2.

The above-described method with subgradient selection requires storing N+3 past subgradients. Also much work may be required by the solution of subproblem (2.14) (or (2.15)) if N is large. Therefore, we shall now use the subgradient aggregation strategy of Chapter 5 to derive a method in which storage and work per iteration can be controlled by the user.

At the k-th iteration of the method with subgradient aggregation we have some past subgradients (g_f^j, f_j^k, s_j^k), $j \in J_f^k$, and (g_F^j, F_j^k, s_j^k), $j \in J_F^k$, and two aggregate subgradients

$$(p_f^{k-1}, f_p^k, s_{f,p}^k) \in \operatorname{conv}\{(g_f^j, f_j^k, s_j^k) : j=1,\dots,k-1\},$$
$$(p_F^{k-1}, F_p^k, s_{F,p}^k) \in \operatorname{conv}\{(g_F^j, F_j^k, s_j^k) : j=1,\dots,k-1\}, \tag{2.17}$$

which were computed at the (k-1)-st iteration. The aggregate subgradients are characterized, similarly to (2.8), by the following aggregate subgradient locality measures

$$\alpha_{f,p}^k = \max\{|f(x^k) - F_p^k|, \gamma_f (s_f^k)^2\},$$
$$\alpha_{F,p}^k = \max\{|F_p^k|, \gamma_F (s_F^k)^2\}. \tag{2.18}$$

The value of $\alpha_{f,p}^k$ $(\alpha_{F,p}^k)$ indicates how far p_f^{k-1} (p_F^{k-1}) is from $\partial H^k(x^k)$.

We recall from Chapter 5 that in the convex case such aggregate subgradients define the (k-1)-st aggregate linearizations

$$\tilde f^{k-1}(x) = f_p^k + \langle p_f^{k-1}, x - x^k \rangle \quad \text{for all } x,$$

$$\tilde F^{k-1}(x) = F_p^k + \langle p_F^{k-1}, x - x^k \rangle \quad \text{for all } x,$$

which are convex combinations of the linearizations f_j and F_j, j=1,...,k-1, respectively. For this reason, in the convex case we defined the following aggregate polyhedral approximations

$$\hat H_a^k(x) = \max\{\hat f_a^k(x) - f(x^k), \hat F_a^k(x)\},$$

$$\hat f_a^k(x) = \max\{\tilde f^{k-1}(x), f_j(x) : j \in J_f^k\},$$

$$\hat F_a^k(x) = \max\{\tilde F^{k-1}(x), F_j(x) : j \in J_F^k\}$$

to H^k, f and F, respectively, and used the following search direction finding subproblem

$$\text{minimize} \quad \hat{H}_a^k(x^k+d)+\tfrac{1}{2}|d|^2 \quad \text{over all} \quad d \in R^N. \tag{2.19}$$

Reasoning as in the transition from (2.11) to (2.14), one can show that in the convex case $\hat{H}_a^k$ can be expressed in terms of subgradient locality measures as

$$\begin{aligned}\hat{H}_a^k(x) = \max\,\big[&\max\{-\alpha_{f,j}^k+\langle g_f^j,x-x^k\rangle : j\in J_f^k\},\\ &-\alpha_{f,p}^k+\langle p_f^{k-1},x-x^k\rangle,\\ &\max\{-\alpha_{F,p}^k+\langle g_F^j,x-x^k\rangle : j\in J_F^k\},\\ &-\alpha_{F,p}^k+\langle p_F^{k-1},x-x^k\rangle\big],\end{aligned} \tag{2.20}$$

while subproblem (2.19) may be solved by finding $(d^k,\hat{v}^k)$ to

$$\begin{aligned}&\underset{(d,\hat{v})\in R^{N+1}}{\text{minimize}} \quad \tfrac{1}{2}|d|^2+\hat{v},\\ &\text{subject to} \quad -\alpha_{f,j}^k+\langle g_f^j,d\rangle \le \hat{v},\ j\in J_f^k,\\ &\qquad -\alpha_{f,p}^k+\langle p_f^{k-1},d\rangle \le \hat{v},\\ &\qquad -\alpha_{F,j}^k+\langle g_F^j,d\rangle \le \hat{v},\ j\in J_F^k,\\ &\qquad -\alpha_{F,p}^k+\langle p_F^{k-1},d\rangle \le \hat{v}.\end{aligned} \tag{2.21}$$

Therefore we shall also use the search direction finding subproblem (2.21) in the nonconvex case. In fact, we shall use the following modified version of (2.21):

$$\begin{aligned}&\underset{(d,\hat{v})\in R^{N+1}}{\text{minimize}} \quad \tfrac{1}{2}|d|^2+\hat{v},\\ &\text{subject to} \quad -\alpha_{f,j}^k+\langle g_f^j,d\rangle \le \hat{v},\ j\in J_f^k,\\ &\qquad -\alpha_{f,p}^k+\langle p_f^{k-1},d\rangle \le \hat{v}, \quad \text{if} \quad r_a^k=0,\\ &\qquad -\alpha_{F,j}^k+\langle g_F^j,d\rangle \le \hat{v},\ j\in J_F^k,\\ &\qquad -\alpha_{F,p}^k+\langle p_F^{k-1},d\rangle \le \hat{v} \quad \text{if} \quad r_a^k=0,\end{aligned} \tag{2.22}$$

where the value of $r_a^k \in \{0,1\}$ indicates whether the (k-1)-st aggregate subgradients are dropped at the k-th iteration, when a so-called distance reset $(r_a^k=1)$ occurs (see Section 3.2). As in the method with subgradient selection, our resetting strategy will ensure locally uniform boundedness of accumulated subgradients.

For updating the aggregate subgradients we may use the rules of Chapter 5, which are applicable to subproblems of the form (2.22) (see Lemma 5.2.4). To this end, let λ_j^k, $j\in J_f^k$, λ_p^k, μ_j^k, $j\in J_F^k$, and μ_p^k denote any Lagrange multipliers of (2.22), where we set $\lambda_p^k=\mu_p^k=0$ if $r_a^k=1$. Similarly to (2.43)-(2.45), we have

$$\lambda_j^k\geq 0,\ j\in J_f^k,\ \lambda_p^k\geq 0,\ \mu_j^k\geq 0,\ j\in J_F^k,\ \mu_p^k\geq 0,$$

$$\sum_{j\in J_f^k}\lambda_j^k+\lambda_p^k+\sum_{j\in J_F^k}\mu_j^k+\mu_p^k=1,$$

hence we may calculate scaled multipliers $\tilde{\lambda}$ and $\tilde{\mu}$ satisfying

$$\nu_f^k=\sum_{j\in J_f^k}\lambda_j^k+\lambda_p^k,\lambda_j^k=\nu_f^k\tilde{\lambda}_j^k,\ j\in J_f^k,\ \lambda_p^k=\nu_f^k\tilde{\lambda}_p^k, \tag{2.23a}$$

$$\nu_F^k=\sum_{j\in J_F^k}\mu_j^k+\mu_p^k,\mu_j^k=\nu_F^k\tilde{\mu}_j^k,\ j\in J_F^k,\ \mu_p^k=\nu_F^k\tilde{\mu}_p^k, \tag{2.23b}$$

$$\tilde{\lambda}_j^k\geq 0,\ j\in J_f^k,\ \tilde{\lambda}_p^k\geq 0,\quad \sum_{j\in J_f^k}\tilde{\lambda}_j^k+\tilde{\lambda}_p^k=1, \tag{2.23c}$$

$$\tilde{\mu}_j^k\geq 0,\ j\in J_F^k,\ \tilde{\mu}_p^k\geq 0,\quad \sum_{j\in J_F^k}\tilde{\mu}_j^k+\tilde{\mu}_p^k=1, \tag{2.23d}$$

and use them for computing the current aggregate subgradients (cf. (3.3.4))

$$\begin{aligned}(p_f^k,\tilde{f}_p^k,\tilde{s}_f^k)&=\sum_{j\in J_f^k}\tilde{\lambda}_j^k(g_f^j,f_j^k,s_j^k)+\tilde{\lambda}_p^k(p_f^{k-1},f_p^k,s_f^k),\\(p_F^k,\tilde{F}_p^k,\tilde{s}_F^k)&=\sum_{j\in J_F^k}\tilde{\mu}_j^k(g_F^j,F_j^k,s_j^k)+\tilde{\mu}_p^k(p_F^{k-1},F_p^k,s_F^k).\end{aligned} \tag{2.24}$$

We recall from Section 5.2 that

$$\nu_f^k\geq 0\ ,\ \nu_F^k\geq 0\ ,\ \nu_f^k+\nu_F^k=1, \tag{2.25}$$

and that

$$\tilde{\lambda}_j^k=\lambda_j^k/\nu_f^k,\ j\in J_p^k,\ \tilde{\lambda}_p^k=\lambda_p^k/\nu_f^k,$$

$$\tilde{\mu}_j^k = \mu_j^k/\nu_F^k, \; j \quad J_F^k, \; \tilde{\mu}_p^k = \mu_p^k/\nu_F^k$$

if $\nu_f^k \neq 0$ and $\nu_F^k \neq 0$. If $\nu_f^k=0$ ($\nu_F^k=0$) then one may pick any numbers satisfying (2.23c) ((2.23d)). We also have

$$d^k = -p^k,$$

$$p^k = \nu_f^k p_f^k + \nu_F^k p_F^k, \tag{2.26}$$

and the k-th aggregate subgradients (2.24) embody, in the sense of Lemma 5.2.4, all that part of the past subgradient information that was active at the k-th search direction finding. Therefore, in the method with subgradient aggregation one has much freedom in the choice of J_f^{k+1} and J_F^{k+1}, subject only to the requirement that $k+1 \in J_f^{k+1} \cup J_F^{k+1}$. For instance, one may set $J_f^k=J_F^k=\{k\}$ for all k, although this will lead to slow convergence, which is enhanced if more subgradients are used for search direction finding.

Having computed $(p_f^k, \tilde{f}_p^k, \tilde{s}_f^k)$ and $(p_F^k, \tilde{F}_p^k, \tilde{s}_F^k)$ and the next point x^{k+1}, one can obtain (f_p^{k+1}, s_f^{k+1}) and (F_p^{k+1}, s_F^{k+1}) by the updating rules of Section 3.2. In particular, we may define the k-th aggregate linearizations

$$\tilde{f}^k(x) = \tilde{f}_p^k + \langle p_f^k, x-x^k \rangle \quad \text{for all } x,$$

$$\tilde{F}^k(x) = \tilde{F}_p^k + \langle p_F^k, x-x^k \rangle \quad \text{for all } x$$

and calculate $f_p^{k+1}=\tilde{f}^k(x^{k+1})$ and $F_p^{k+1}=\tilde{F}^k(x^{k+1})$. This ends the k-th iteration.

<u>Remark 2.3</u>. For convenience, in the two methods described above we have assumed that one calculates

$$g_f^{k+1} = g_f(y^{k+1}), \; f_{k+1}^{k+1} = f(y^{k+1}) + \langle g_f^{k+1}, x^{k+1}-y^{k+1} \rangle, \tag{2.27a}$$

$$g_F^{k+1} = g_F(y^{k+1}), \; F_{k+1}^{k+1} = F(y^{k+1}) + \langle g_F^{k+1}, x^{k+1}-y^{k+1} \rangle, \tag{2.27b}$$

and chooses sets of the form

$$J_f^{k+1} = \hat{J}_f^k \cup \{k+1\}, \; \hat{J}_f^k \subset J_f^k, \tag{2.28a}$$

$$J_F^{k+1} = \hat{J}_F^k \cup \{k+1\}, \ \hat{J}_F^k \subset J_F^k, \tag{2.28b}$$

for all $k \geq 1$, and that the methods are initialized by setting $y^1=x^1$ and

$$J_f^1 = \{1\}, \ g_f^1 = g_f(y^1), \ f_1^1 = f(y^1), \tag{2.29a}$$

$$J_F^1 = \{1\}, \ g_F^1 = g_F(y^1), \ F_1^1 = F(y^1). \tag{2.29b}$$

If f and g_f, or F and g_F, cannot be evaluated at each $y \in R^N$, then the following modifications are necessary. We replace (2.28a) by the following requirement

$$J_f^{k+1} = \begin{cases} \hat{J}_f^k \cup \{k+1\} & \text{if } y^{k+1} \in S, \\ \hat{J}_f^k & \text{if } y^{k+1} \notin S, \end{cases} \tag{2.30a}$$

where $\hat{J}_f^k \subset J_f^k$, for all k. Then there is no need for (2.27a) if y^{k+1} is infeasible (Another possibility is to use (2.28a) with (2.27a) if $y^{k+1} \in S$, and with $g_f^{k+1}=g_f(x^{k+1})$ and $f_{k+1}^{k+1}=f(x^{k+1})$ if $y^{k+1} \notin S$). If F and g_F cannot be evaluated at feasible points then we set $J_F^1=\emptyset$ and replace (2.28b) by

$$J_F^{k+1} = \begin{cases} \hat{J}_F^k & \text{if } y^{k+1} \in S, \\ \hat{J}_F^k \cup \{k+1\} & \text{if } y^{k+1} \notin S, \end{cases} \tag{2.30b}$$

where $\hat{J}_F^k \subset J_F^k$, for all k. Then (2.27b) need not be used if y^{k+1} is feasible. In this case the last constraint (2.22) should be dropped for all k such that $y^j \in S$ for $j=1,\ldots,k-1$, i.e. we do not use the constraint subgradients until the first infeasible trial point is found. It will be seen that all the subsequent proofs need only minor changes to cover the rules (2.30), while the rules (2.28) require simpler notation. Specific techniques for dealing with the rules (2.30) will be described in Section 7.

We shall now consider versions of the above-described methods that are obtained if one uses subgradient locality measures different from (2.8). To this end, for any x and y in R^N define the linearizations

$$\bar{f}(x;y) = f(y) + \langle g_f(y), x-y \rangle, \tag{2.31}$$

$$\bar F(x;y) = F(y) + \langle g_F(y), x-y \rangle$$

and the following subgradient locality measures

$$\alpha_f(x,y) = \max\{|f(x)-\bar f(x;y)|,\ \gamma_f|x-y|^2\},$$
$$\alpha_F(x,y) = \max\{|\bar F(x;y)|,\ \gamma_F|x-y|^2\}, \tag{2.32}$$

which indicate how far $g_f(y)$ and $g_F(y)$ are from $\partial H(x;x)$, respectively. Since

$$f_j^k = \bar f(x^k;y^j) \quad \text{and} \quad F_j^k = \bar F(x^k;y^j),$$

$$\alpha_f(x^k,y^j) = \max\{|f(x^k)-f_j^k|,\ \gamma_f|x^k-y^j|^2\},$$

$$\alpha_F(x^k,y^j) = \max\{|F_j^k|,\ \gamma_F|x^k-y^j|^2\},$$

we see that (2.8) differs from (2.32) by using the distance measures s_j^k instead of $|x^k-y^j|$. This enables us not to store the points y^j.

In fact, one may use $\alpha_f(x^k,y^j)$ and $\alpha_F(x^k,y^j)$ instead of $\alpha_{f,j}^k$ and $\alpha_{F,j}^k$ in the search direction finding subproblems (2.14) and (2.22). Then the method with subgradient selection has subproblems of the form

$$\begin{aligned}
&\underset{(d,\hat v)\in R^{N+1}}{\text{minimize}} && \tfrac{1}{2}|d|^2+\hat v,\\
&\text{subject to} && -\alpha_f(x^k,y^j) + \langle g_f^j, d\rangle \le \hat v,\ j\in J_f^k,\\
& && -\alpha_F(x^k,y^j) + \langle g_F^j, d\rangle \le \hat v,\ j\in J_F^k
\end{aligned} \tag{2.33}$$

while the k-th iteration of the method with subgradient aggregation uses the subproblem

$$\begin{aligned}
&\underset{(d,\hat v)\in R^{N+1}}{\text{minimize}} && \tfrac{1}{2}|d|^2+\hat v,\\
&\text{subject to} && -\alpha_f(x^k,y^j) + \langle g_f^j, d\rangle \le \hat v,\ j\in J_f^k,\\
& && -\alpha_{f,p}^k + \langle p_f^{k-1}, d\rangle \le \hat v \quad \text{if} \quad r_a^k=0,\\
& && -\alpha_F(x^k,y^j) + \langle g_F^j, d\rangle \le \hat v,\ j\in J_F^k,
\end{aligned} \tag{2.34}$$

$$-\alpha_{F,p}^{k} + \langle p_F^{k-1}, d \rangle \le \hat{v} \quad \text{if} \quad r_a^k = 0.$$

In this case the aggregate subgradient updating rules (2.24) should be replaced by the following

$$\begin{aligned}(p_f^k, \tilde f_p^k, \tilde s_f^k) &= \sum_{j \in J_f^k} \tilde\lambda_j^k (g_f^j, f_j^k, |x^k - y^j|) + \tilde\lambda_p^k (p_f^{k-1}, f_p^k, s_f^k),\\ (p_F^k, \tilde F_p^k, \tilde s_F^k) &= \sum_{j \in J_F^k} \tilde\mu_j^k (g_F^j, F_j^k, |x^k - y^j|) + \tilde\mu_p^k (p_F^{k-1}, F_p^k, s_F^k).\end{aligned} \tag{2.35}$$

Other versions of the methods are obtained if we replace (2.32) by the following definition of Mifflin (1982)

$$\begin{aligned}\alpha_f(x,y) &= \max\{f(x) - \overline{f}(x;y),\ \gamma_f |x-y|^2\},\\ \alpha_F(x,y) &= \max\{-F(x;y),\ \gamma_F |x-y|^2\}\end{aligned} \tag{2.36}$$

and (2.8) and (2.18) by

$$\begin{aligned}\alpha_{f,j}^k &= \max\{f(x^k) - f_j^k,\ \gamma_f (s_j^k)^2\},\\ \alpha_{F,j}^k &= \max\{-F_j^k,\ \gamma_F (s_j^k)^2\},\\ \alpha_{f,p}^k &= \max\{f(x^k) - f_p^k,\ \gamma_f (s_f^k)^2\},\\ \alpha_{F,p}^k &= \max\{-F_p^k,\ \gamma_F (s_F^k)^2\}.\end{aligned} \tag{2.37}$$

We note that in the convex case $(\gamma_f = \gamma_F = 0)$ the values of the subgradient locality measures (2.32), (2.8) and (2.18) coincide with those given by (2.36)-(2.37), respectively.

To compare our method with subgradient selection with the Mifflin (1982) algorithm, we shall need the following notation. Define the subgradient mapping

$$g(x) = \begin{cases} g_f(x) & \text{if} \quad x \in S,\\ g_F(x) & \text{if} \quad x \notin S,\end{cases} \tag{2.38}$$

and the corresponding subgradient locality measure

$$\alpha(x,y) = \begin{cases} \alpha_f(x,y) & \text{if} \quad y \in S\\ \alpha_F(x,y) & \text{if} \quad y \notin S.\end{cases} \tag{2.39}$$

Suppose that the rules for choosing J_f^k and J_F^k satisfy (2.30) for all k, with $J_f^1=\{1\}$ and $J_F^1=\emptyset$. Then we have

$$J_f^k \cap J_F^k = \emptyset$$

and

$$g_f^j = g(y^j) \quad \text{and} \quad \alpha_f(x^k,y^j)=\alpha(x^k,y^j) \quad \text{if} \quad y^j \in S,$$

$$g_F^j = g(y^j) \quad \text{and} \quad \alpha_F(x^k,y^k)=\alpha(x^k,y^k) \quad \text{if} \quad y^j \notin S,$$

for all k. Hence, letting

$$J^k = J_f^k \cup J_F^k \qquad \text{for all } k,$$

we conclude that the search direction finding subproblem (2.33) can be formulated as follows

$$\begin{aligned} &\underset{(d,\hat{v}) \in R^{N+1}}{\text{minimize}} && \tfrac{1}{2}|d|^2+\hat{v}, \\ &\text{subject to} && -\alpha(x^k,y^j) + \langle g(y^j),d \rangle \le \hat{v}, \; j \in J^k. \end{aligned} \tag{2.40}$$

Subproblem (2.40) is the search direction finding subproblem of the Mifflin (1982) algorithm if $J^k=\{1,...,k\}$ and $\alpha(x^k,y^j)$ is defined via (2.39) and (2.36). Thus this algorithm uses total subgradient accumulation, coresponding to the choice $\hat{J}_f^k=J_f^k$ and $\hat{J}_F^k=J_F^k$ in (2.30), for all k. Moreover, this choice of J_f^k and J_F^k combined with the subgradient locality measures (2,37) and the search direction finding subproblems (2.14) leads to a version of the Mifflin (1982) algorithm that does not need storing the points y^j.

To sum up, we shall now comment on relations of the above-described methods with other algorithms. If we neglect the variables corresponding to the constraint function F then the methods reduce to the algorithms for unconstrained minimization from Chapter 3. In the convex case we automatically obtain the search direction finding subproblems studied in Chapter 5. Thus the methods generalize the method of centers for inequality constrained minimax problems (Kiwiel, 1981a), which in turn extends the Pironneau and Polak method of centers and method of feasible directions for smooth problems.

3. The Algorithm with Subgradient Aggregation

We now state an algorithm procedure for solving problem (1.1). Its line searches are discussed below.

Algorithm 3.1.

Step 0 (Initialization). Select the starting point $x^1 \in S$ and a final accuracy tolerance $\varepsilon_s \geq 0$. Choose fixed positive line search parameters m_L, m_R, $\bar{a}$ and $\bar{t}$, $\bar{t} \leq 1$ and $0 < m_L < m_R < 1$, and distance measure parameters $\gamma_f > 0$ and $\gamma_F > 0$ ($\gamma_f = 0$ if f is convex; $\gamma_F = 0$ if F is convex). Set $y^1 = x^1$, $s_1^1 = s_f^1 = s_F^1 = 0$ and

$$J_f^1 = \{1\},\ g_f^1 = p_f^0 = g_f(y^1),\ f_1^1 = f_p^1(y^1),$$

$$J_F^1 = \{1\},\ g_F^1 = p_F^0 = g_F(y^1),\ F_1^1 = F_p^1(y^1),$$

and the reset indicator $r_a^1 = 1$. Set the counter $k=1$.

Step 1 (Direction finding). Find multipliers λ_j^k, $j \in J_f^k$, λ_p^k, μ_j^k, $j \in J_F^k$, and μ_p^k that solve the following k-th dual search direction finding subproblem

$$\underset{\lambda,\mu}{\text{minimize}}\ \frac{1}{2}\Big|\sum_{j \in J_f^k} \lambda_j g_f^j + \lambda_p p_f^{k-1} + \sum_{j \in J_F^k} \mu_j g_F^j + \mu_p p_F^{k-1}\Big|^2 +$$

$$+ \sum_{j \in J_f^k} \lambda_j \alpha_{f,j}^k + \lambda_p \alpha_{f,p}^k + \sum_{j \in J_F^k} \mu_j \alpha_{F,j}^k + \mu_p \alpha_{F,p}^k,$$

$$\text{subject to}\ \lambda_j \geq 0,\ j \in J_f^k,\ \lambda_p \geq 0,\ \mu_j \geq 0,\ j \in J_F^k,\ \mu_p \geq 0, \tag{3.1}$$

$$\sum_{j \in J_f^k} \lambda_j + \lambda_p + \sum_{j \in J_F^k} \mu_j + \mu_p = 1,$$

$$\lambda_p = \mu_p = 0 \quad \text{if} \quad r_a^k = 1,$$

where

$$\alpha_{f,j}^k = \max\{|f(x^k) - f_j^k|,\ \gamma_f(s_j^k)^2\},\quad \alpha_{F,j}^k = \max\{|F_j^k|,\ \gamma_F(s_j^k)^2\}, \tag{3.2a}$$

$$\alpha_{f,p}^k = \max\{|f(x^k) - f_p^k|,\ \gamma_f(s_f^k)^2\},\quad \alpha_{F,p}^k = \max\{|F_p^k|,\ \gamma_f(s_F^k)^2\}. \tag{3.2b}$$

Compute

$$\nu_f^k = \sum_{j\in J_f^k}\lambda_j^k + \lambda_p^k \quad \text{and} \quad \nu_F^k = \sum_{j\in J_F^k}\mu_j^k + \mu_p^k. \tag{3.3}$$

Set

$$\begin{aligned}
&\tilde\lambda_j^k = \lambda_j^k/\nu_f^k \quad \text{for} \quad j\in J_f^k \quad \text{and} \quad \tilde\lambda_p^k = \lambda_p^k/\nu_f^k \quad \text{if} \quad \nu_f^k \neq 0,\\
&\tilde\lambda_k^k = 1,\ \tilde\lambda_j^k = 0,\ j\in J_f^k\setminus\{k\},\ \tilde\lambda_p^k = 0 \quad \text{if} \quad \nu_f^k = 0,\\
&\tilde\mu_j^k = \mu_j^k/\nu_F^k \quad \text{for} \quad j\in J_F^k \quad \text{and} \quad \tilde\mu_p^k = \mu_p^k/\nu_F^k \quad \text{if} \quad \nu_F^k \neq 0,\\
&\tilde\mu_k^k = 1,\ \tilde\mu_j^k = 0,\ j\in J_F^k\setminus\{k\},\ \tilde\lambda_p^k = 0 \quad \text{if} \quad \nu_F^k = 0.
\end{aligned} \tag{3.4}$$

Calculate $a^k=\max\{s_j^k : j\in J_f^k\cup J_F^k\}$ if $\lambda_p^k=\mu_p^k=0$. Set

$$\begin{aligned}
(p_f^k,\tilde f_p^k,\tilde s_f^k) &= \sum_{j\in J_f^k}\tilde\lambda_j^k(g_f^j,f_j^k,s_j^k) + \tilde\lambda_p^k(p_f^{k-1},f_p^k,s_f^k),\\
(p_F^k,\tilde F_p^k,\tilde s_F^k) &= \sum_{j\in J_F^k}\tilde\mu_j^k(g_F^j,F_j^k,s_j^k) + \tilde\mu_p^k(p_f^{k-1},F_p^k,s_F^k),
\end{aligned} \tag{3.5}$$

$$p^k = \nu_f^k p_f^k + \nu_F^k p_F^k, \tag{3.6}$$

$$d^k = -p^k, \tag{3.7}$$

$$\tilde\alpha_{f,p}^k = \max\{|f(x^k)-\tilde f_p^k|,\ \gamma_f(\tilde s_f^k)^2\}, \tag{3.8a}$$

$$\tilde\alpha_{F,p}^k = \max\{|\tilde F_p^k|,\ \gamma_F(\tilde s_F^k)^2\}, \tag{3.8b}$$

$$v^k = -\{|p^k|^2+\nu_f^k\tilde\alpha_{f,p}^k + \nu_F^k\tilde\alpha_{F,p}^k\}. \tag{3.8c}$$

<u>Step 2 (Stopping criterion)</u>. Set

$$w^k = \tfrac{1}{2}|p^k|^2+\nu_f^k\tilde\alpha_{f,p}^k+\nu_F^k\tilde\alpha_{F,p}^k. \tag{3.10}$$

If $w^k \le \varepsilon_s$ terminate; otherwise, go to Step 3.

<u>Step 3 (Line search)</u>. By a line search procedure as given below, find two stepsizes t_L^k and t_R^k such that $0\le t_L^k\le t_R^k$ and such that the two corresponding points defined by

$$x^{k+1} = x^k+t_L^k d^k \quad \text{and} \quad y^{k+1} = x^k+t_R^k d^k$$

satisfy $t_L^k \le 1$ and

$$f(x^{k+1}) \le f(x^k) + m_L t_L^k v^k, \tag{3.11a}$$

$$F(x^{k+1}) \le 0, \tag{3.11b}$$

$$t_R^k = t_L^k \quad \text{if} \quad t_L^k \ge \bar{t}, \tag{3.11c}$$

$$-\alpha(x^{k+1},y^{k+1}) + \langle g(y^{k+1}),d^k \rangle \ge m_R v^k \quad \text{if} \quad t_L^k < \bar{t} \tag{3.11d}$$

$$|y^{k+1}-x^{k+1}| \le \bar{a}/2, \tag{3,11e}$$

where

$$g(y) = g_f(y) \quad \text{and} \quad \alpha(x,y)=\max\{|f(x)-\bar{f}(x;y)|,\gamma_f|x-y|^2\} \quad \text{if} \quad F(y) \le 0,$$

$$\tag{3.12}$$

$$g(y) = g_F(y) \quad \text{and} \quad \alpha(x,y)=\max\{|\bar{F}(x;y)|,\gamma_F|x-y|^2\} \quad \text{if} \quad F(y) > 0.$$

Step 4 (Subgradient updating). Select sets $\hat{J}_f^k \subset J_f^k$ and $\hat{J}_F^k \subset J_F^k$, and set

$$J_f^{k+1} = \hat{J}_f^k \cup \{k+1\} \quad \text{and} \quad J_F^{k+1} = \hat{J}_F^k \cup \{k+1\}. \tag{3.13}$$

Set $g_f^{k+1}=g_f(y^{k+1})$, $g_F^{k+1}=g_F(y^{k+1})$ and

$$f_{k+1}^{k+1} = f(y^{k+1}) + \langle g_f^{k+1},x^{k+1}-y^{k+1} \rangle,$$

$$f_j^{k+1} = f_j^k + \langle g_f^j,x^{k+1}-x^k \rangle \quad \text{for} \quad j \in \hat{J}_f^k,$$

$$f_p^{k+1} = \tilde{f}_p^k + \langle p_f^k,x^{k+1}-x^k \rangle,$$

$$F_{k+1}^{k+1} = F(y^{k+1}) + \langle g_F^{k+1},x^{k+1}-y^{k+1} \rangle,$$

$$F_j^{k+1} = F_j^k + \langle g_F^j,x^{k+1}-x^k \rangle \quad \text{for} \quad j \in \hat{J}_F^k, \tag{3.14}$$

$$F_p^{k+1} = \tilde{F}_p^k + \langle p_F^k,x^{k+1}-x^k \rangle,$$

$$s_{k+1}^{k+1} = |y^{k+1}-x^{k+1}|,$$

$$s_j^{k+1} = s_j^k + |x^{k+1}-x^k| \quad \text{for} \quad j \in \hat{J}_f^k \cup \hat{J}_F^k,$$

$$s_f^{k+1} = \tilde{s}_f^k + |x^{k+1}-x^k|,$$

$$s_F^{k+1} = \tilde{s}_F^{k+1} + |x^{k+1}-x^k|.$$

Step 5 (Distance resetting test). Set

$$a^{k+1} = \max\{a^k+|x^{k+1}-x^k|, s_{k+1}^{k+1}\}.$$

If $a^{k+1} \le \bar{a}$ then set $r_a^{k+1}=0$ and go to Step 7. Otherwise, set $r_a^{k+1}=1$ and go to Step 6.

Step 6 (Distance resetting). Keep deleting from J_f^{k+1} and J_F^{k+1} indices with the smallest values until the reset value of a^{k+1} satisfies

$$a^{k+1} = \max\{s_j^{k+1} : j \in J_f^{k+1} \cup J_F^{k+1}\} \le \bar{a}/2.$$

Step 7. Increase k by 1 and go to Step 1.

A few remarks on the algorithm are in order.

By Lemma 5.2.3, subproblem (3.1) is the dual of the k-th primal search direction finding subproblem (2.22), and λ_j^k, $j \in J_f^k$, λ_p^k, μ_j^k, $j \in J_F^k$ and μ_p^k are the corresponding Lagrange multipliers. Relations (3.3)-(3.4) satisfy (2.23), hence (see Section 5.2) one can calculate $(d^k,\hat{v}^k)$, the solution of (2.22), via (3.5)-(3.7) and

$$\hat{v}^k = -\{|p^k|^2 + \sum_{j \in J_f^k} \tilde{\lambda}_j^k \alpha_{f,j}^k + \tilde{\lambda}_p^k \alpha_{f,p}^k + \sum_{j \in J_F^k} \tilde{\mu}_j^k \alpha_{F,j}^k + \tilde{\mu}_p^k \alpha_{F,p}^k\}. \tag{3.15}$$

One may, of course, solve the k-th primal search direction finding subproblem (2.22) in Step 1 of the method.

The stopping criterion of Step 2 admits of the following interpretation. The values of the aggregate subgradient locality measures $\tilde{\alpha}_{f,p}^k$ and $\tilde{\alpha}_{F,p}^k$ given by (3.8) indicate how far p_f^k and p_F^k, respectively, are from $\hat{M}(x^k)$. At the same time, the value of the following subgradient locality measure

$$\tilde{\alpha}_p^k = \nu_f^k \tilde{\alpha}_{f,p}^k + \nu_F^k \tilde{\alpha}_{F,p}^k \tag{3.16}$$

indicates how much the aggregate subgradient $p^k=\nu_f^k p_f^k+\nu_F^k p_F^k$ differs from being an element of $\hat{M}(x^k)$, since $\nu_f^k \ge 0$, $\nu_F^k \ge 0$ and $\nu_f^k+\nu_F^k=1$. In particular, in the convex case we have $\hat{M}(x^k)=\partial H(x^k,x^k)$ and

$$p^k \in \partial_\varepsilon H(x^k;x^k) \quad \text{for} \quad \varepsilon=\tilde{\alpha}_p^k,$$

see Lemma 5.4.2. By (3.10) and (3.16),

$$w^k = \tfrac{1}{2}|p^k|^2+\tilde{\alpha}_p^k. \tag{3.17}$$

Therefore, a small value of w^k indicates that both $|p^k|$ is small and that p^k is close to $\hat{M}(x^k)$, i.e. the null vector is close to $\hat{M}(x^k)$,so that x^k is approximately stationary (stationary points $\bar{x}$ satisfy $0 \in \hat{M}(\bar{x})$). Thus w^k may be called the stationarity measure of x^k. On the other hand, since p^k is a convex combination of approximate subgradients p_f^k and p_F^k of f and F at x^k, respectively, we may regard p^k as an approximate subgradient of some Lagrangian function of problem (1.1)

$$L(x,\nu) = \nu_f f(x)+\nu_F F(x),$$

i.e. p^k is close to $\nu_f^k \partial f(x^k)+\nu_F^k \partial F(x^k)$ if the value of $\tilde{\alpha}_p^k$ is small. Thus our stopping criterion generalizes the usual criterion of a small value of the gradient of the Lagrangian, which is frequently employed in algorithms for smooth problems.

Our line search rules (3.11) extend the rules (3.3.8)-(3.3.11) to the constrained case. As in Algorithm 5.3.1, v^k approximates the directional derivative of H^k at x^k in the direction d^k, and the line search is entered with

$$v^k = \hat{H}_a^k(x^k+d^k)-H^k(x^k) < 0.$$

The criteria (3.11a)-(3.11b) ensure monotonicity in the objective value and feasibility, i.e. $f(x^{k+1}) \le f(x^k)$ and $x^{k+1} \in S$ for all k. The rule (3.11c) means that we do not pose any demands on new subgradients $g^{k+1} = g_f(y^{k+1})$ and $g_F^{k+1}=g_F(y^{k+1})$ if the algorithm makes sufficient progress, i.e. $f(x^{k+1})$ is significantly smaller than $f(x^k)$. On the other hand, the criterion (3.11d), yielding either

$$-\alpha_{f,k+1}^{k+1} + \langle g_f^{k+1},d^k \rangle \ge m_R v^k \quad \text{if} \quad t_L^k < \bar{t} \quad \text{and} \quad y^{k+1} \in S$$

or

$$-\alpha_{F,k+1}^{k+1} + \langle g_F^{k+1},d^k \rangle \ge m_R v^k \quad \text{if} \quad t_L^k < \bar{t} \quad \text{and} \quad y^{k+1} \notin S,$$

ensures that at least one of the two new subgradients will significantly modify the next polyhedral approximation to H^{k+1} after a null step or a short serious step. This prevents the algorithm from jamming at non-stationary points. The criterion (3.11e) is connected with the distance resetting strategy discussed below.

The line search rules (3.11) are general enough to allow for constructing many efficient procedures for executing Step 3 (Mifflin, 1982 and 1983). For completeness, we give below a simple extension of Line Search Procedure 3.3.1 for finding stepsizes $t_L=t_L^k$ and $t_R=t_R^k$. In this procedure ζ is a fixed parameter satisfying $\zeta\in(0,0.5)$, $x=x^k$, $d=d^k$ and $v=v^k<0$.

Line Search Procedure 3.2.

(i) Set $t_L=0$ and $t=t_U=1$.

(ii) If $f(x+td)\le f(x)+m_L tv$ and $F(x+td)\le 0$ set $t_L=t$; otherwise set $t_U=t$.

(iii) If $t_L\ge\bar{t}$ set $t_R=t_L$ and return.

(iv) If $-\alpha(x+t_Ld,x+td)+\langle g(x+td),d\rangle\ge m_Rv$ and $t_L<\bar{t}$ and $(t-t_L)|d|\le\bar{a}/2$ set $t_R=t$ and return.

(v) Choose $t\in[t_L+\zeta(t_U-t_L),\ t_U-\zeta(t_U-t_L)]$ by some interpolation procedure and go to (ii).

To study convergence of the above procedure, consider the following "semismoothness" hypothesis:

for any $x\in R^N, d\in R^N$ and sequences $\{\bar{g}^i\}\subset R^N$ and $\{t^i\}\subset R_+$

satisfying $\bar{g}^i\in\partial F(x+t^id), F(x+t^id)>0, F(x)=0$ and $t^i\downarrow 0$, one has (3.18)

$$\limsup_{i\to\infty}\langle\bar{g}^i,d\rangle\ge\liminf_{i\to\infty}[F(x+t^id)-F(x)]/t^i.$$

Lemma 3.3. If f and F are semismooth in the sense of (3.3.23) and (3.18) then Line Search Procedure 3.2 terminates with $t_L^k=t_L$ and $t_R^k=t_R$ satisfying (3.11).

Proof. We shall use the following modification of the proof of Lemma 3.3.3. Let

$$TL=\{t\ge 0: f(x+td)\le f(x)+m_L tv \text{ and } F(x+td)\le 0\}$$

and observe that we now have $\{\tilde{t}_L^i\} \subset TL$, $\tilde{t}_L^i \uparrow \tilde{t}$ and $\tilde{t} \in TL$, because both f and F are continuous, so we have $F(x+\tilde{t}d) \le 0$ in addition to (3. 3.24a). Using (2.31), (3.12), the continuity of f and F, and the local boundedness of g_f and g_F, we obtain

$$\alpha(x+\tilde{t}_L^i d, x+t^i d) \longrightarrow 0,$$

hence the rules of the procedure yield, as before, that

$$\limsup_{i\to\infty} \langle \bar{g}^i, d \rangle \le m_R v, \tag{3.19}$$

where $\bar{g}^i = g(x+t^i d)$ for all i. Since $t_U^i \downarrow \tilde{t}$, $\tilde{t} \in LT$ and $t_U^i \in LT$ if $t_U^i = t^i$, there exists an infinite set $\tilde{I} \subset \{1,2,\ldots\}$ such that $t_U^i = t^i > \tilde{t}$ and either $f(x+t^i d) > f(x)+m_L t^i v$ or $F(x+t^i d) > 0$ for all $i \in \tilde{I}$. If $\tilde{I}$ contained an infinite set I satisfying (3.3.24b), we would obtain a contradiction with (3.3.23) as before. Therefore we may suppose that

$$F(x+t^i d) > 0 \quad \text{for all} \quad i \in \tilde{I}, \tag{3.20a}$$

$$\bar{g}^i = g_F(x+t^i d) \in \partial F(x+\tilde{t}d+(t^i-\tilde{t})d) \quad \text{for all} \quad i \in \tilde{I}. \tag{3.20b}$$

Then, since $t^i \downarrow \tilde{t}$, $F(x+\tilde{t}d) \le 0$ and F is continuous, (3.20a) yields

$$F(x+\tilde{t}d) = 0, \tag{3.20c}$$

$$\liminf_{i\to\infty, i\in\tilde{I}} \left[F(x+\tilde{t}d+(t^i-\tilde{t})d)-F(x+\tilde{t}d)\right]/(t^i-\tilde{t}) \ge 0 \ge m_L v, \tag{3.20d}$$

because $m_L v < 0$. Since $m_R v < m_L v$, (3.19)-(3.20) contradict (3.18). Therefore the search terminates. □

<u>Remark 3.4</u>. As in Remark 3.3.4, we observe that (3.18) holds if F is weakly upper semismooth in the sense of Mifflin (1982), e.g. if F is convex.

The subgradient deletion rules of Step 5 and Step 6 are taken from Algorithm 3.3.1. Therefore, similarly to (3.3.29), we have

$$k \in J_f^k \cap J_F^k \quad \text{for all } k, \tag{3.21a}$$

while the use of (2.30) instead of (3.13) yields

$$k \in J_f^k \;\text{ if }\; y^k \in S, \text{ and } \; k \in J_F^k \;\text{ if }\; y^k \notin S. \tag{3.21b}$$

4. Convergence

In this section we shall establish global convergence of Algorithm 3.1. We suppose that each execution of Line Search Procedure 3.2 is finite, e.g. f and F have the additional semismoothness properties (3.3.23) and (3.18). Moreover, convergence results assume that the final accuracy tolerance ε_s is set to zero. In the absence of convexity, we will content ourselves with finding stationary points for f on S. Our principal result states that the algorithm either terminates at a stationary point or generates an infinite sequence $\{x^k\}$ whose accumulation points are stationary. If problem (1.1) is convex and satisfies the Slater constraint qualification, then $\{x^k\}$ is a minimizing sequence for f on S, which converges to a solution of (1.1) whenever f attains its infimum on S.

Since Algorithm 3.1 is a combination of Algorithm 3.3.1 and Algorithm 5.3.1, our convergence analysis will consist in modifying the proofs of Section 3.4 with the help of the results of Section 5.4.

We start by observing that the rules of Algorithm 3.1 for aggregating subgradients of each of the problem functions are similar to those of Algorithm 3.3.1. Therefore, the reader can establish the following result on convex representations of the aggregate subgradients analogously to Lemma 3.4.1 through Lemma 3.4.3, and Lemma 5.4.1.

Lemma 4.1. Suppose $k \geq 1$ is such that Algorithm 3.1 did not stop before k-th iteration, and let $M=N+3$. Then there exist numbers $\hat{\lambda}_i^k$ and $\hat{\mu}_i^k$, and vectors $(y_f^{k,i}, f^{k,i}, s_f^{k,i}) \in R^N \times R \times R$ and $(g_F^{k,i}, F^{k,i}, s_F^{k,i}) \in R^N \times R \times R$, $i=1,\ldots,M$, satisfying

$$(p_f^k, \tilde{f}_p^k, \tilde{s}_f^k) = \sum_{i=1}^{M} \hat{\lambda}_i^k (g_f(y_f^{k,i}), f^{k,i}, s_f^{k,i}), \tag{4.1a}$$

$$\hat{\lambda}_i^k \geq 0, \ i=1,\ldots,M, \quad \sum_{i=1}^{M} \hat{\lambda}_i^k = 1, \tag{4.1b}$$

$$(g_f(y_f^{k,i}), f^{k,i}, s_f^{k,i}) \in \{(g_f(y^j), f_j^k, s_j^k) : j=1,\ldots,k\}, \ i=1,\ldots,M, \tag{4.1c}$$

$$|y_f^{k,i} - x^k| \leq s_f^{k,i}, \ i=1,\ldots,M, \tag{4.1d}$$

$$\max\{s_f^{k,i} : i=1,\ldots,M\} \leq a^k \leq \bar{a}, \tag{4.1e}$$

and

$$(p_F^k, \tilde{F}_p^k, \tilde{s}_F^k) = \sum_{i=1}^{M} \hat{\mu}_i^k (g_f(y_F^{k,i}), F^{k,i}, s_F^{k,i}),$$

$$\hat{\mu}_i^k \geq 0,\ i=1,\dots,M,\quad \sum_{i=1}^{M} \hat{\mu}_i^k = 1,$$

$$(g_F(y_F^{k,i}), F^{k,i}, s_F^{k,i}) \in \{(g_F(y^j), F_j^k, s_j^k) : j=1,\dots,k\},\ i=1,\dots,M, \quad (4.2)$$

$$|y_F^{k,i} - x^k| \leq s_F^{k,i},\ i=1,\dots,M,$$

$$\max\{s_F^{k,i} : i=1,\dots,M\} \leq a^k \leq \bar{a}\ .$$

If additionally f is convex then

$$p_f^k \in \partial_\varepsilon f(x^k) \quad \text{for} \quad \varepsilon = f(x^k) - \tilde{f}_p^k = \tilde{\alpha}_{f,p}^k, \tag{4.3a}$$

while if F is convex then

$$p_F^k \in \partial_\varepsilon F(x^k) \quad \text{for} \quad \varepsilon = F(x^k) - \tilde{F}_p^k \leq \tilde{\alpha}_{F,p}^k. \tag{4.3b}$$

The following lemma, which generalizes Lemma 3.4.4, is useful in deriving asymptotic results from the representations (4.1)-(4.2). We recall that $\gamma_f=0$ ($\gamma_F=0$) only if f (F) is convex; otherwise $\gamma_f > 0$ ($\gamma_F > 0$).

Lemma 4.2. (i) Suppose that a point $\bar{x} \in R^N$, N-vectors $\bar{p}_f, \bar{y}_f^i$ and $\bar{g}_f^i$, and numbers $\bar{f}_p, \bar{s}_f, \bar{f}^i$ and $\bar{s}_f^i$, $i=1,\dots,M=N+3$, satisfy

$$(\bar{p}_f, \bar{f}_p, \bar{s}_f) = \sum_{i=1}^{M} \bar{\lambda}_i (\bar{g}_f^i, \bar{f}^i, \bar{s}_f^i),$$

$$\bar{\lambda}_i \geq 0,\ i=1,\dots,M,\quad \sum_{i=1}^{M} \bar{\lambda}_i = 1,$$

$$\bar{g}_f^i \in \partial f(\bar{y}_f^i),\ i=1,\dots,M,$$

$$\bar{f}^i = f(\bar{y}_f^i) + \langle \bar{g}_f^i, \bar{x} - \bar{y}_f^i \rangle,\ i=1,\dots,M, \tag{4.4}$$

$$|\bar{y}_f^i - \bar{x}| \leq \bar{s}_f^i,\ i=1,\dots,M,$$

$$f(\bar{x}) = \bar{f}_p,$$

$$\gamma_f \bar{s}_f = 0.$$

Then $p_f \in \partial f(\bar{x})$.

(ii) Suppose that a point $\bar{x} \in R^N$, N-vectors $\bar{p}_F, \bar{y}_F^i$ and $\bar{g}_F^i$, and numbers $\bar{F}_p, \bar{s}_F, \bar{F}^i$ and $\bar{s}_F^i$, i=1,...,M, satisfy

$$(\bar{p}_F, \bar{F}_p, \bar{s}_F) = \sum_{i=1}^{M} \bar{\mu}_i (\bar{g}_F^i, \bar{F}^i, \bar{s}_F^i), \tag{4.5a}$$

$$\bar{\mu}_i \geq 0, \ i=1,\ldots,M, \quad \sum_{i=1}^{M} \bar{\mu}_i = 1, \tag{4.5b}$$

$$\bar{g}_F^i \in \partial F(\bar{y}_F^i), \ i=1,\ldots,M, \tag{4.5c}$$

$$\bar{F}^i = F(\bar{y}_F^i) + \langle \bar{g}_F^i, \bar{x}-\bar{y}_F^i \rangle, \ i=1,\ldots,M, \tag{4.5d}$$

$$|\bar{y}_F^i - \bar{x}| \leq \bar{s}_F^i, \ i=1,\ldots,M, \tag{4.5e}$$

$$F(\bar{x})_+ = \bar{F}_p, \tag{4.5f}$$

$$\gamma_F \bar{s}_F = 0. \tag{4.5g}$$

Then $\bar{p}_F \in \partial F(\bar{x})$ and $F(\bar{x}) \geq 0$.

Proof. We shall only prove part (ii) of the lemma, since part (i) follows from Lemma 3.4.4.

(a) First, suppose that $\gamma_F > 0$. Then $\bar{s}_F=0$ by (4.5g), so (4.5a,b,e) yields

$$\bar{y}_F^i = \bar{x} \quad \text{if} \quad \bar{\mu}_i \neq 0 \tag{4.6}$$

and we have $\bar{p}_F \in \partial F(\bar{x})$ from (4.5a,b,c), (4.6) and the convexity of $\partial F(\bar{x})$. By (4.5a,b,d,f) and (4.6)

$$0 = F(x)_+ - \bar{F}_p = \sum_{i=1}^{M} \bar{\mu}_i [F(\bar{x})_+ - \bar{F}^i] =$$

$$= \sum_{i=1}^{M} \bar{\mu}_i [F(\bar{x})_+ - F(\bar{y}_F^i) - \langle \bar{g}_F^i, \bar{x}-\bar{y}_F^i \rangle] =$$

$$= \sum_{\bar{\mu}_i \neq 0} \bar{\mu}_i [F(\bar{x})_+ - F(\bar{x})] = F(\bar{x})_+ - F(\bar{x}).$$

Thus $F(\bar{x}) = F(\bar{x})_+ \geq 0$.

(b) Next, suppose that $\gamma_F=0$. Then F is convex and (4.5c,d) yield

$$F(\bar{x}) \ge F(\bar{y}_F^i) + \langle \bar{g}_F^i, \bar{x}-\bar{y}_F^i \rangle = \bar{F}^i \quad \text{for} \quad i=1,\dots,M.$$

Multiplying the above inequality by $\bar{\mu}_i$ and summing, we obtain $F(\bar{x}) \ge \bar{F}_p$ from (4.5a,b). Therefore, by (4.5f), $F(\bar{x}) \ge \bar{F}_p = F(x)_+ \ge 0$.

We shall now consider the case when the algorithm terminates.

Lemma 4.3. If Algorithm 3.1 terminates at the k-th iteration then the point $\bar{x}=x^k$ is stationary for f on S.

Proof. Suppose the algorithm terminates at Step 2 due to $w^k \le \varepsilon_s = 0$, and let $\bar{x}=x^k$. We have

$$w^k = \tfrac{1}{2}|p^k|^2 + \tilde{\alpha}_p^k,$$

$$p^k = \nu_f^k p_f^k + \nu_F^k p_F^k,$$

$$\tilde{\alpha}_p^k = \nu_f^k \max\{|f(\bar{x})-\tilde{f}_p^k|, \gamma_f(\tilde{s}_f^k)^2\} + \nu_F^k \max\{|\tilde{F}_p^k|, \gamma_F(\tilde{s}_F^k)^2\},$$

$$\nu_f^k \ge 0,\ \nu_F^k \ge 0,\ \nu_f^k + \nu_F^k = 1 \tag{4.7a}$$

from (3.17), (3.6), (3.16), (3.8) and (2.25). Therefore $w^k=0$ and

$$\nu_f^k p_f^k + \nu_F^k p_F^k = 0, \tag{4.7b}$$

$$\nu_f^k[f(\bar{x})-\tilde{f}_p^k] = 0, \quad \nu_f^k \gamma_f \tilde{s}_f^k = 0, \tag{4.8a}$$

$$\nu_F^k[F(\bar{x})_+ - \tilde{F}_p^k] = 0, \quad \nu_F^k \gamma_F \tilde{s}_F^k = 0, \tag{4.8b}$$

where $F(\bar{x})_+ = 0$, because $F(\bar{x}) = F(x^k) \le 0$. Suppose that $\nu_f^k \ne 0$. Then (4.8a), Lemma 4.1 and Lemma 4.2 imply $p_f^k \in \partial f(\bar{x})$, i.e.

$$p_f^k \in \partial f(\bar{x}) \quad \text{if} \quad \nu_f^k \ne 0. \tag{4.9a}$$

Next, if $\nu_F^k \ne 0$ then (4.8b), Lemma 4.1 amd Lemma 4.2 yield $p_F^k \in \partial F(\bar{x})$ and $F(\bar{x}) \ge 0$, so, because $F(\bar{x}) \le 0$, we have

$$p_F^k \in \partial F(\bar{x}) \quad \text{and} \quad F(\bar{x}) = 0 \quad \text{if} \quad \nu_F^k \ne 0. \tag{4.9b}$$

Since $F(\bar{x}) \le 0$, (4.7) and (4.9) imply $0 \in \hat{M}(\bar{x})$ (see (2.2)) and $\bar{x} \in S$. □

From now on we suppose that the algorithm generates an infinite sequence $\{x^k\}$, i.e. $w^k > 0$ for all k.

The following lemma, which generalizes Lemma 3.4.6, states useful asymptotic properties of the aggregate subgradients.

Lemma 4.4. Suppose that there exist a point $\bar{x} \in R^N$ and an infinite set $K \subset \{1,2,...\}$ satisfying $x^k \xrightarrow{K} \bar{x}$. Then there exists an infinite set $\bar{K} \subset K$ and two N-vectors $\bar{p}_f$ and $\bar{p}_F$ such that

$$p_f^k \xrightarrow{K} \bar{p}_f \quad \text{and} \quad p_F^k \xrightarrow{K} \bar{p}_F.$$

If additionally $\tilde{\alpha}_{f,p}^k \xrightarrow{K} 0$ then $\bar{p}_f \in \partial f(\bar{x})$, while if $\tilde{\alpha}_{F,p}^k \xrightarrow{K} 0$ then $\bar{p}_F \in \partial F(\bar{x})$ and $F(\bar{x}) \geq 0$.

Proof. Use Lemma 4.1, Lemma 4.2 and proceed as in the proof of Lemma 3.4.6. □

Our next result, which extends Lemma 3.4.7, establishes a crucial property of the stationarity measures $\{w^k\}$.

Lemma 4.5. Suppose that for some $\bar{x} \in R^N$ we have

$$\liminf_{k\to\infty} \max\{w^k, |\bar{x}-x^k|\} = 0, \tag{4.10}$$

or equivalently

there exists an infinite set $K\subset\{1,2,...\}$ such that

$$x^k \xrightarrow{K} \bar{x} \quad \text{and} \quad w^k \xrightarrow{K} 0. \tag{4.11}$$

Then $0 \in \hat{M}(\bar{x})$ and $\bar{x} \in S$.

Proof. Suppose that (4.11) holds. Since $w^k = \frac{1}{2}|p^k| + \nu_f^k \tilde{\alpha}_{f,p}^k + \nu_F^k \tilde{\alpha}_{F,p}^k$ with $\nu_f^k \tilde{\alpha}_{f,p}^k \geq 0$ and $\nu_F^k \tilde{\alpha}_{F,p}^k \geq 0$ for all k by (2.25) and (3.8), we have $|p^k| \xrightarrow{K} 0$ and

$$\nu_f^k \tilde{\alpha}_{f,p}^k \xrightarrow{K} 0 \quad \text{and} \quad \nu_F^k \tilde{\alpha}_{F,p}^k \xrightarrow{K} 0. \tag{4.12}$$

Since $x^k \xrightarrow{K} \bar{x}$ by assumption and $|p^k| \xrightarrow{K} 0$, we may use (2.25),(3.6) and Lemma 4.4 to deduce the existence of an infinite set $\bar{K} \subset K$, numbers $\bar{\nu}_f$ and $\bar{\nu}_F$, and N-vectors $\bar{p}_f$ and $\bar{p}_F$ such that

$$\nu_f^k \xrightarrow{\bar{K}} \bar{\nu}_f, \quad \nu_F^k \xrightarrow{\bar{K}} \bar{\nu}_F,$$

$$p_f^k \xrightarrow{\bar{K}} \bar{p}_f, \quad p_F^k \xrightarrow{\bar{K}} \bar{p}_F,$$

$$\bar{\nu}_f \geq 0, \ \bar{\nu}_F \geq 0, \ \bar{\nu}_f + \bar{\nu}_F = 1, \tag{4.13a}$$

$$\bar{\nu}_f \bar{p}_f + \bar{\nu}_F \bar{p}_F = 0. \tag{4.13b}$$

Suppose that $\bar{\nu}_f \neq 0$. Then (4.12) yields $\tilde{\alpha}_{f,p}^k \xrightarrow{\bar{K}} 0$, so $\bar{p}_f \in \partial f(\bar{x})$ by Lemma 4.4. Thus

$$\bar{p}_f \in \partial f(\bar{x}) \quad \text{if} \quad \bar{\nu}_f \neq 0. \tag{4.13c}$$

Similarly, (4.12) and Lemma 4.4 imply

$$\bar{p}_F \in \partial F(\bar{x}) \quad \text{and} \quad F(\bar{x}) \geq 0 \quad \text{if} \quad \bar{\nu}_F \neq 0. \tag{4.13d}$$

Since $F(x^k) \leq 0$ and $x^k \xrightarrow{K} \bar{x}$, the continuity of F yields $F(\bar{x}) \leq 0$. Combining this with (4.13) we obtain $0 \in \hat{M}(\bar{x})$ and $\bar{x} \in S$. The equivalence of (4.10) and (4.11) follows from the nonnegativity of w^k-s. $\square$

Proceeding as in Section 3.4, we shall now relate the stationarity measures w^k with the dual search direction finding subproblems.

Let $\hat{w}^k$ denote the optimal value of the dual search direction finding subproblem (3.1), for all k. A useful relation between $\hat{w}^k$ and w^k is established in the following lemma, which generalizes Lemma 3.4.8.

<u>Lemma 4.6</u>. At the k-th iteration of Algorithm 3.1, one has

$$\hat{w}^k = \tfrac{1}{2}|p^k|^2 + \hat{\alpha}_p^k, \tag{4.14a}$$

$$\hat{\alpha}_p^k = \nu_f^k \hat{\alpha}_{f,p}^k + \nu_F^k \hat{\alpha}_{F,p}^k, \tag{4.14b}$$

$$\hat{\alpha}_{f,p}^k = \sum_{j \in J_f^k} \tilde{\lambda}_j^k \alpha_{f,j}^k + \tilde{\lambda}_p^k \alpha_{f,p}^k, \tag{4.14c}$$

$$\hat{\alpha}_{F,p}^k = \sum_{j \in J_F^k} \tilde{\mu}_j^k \alpha_{F,p}^k + \tilde{\mu}_p^k \alpha_{F,p}^k, \tag{4.14d}$$

$$\tilde{\alpha}_{f,p}^k \leq \hat{\alpha}_{f,p}^k \quad \text{and} \quad \tilde{\alpha}_{F,p}^k \leq \hat{\alpha}_{F,p}^k \tag{4.15}$$

$$w^k \le \hat{w}^k, \tag{4.16}$$

$$v^k = -\{|p^k|^2 + \tilde{\alpha}_p^k\} \le -w^k \le 0, \tag{4.17}$$

$$\hat{v}^k \le v^k. \tag{4.18}$$

Moreover, if f and F are convex then $\tilde{\alpha}_p^k = \hat{\alpha}_p^k$, $w^k = \hat{w}^k$ and $v^k = \hat{v}^k$.

<u>Proof</u>.(i) By (3.6), (3.5) and (2.23a,b),

$$p^k = \sum_{j \in J_f^k} \lambda_j^k g_f^j + \lambda_p^k p_f^{k-1} + \sum_{j \in J_F^k} \mu_j^k g_F^j + \mu_p^k p_F^{k-1},$$

while (4.14b,c,d) and (2.23a,b) yield

$$\hat{\alpha}_p^k = \sum_{j \in J_f^k} \lambda_j^k \alpha_{f,j}^k + \lambda_p^k \alpha_{f,p}^k + \sum_{j \in J_F^k} \mu_j^k \alpha_{F,j}^k + \mu_p^k \alpha_{F,p}^k . \tag{4.19}$$

The above two equalities imply (4.14a).

(ii) One can establish (4.15) by using (4.14c,d), (3.8), (3.5) and (2.23c,d) as in the proof of Lemma 4.8. Since ν_f^k and ν_F^k are nonnegative, we obtain

$$\tilde{\alpha}_p^k = \nu_f^k \tilde{\alpha}_{f,p}^k + \nu_F^k \tilde{\alpha}_{F,p}^k \le \nu_f^k \hat{\alpha}_{f,p}^k + \nu_F^k \hat{\alpha}_{F,p}^k = \hat{\alpha}_p^k$$

from (3.16), (4.15) and (4.14b). Hence

$$w^k = \tfrac{1}{2}|p^k|^2 + \tilde{\alpha}_p^k \le \tfrac{1}{2}|p^k|^2 + \hat{\alpha}_p^k = \hat{w}^k,$$

$$v^k = -\{|p^k|^2 + \tilde{\alpha}_p^k\} \le -\{\tfrac{1}{2}|p^k|^2 + \tilde{\alpha}_p^k\} = -w^k$$

$$\hat{v}^k = -\{|p^k|^2 + \hat{\alpha}_p^k\} < -\{|p^k|^2 + \tilde{\alpha}_p^k\} = v^k,$$

from (3.17), (4.14a), (3.9), (3.16), (3.15) and (4.19). This proves (4.16)-(4.18).

(iii) If and F are convex then $f(x^k) - f_j^k \ge 0$, $j \in J_f^k$, $f(x^k) - f_p^k \ge 0$, $f(x^k) \ge \tilde{f}_p^k$, $-F_j^k \ge 0$, $j \in J_F^k$, $-F_p^k \ge 0$, and $-\tilde{F}_p^k \ge 0$ (see Lemma 5.4.2), hence we obtain $\tilde{\alpha}_{f,p}^k = \hat{\alpha}_{f,p}^k$ and $\tilde{\alpha}_{F,p}^k = \hat{\alpha}_{F,p}^k$ as in the proof of Lemma 3.4.8. Then part (ii) above yields $\tilde{\alpha}_p^k = \hat{\alpha}_p^k$, $w^k = \hat{w}^k$ and $v^k = \hat{v}^k$. □

The reader can easily establish analogues of Lemma 3.4.9 and Corollary 3.4.10 for Algorithm 3.1.

We shall now prove the following extension of Lemma 3.4.11. Let

$$g^k = g_f(y^k) \text{ and } \alpha^k = \alpha_f(x^k,y^k) \text{ if } y^k \in S,$$
$$g^k = g_F(y^k) \text{ and } \alpha^k = \alpha_F(x^k,y^k) \text{ if } y^k \notin S. \tag{4.20}$$

Lemma 4.7. Suppose that $t_L^{k-1} < \bar{t}$ and $r_a^k=0$ for some $k > 1$. Then

$$w^k \le \hat{w}^k \le \phi_C(w^{k-1})+|\alpha_p^k-\tilde{\alpha}_p^{k-1}|, \tag{4.21}$$

where the function ϕ_C is defined by

$$\phi_C(t) = t-(1-m_R)^2t^2/(8C^2),$$

C is any number satisfying

$$\max\{|p^{k-1}|,|g^k|, \tilde{\alpha}_p^{k-1},1\} \le C, \tag{4.22}$$

and

$$\alpha_p^k = \nu_f^{k-1}\,\alpha_{f,p}^k + \nu_F^{k-1}\,\alpha_{F,p}^k. \tag{4.23}$$

Proof. (i) Observe that $k > 1$, $t_L^{k-1} < \bar{t}$ and the line search rule (3.11d) yield

$$-\alpha(x^k,y^k) + \langle g(y^k),d^{k-1}\rangle \ge m_R v^{k-1},$$

so we have

$$-\alpha^k + \langle g^k,d^{k-1}\rangle \ge m_R v^{k-1} \tag{4.24}$$

from (4.20), (2.38) and (2.36).

(ii) Let $\nu \in [0,1]$ and define the multipliers

$$\lambda_k(\nu)=\nu,\ \lambda_j(\nu)=0 \text{ for } j \in J_f^k\setminus\{k\},\ \lambda_p(\nu)=(1-\nu)\nu_f^{k-1},$$
$$\mu_j(\nu)=0 \text{ for } j \in J_F^k,\ \mu_p(\nu)=(1-\nu)\nu_F^{k-1} \tag{4.25a}$$

if $y^k \in S$, and

$$\lambda_j(\nu)=0 \text{ for } j \in J^k,\ \lambda_p(\nu)=(1-\nu)\nu^{k-1},$$

$$\mu_k(\nu)=\nu,\ \mu_j(\nu)=0 \quad \text{for} \quad j\in J_F^k\setminus\{k\},\ \mu_p(\nu)=(1-\nu)\nu_F^{k-1} \tag{4.25b}$$

if $y^k \notin S$. Since $p^{k-1}=\nu_f^{k-1}p_f^{k-1}+\nu_F^{k-1}p_F^{k-1}$ by (3.6), we obtain

$$\sum_{j\in J_f^k}\lambda_j(\nu)g_f^j+\lambda_p(\nu)p_f^{k-1}+\sum_{j\in J_F^k}\mu_j(\nu)g_F^j+\mu_p(\nu)p_F^{k-1}=(1-\nu)p^{k-1}+\nu g^k \tag{4.26a}$$

from (4.25), (4.20) and the fact that $g_f^k=g_f(y^k)$ and $g_F^k=g_F(y^k)$. Similarly, (4.25), (4.20), (4.23) and the fact that $\alpha_{f,k}^k=\alpha_f(x^k,y^k)$ and $\alpha_{F,k}^k=\alpha_F(x^k,y^k)$ yield

$$\sum_{j\in J_f^k}\lambda_j(\nu)\alpha_{f,j}^k+\lambda_p(\nu)\alpha_{f,p}^k+\sum_{j\in J_F^k}\mu_j(\nu)\alpha_{F,j}^k+\mu_p(\nu)\alpha_{F,p}^k=(1-\nu)\alpha_p^k+\nu\alpha^k. \tag{4.26b}$$

By (2.25), $\nu_f^{k-1}\geq 0$, $\nu_F^{k-1}\geq 0$ and $\nu_f^{k-1}+\nu_F^{k-1}=1$, hence (4.25) yields

$$\lambda_j(\nu)\geq 0,\ j\in J_f^k,\ \lambda_p(\nu)\geq 0,\ \mu_j(\nu)\geq 0,\ j\in J_F^k,\ \mu_p(\nu)\geq 0,$$

$$\sum_{j\in J_f^k}\lambda_j(\nu)+\lambda_p(\nu)+\sum_{j\in J_F^k}\mu_j(\nu)+\lambda_p(\nu)=$$

$$=\nu+(1-\nu)\nu_f^{k-1}+(1-\nu)\nu_F^{k-1}=1$$

for all $\nu\in[0,1]$. Combining this with our assumption that $r_a^k=0$ and with (3.21b), we deduce that the multipliers (4.25) are feasible for subproblem (3.1) for all $\nu\in[0,1]$. Therefore $\hat{w}^k$ (the optimal value of subproblem (3.1)) satisfies

$$\hat{w}^k\leq\min\{\tfrac{1}{2}|(1-\nu)p^{k-1}+\nu g^k|^2+(1-\nu)\alpha_p^k+\nu\alpha^k:\nu\in[0,1]\}\leq$$

$$\leq\min\{\tfrac{1}{2}|(1-\nu)p^{k-1}+\nu g^k|^2+(1-\nu)\tilde{\alpha}_p^{k-1}+\nu\alpha^k:\nu\in[0,1]\}+|\alpha_p^k-\tilde{\alpha}_p^{k-1}|. \tag{4.27}$$

Using Lemma 2.4.10 and relations (3.7), (3.17), (4.17), (4.24) and (4.22), we obtain

$$\min\{\tfrac{1}{2}|(1-\nu)p^{-1}+\nu g^k|^2+(1-\nu)\tilde{\alpha}_p^{k-1}+\nu\alpha^k:\nu\in[0,1]\}\leq\phi_C(w^{k-1}),$$

hence (4.27) and (4.16) imply (4.21). □

To obtain locally uniform bounds of the form (4.22), we shall need the following generalization of Lemma 3.4.13.

Lemma 4.8. (i) For each $k \geq 1$

$$\max\{|p^k|,\tilde{\alpha}_p^k\} \leq \max\{\tfrac{1}{2}|g^k|^2+\alpha^k,(|g^k|^2+2\alpha^k)^{1/2}\}. \tag{4.28}$$

(ii) Suppose that $\bar{x} \in R^N$, $B=\{y \in R^N : |\bar{x}-y| \leq 2\bar{a}\}$, where $\bar{a}$ is the line search parameter involved in (3.11e), and let

$$C_g = \sup\{|g(y)| : y \in B\}, \tag{4.29a}$$

$$C_\alpha = \sup\{\alpha(x,y) : x \in B,\ y \in B\}, \tag{4.29b}$$

$$C = \max\{\tfrac{1}{2}C_g^2 + C_\alpha,(C_g^2 + 2C_\alpha)^{1/2},1\}. \tag{4.29c}$$

Then C is finite and

$$\max\{|p^k|,|g^k|,\tilde{\alpha}_p^k,1\} \leq C \quad \text{if} \quad |x^k-\bar{x}| \leq \bar{a}. \tag{4.30}$$

Proof.(i) Let $k \geq 1$ be fixed and define the multipliers

$$\lambda_k = 1,\ \lambda_j = 0 \quad \text{for} \quad j \in J_f^k \setminus \{k\},\ \lambda_p = 0,$$

$$\mu_j = 0, \quad \text{for} \quad j \in J_F^k,\ \mu_p = 0$$

if $y^k \in S$, and

$$\lambda_j = 0 \quad \text{for} \quad j \in J_f^k,\ \lambda_p = 0,$$

$$\mu_k = 1,\ \mu_j = 0 \quad \text{for} \quad j \in J_F^k \setminus \{k\},\ \mu_p = 0$$

if $y^k \notin S$. Since $k \in J_f^k \cap J_F^k$ by (3.21a), the above multipliers are feasible for the k-th dual subproblem (3.1). Therefore $\hat{w}^k$, the optimal value of (3.1), satisfies $\hat{w}^k \leq \frac{1}{2}|g^k|^2+\alpha^k$ from (4.20), and we have

$$\tfrac{1}{2}|p^k|^2+\tilde{\alpha}_p^k = w^k \leq \hat{w}^k \leq \tfrac{1}{2}|g^k|^2+\alpha^k$$

by (3.17) and (4.16). The above inequality and the fact that $\tilde{\alpha}_p^k \geq 0$ yield (4.28).

(ii) Use the local boundedness of g_f and g_F (Lemma 1.2.2) and the continuity of f and F to deduce from (2.31) and (3.12) that the mapping $g(\circ)$ is bounded on the bounded set B, while $\alpha(\cdot,\cdot)$ is bounded on $B\times B$. Therefore the constants defined by (4.29) are finite. (4.30) follows from (4.28)-(4.29) and the fact that $|y^k-x^k| \le \bar{a}$ by (3.11e), while $g^k=g(y^k)$ and $\alpha^k=\alpha(x^k,y^k)$ by (4.20) and (3.12). □

To verify that Lemma 3.4.15 holds for Algorithm 3.1, we shall need the following result.

<u>Lemma 4.9</u>. Suppose that there exist a point $\bar{x} \in R^N$ and an infinite set $K \subset \{1,2,...\}$ such that $x^k \xrightarrow{K} \bar{x}$ and $|x^{k+1}-x^k| \longrightarrow 0$ as $k\to\infty$, $k \in K$. Then the sequences $\{p_f^k\}_{k\in K}$ and $\{p_F^k\}_{k\in K}$ are bounded, and

$$|\alpha_{f,p}^{k+1} - \tilde{\alpha}_{f,p}^k| \longrightarrow 0 \quad \text{as } k\to\infty,\ k\in K, \tag{4.31a}$$

$$|\alpha_{F,p}^{k+1} - \tilde{\alpha}_{F,p}^k| \longrightarrow 0 \quad \text{as } k\to\infty,\ k\in K, \tag{4.31b}$$

$$|\alpha_p^{k+1} - \tilde{\alpha}_p^k| \longrightarrow 0 \quad \text{as } k\to\infty,\ k\in K. \tag{4.32}$$

<u>Proof</u>. Suppose $x^k \xrightarrow{K} \bar{x}$ and let $B=\{y \in R^N : |\bar{x}-y| \le 2\bar{a}\}$. Since $\bar{a} > 0$ and $x^k \xrightarrow{K} \bar{x}$, we deduce from (4.1d,e) the existence of a number $\bar{k}$ such that $y_f^{k,i} \in B$ for $i=1,...,M$ and all $k \ge \bar{k}$, $k \in K$. Then (4.1a,b) and the boundedness of g_f on B imply that $\{p_f^k\}_{k\in K}$ is bounded. In a similar way we deduce the boundedness of $\{p_F^k\}_{k\in K}$ from (4.2). Next,

$$\begin{aligned} ||f(x^{k+1})-f_p^{k+1}|-|f(x^k)-\tilde{f}_p^k|| &\le |f(x^{k+1})-f_p^{k+1}-f(x^k)+\tilde{f}_p^k| \le \\ &\le |f(x^{k+1})-f(x^k)|+|f_p^{k+1}-\tilde{f}_p^k| = \\ &= |f(x^{k+1})-f(x^k)|+|\langle p_f^k, x^{k+1}-x^k\rangle| \le \\ &\le |f(x^{k+1})-f(x^k)|+|p_f^k||x^{k+1}-x^k| \xrightarrow{K} 0, \end{aligned} \tag{4.33a}$$

since $x^k \xrightarrow{K} \bar{x}$, $|x^{k+1}-x^k| \xrightarrow{K} 0$, f is continuous and $\{p_f^k\}_{k\in K}$ is bounded. A similar argument yields

$$|F_p^{k+1}-\tilde{F}_p^k| = |\langle p_F^k, x^{k+1}-x^k\rangle| \le |p_F^k||x^{k+1}-x^k| \xrightarrow{K} 0. \tag{4.33b}$$

Analogously to (3.4.37b), we obtain for large $k \in K$

$$|\gamma_f(s_f^{k+1})^2 - \gamma_f(\tilde{s}_f^k)| \le |x^{k+1}-x^k|(2(\gamma_f C)^{1/2} + \gamma_f |x^{k+1}-x^k|),$$

$$|\gamma_F(s_F^{k+1}) - \gamma_F(s_F^k)| \le |x^{k+1}-x^k|(2(\gamma_F C)^{1/2} + \gamma_F |x^{k+1}-x^k|).$$

Combining this with our assumption that $|x^{k+1}-x^k| \xrightarrow{K} 0$ and with (4.33), we obtain (4.31) from (3.2b) and (3.8). By (4.23), (3.15) and (2.45),

$$|\alpha_p^{k+1} - \tilde{\alpha}_p^k| = |\nu_f^k(\alpha_{f,p}^{k+1} - \tilde{\alpha}_{f,p}^k) + \nu_F^k(\alpha_{F,p}^{k+1} - \tilde{\alpha}_{F,p}^k)| \le$$

$$\le \max\{|\alpha_{f,p}^{k+1} - \tilde{\alpha}_{f,p}^k|, |\alpha_{F,p}^{k+1} - \tilde{\alpha}_{F,p}^k|\} \xrightarrow{K} 0$$

from (4.31). □

Using the above lemma, one can easily modify the proof of Lemma 3. 4.15 for Algorithm 3.1. Then, since the proof of Lemma 3.4.16 requires no modifications, we obtain

<u>Lemma 4.10</u>. If $x^k \xrightarrow{K} \bar{x}$ then (4.10) holds.

Lemma 4.5 and Lemma 4.10 yield

<u>Theorem 4.11</u>. Every accumulation point of a sequence $\{x^k\}$ generated by Algorithm 3.1 is stationary for f on S.

In the convex case, the above result can be strengthened as follows.

<u>Theorem 4.12</u>. If f and F are convex and $F(\tilde{x}) < 0$ for some $\tilde{x} \in R^N$, then Algorithm 3.1 generates a minimizing sequence $\{x^k\}$:

$$\{x^k\} \subset S \quad \text{and} \quad f(x^k) \downarrow \inf\{f(x) : x \in S\}.$$

If additionally problem (1.1) admits of a solution, then $\{x^k\}$ converges to a solution of problem (1.1).

<u>Proof</u>. One can use the proofs of Lemma 5.4.14, Theorem 5.4.15 and Theorem 5.4.16 to obtain the desired conclusion. □

The folowing result can be established similarly to Lemma 3.4.19.

Corollary 4.13. If the level set $\{x \in S : f(x) \le f(x^1)\}$ is bounded and the final accuracy tolerance ε_s is positive, then Algorithm 3.1 terminates in a finite number of iterations.

Remark 4.14. The results of this section hold for the case of many constraints considered in Remark 2.1 and Remark 2.2. This follows from the fact that the mapping $\tilde{\partial} F$ has essentially the same properties as the subdifferential ∂F, i.e. is locally bounded and upper semicontinuous.

5. The Algorithm with Subgradient Selection

In this section we state in detail and analyze the method with subgradient selection introduced in Section 2.

Algorithm 5.1.

Step 0 (Initialization). Select the starting point $x^1 \in S$ and a final accuracy tolerance $\varepsilon_s \ge 0$. Choose fixed positive line search parameters m_L, m_R, $\bar{a}$ and $\bar{t}$, $\bar{t} < 1$ and $m_L < m_R < 1$, and distance measure parameters $\gamma_f > 0$ and $\gamma_F > 0$ ($\gamma_f = 0$ if f is convex; $\gamma_F = 0$ if F is convex). Set $y^1 = x^1$, $s_1^1 = 0$ and

$$J_f^1 = \{1\},\ g_f^1 = g_f(y^1),\ f_1^1 = f(y^1),$$

$$J_F^1 = \{1\},\ g_F^1 = g_F(y^1),\ F_1^1 = F(y^1).$$

Set the counter $k=1$.

Step 1 (Direction finding). Find multipliers λ_j^k, $j \in J_f^k$, and μ_j^k, $j \in J_F^k$, that solve the following k-th dual search direction finding subproblem

$$\underset{\lambda,\mu}{\text{minimize}}\ \tfrac{1}{2}\Big|\sum_{j\in J_f^k} \lambda_j g_f^j + \sum_{j\in J_F^k} \mu_j g_F^j\Big|^2 + \sum_{j\in J_f^k} \lambda_j \alpha_{f,j}^k + \sum_{j\in J_F^k} \mu_j \alpha_{F,j}^k\ , \tag{5.1}$$

$$\text{subject to}\quad \lambda_j \ge 0,\ j \in J_f^k,\ \mu_j \ge 0,\ j \in J_F^k,\ \sum_{j\in J_f^k} \lambda_j + \sum_{j\in J_F^k} \mu_j = 1,$$

where

$$\alpha_{f,j}^k = \max\{|f(x^k) - f_j^k|, \gamma_f (s_j^k)^2\},\ \alpha_{F,j}^k = \max\{|F_j^k|, \gamma_F (s_j^k)^2\}, \tag{5.2}$$

and sets $\hat{J}_f^k$ and $\hat{J}_F^k$ satisfying

$$\hat{J}_f^k = \{j \in J_f^k : \lambda_j^k \neq 0\} \quad \text{and} \quad \hat{J}_F^k = \{j \in J_F^k : \mu_j^k \neq 0\}, \tag{5.3a}$$

$$|\hat{J}_f^k \cup \hat{J}_F^k| \leq N+1. \tag{5.3b}$$

Compute

$$p^k = \sum_{j \in J_f^k} \lambda_j^k g_f^j + \sum_{j \in J_F^k} \mu_j^k g_F^j \tag{5.4}$$

$$d^k = -p^k, \tag{5.5}$$

$$\hat{\alpha}_p^k = \sum_{j \in J_f^k} \lambda_j^k \alpha_{f,j}^k + \sum_{j \in J_F^k} \mu_j^k \alpha_{F,j}^k, \tag{5.6}$$

$$\hat{v}^k = -\{|p^k|^2 + \hat{\alpha}_p^k\}. \tag{5.7}$$

Step 2 (Stopping criterion). Set

$$\hat{w}^k = \tfrac{1}{2}|p^k|^2 + \hat{\alpha}_p^k. \tag{5.8}$$

If $\hat{w}^k \leq \varepsilon_s$ then terminate; otherwise, go to Step 3.

Step 3 (Line search). By a line search procedure as discussed below, find two stepsizes t_L^k and t_R^k such that $0 \leq t_L^k \leq t_R^k$ and such that the two corresponding points defined by

$$x^{k+1} = x^k + t_L^k d^k \quad \text{and} \quad y^{k+1} = x^k + t_R^k d^k$$

satisfy $t_L^k \leq 1$ and

$$f(x^{k+1}) \leq f(x^k) + m_L t_L^k \hat{v}^k, \tag{5.9a}$$

$$F(x^{k+1}) \leq 0, \tag{5.9b}$$

$$t_R^k = t_L^k \quad \text{if} \quad t_L^k \geq \bar{t}, \tag{5.9c}$$

$$-\alpha(x^{k+1}, y^{k+1}) + \langle g(y^{k+1}), d^k \rangle \geq m_R \hat{v}^k \quad \text{if} \quad t_L^k < \bar{t}, \tag{5.9d}$$

$$|y^{k+1} - x^{k+1}| \leq \bar{a}/2 \tag{5.9e}$$

where the mappings $g(\cdot)$ and $\alpha(\cdot,\cdot)$ are defined by (3.12).

Step 4 (Subgradient updating). Set

$$J_f^{k+1} = \hat{J}_f^k \cup \{k+1\} \quad \text{and} \quad J_F^{k+1} = \hat{J}_F^k \cup \{k+1\}. \tag{5.10}$$

Set $g_f^{k+1}=g_f(y^{k+1})$, $g_F^{k+1}=g_F(y^{k+1})$ and

$$f_{k+1}^{k+1} = f(y^{k+1}) + \langle g_f^{k+1}, x^{k+1}-y^{k+1} \rangle ,$$

$$f_j^{k+1} = f_j^k + \langle g_f^j, x^{k+1}-x^k \rangle \quad \text{for} \quad j \in \hat{J}_f^k,$$

$$F_{k+1}^{k+1} = F(y^{k+1}) + \langle y^{k+1}, x^{k+1}-y^{k+1} \rangle ,$$

$$F_j^{k+1} = F_j^k + \langle g_F^j, x^{k+1}-x^k \rangle \quad \text{for} \quad j \in \hat{J}_F^k,$$

$$s_{k+1}^{k+1} = s_j^k + |x^{k+1}-x^k| \quad \text{for} \quad j \in \hat{J}_f^k \cup \hat{J}_F^k.$$

<u>Step 5 (Distance resetting test)</u>. Set

$$a^{k+1}=\max\{ s_j^{k+1} : j \in J_f^{k+1} \cup J_F^{k+1}\}. \tag{5.11}$$

If $a^{k+1} \le \bar{a}$ then set $r_a^{k+1}=0$ and go to Step 7; otherwise, set $r_a^{k+1}=1$ and go to Step 6.

<u>Step 6 (Distance resetting)</u>. Keep deleting from $J_f^{k+1} \cup J_F^{k+1}$ indices with the smallest values until the reset value of a^{k+1} satisfies

$$a^{k+1}=\max\{ s_j^{k+1} : j \in J_f^{k+1} \cup J_F^{k+1}\} \le \bar{a}/2.$$

<u>Step 7</u>. Increase k by 1 and go to Step 1.

We shall now comment on relations between the above method and Algorithm 3.1.

By Lemma 2.2.1, subproblem (5.1) is the dual of the k-th (primal) search direction finding subproblem (2.14), which has the unique solution $(d^k, \hat{v}^k)$ and (possibly nonunique) Lagrange multipliers λ_j^k, $j \in J_f^k$, μ_j^k, $j \in J_F^k$. We refer the reader to Remark 2.5.2 for a discussion of possible ways of finding the k-th Lagrange multipliers satisfying the requirement (5.3).

The stopping criterion of Step 2 can be interpreted similarly to the termination rule of Algorithm 3.1. To see this, let

$$\nu_f^k = \sum_{j\in J_f^k} \lambda_j^k \quad \text{and} \quad \nu_F^k = \sum_{j\in J_F^k} \mu_j^k,$$

define the scaled multipliers $\tilde\lambda_j^k$ and $\tilde\mu_j^k$ satisfying

$$\lambda_j^k = \nu_f^k \tilde\lambda_j^k \quad \text{for} \quad j \in J_f^k,$$

$$\tilde\lambda_j^k \geq 0,\ j \in J_f^k,\quad \sum_{j\in J_f^k} \tilde\lambda_j^k = 1,$$

$$\mu_j^k = \nu_F^k \tilde\mu_j^k \quad \text{for} \quad j \in J_F^k,$$

$$\tilde\mu_j^k \geq 0,\ j \in J_F^k,\quad \sum_{j\in J_F^k} \tilde\mu_j^k = 1,$$

and let

$$(p_f^k, \tilde f_p^k, \tilde s_f^k) = \sum_{j\in J_f^k} \tilde\lambda_j^k (g_f^j, f_j^k, s_j^k),$$

$$(p_F^k, \tilde F_p^k, \tilde s_F^k) = \sum_{j\in J_F^k} \tilde\mu_j^k (g_F^j, F_j^k, s_j^k),$$

$$\tilde\alpha_{f,p}^k = \max\{|f(x^k) - \tilde f_p^k|,\ \gamma_f (\tilde s_f^k)^2\},$$

$$\tilde\alpha_{F,p}^k = \max\{|\tilde F_p^k|,\ \gamma_F (\tilde s_F^k)^2\},$$

$$\tilde\alpha_p^k = \nu_f^k \tilde\alpha_{f,p}^k + \nu_F^k \tilde\alpha_{F,p}^k,$$

$$w^k = \tfrac{1}{2}|p^k|^2 + \tilde\alpha_p^k,$$

$$v^k = -\{|p^k|^2 + \tilde\alpha_p^k\}.$$

Then one may simply set

$$\lambda_p^k = \mu_p^k = 0$$

in the relevant relations of the preceding sections to see that Lemma 4.6 holds for Algorithm 5.1. In particular, we have $w^k \leq \hat w^k$. Thus both w^k and $\hat w^k$ can be regarded as stationarity measures of x^k; see Section 3.

The line search rules (5.9) differ from the rules (3.11) only in that we now use $\hat v^k$ instead of v^k for estimating the directional derivative of f at x^k in the direction d^k. Note that we always have $\hat v^k < 0$ at Step 3, since $\hat v^k \leq v^k < 0$ by Lemma 4.6. Hence to implement Step 3 one can use

Line Search Procedure 3.2 with v^k replaced by $\hat{v}^k$.

We also note that in Algorithm 5.1 the locality radius a^{k+1} is calculated directly via (5.11), instead of using the recursive formulae of Algorithm 3.1.

We refer the reader to Remark 3.5.2 on the possible use of more than $N+3$ subgradients for each search direction finding.

Let us now pass to convergence analysis. Global convergence of Algorithm 5.1 can be established by modifying the results of Section 4 in the spirit of Section 3.5 and Section 5.5. One easily checks that only Lemma 4.7, Lemma 4.8 and Lemma 4.9 need to be modified.

Lemma 4.7 should be replaced by the following results.

<u>Lemma 5.2</u>. Suppose that $t_L^{k-1} < \bar{t}$ and $r_a^k=0$ for some $k > 1$. Then

$$\hat{w}^k \le \phi_C(\hat{w}^{k-1}) + |\alpha_p^k - \hat{\alpha}_p^{k-1}|,$$

where ϕ_C is defined by

$$\phi_C(t) = t-(1-m_R)^2t^2/(8C^2),$$

C is any number satisfying

$$\max\{|p^{k-1}|,|g^k|, \hat{\alpha}_p^{k-1},1\} \le C,$$

and

$$\alpha_p^k = \sum_{j\in \hat{J}_f^{k-1}} \lambda_j^{k-1} \alpha_{f,j}^k + \sum_{j\in \hat{J}_F^{k-1}} \mu_j^{k-1} \alpha_{F,j}^k.$$

<u>Proof</u>. In the proof of Lemma 4.7, replace (4.25a) by

$$\lambda_k(\nu)=\nu, \quad \lambda_j(\nu)=(1-\nu)\lambda_j^{k-1}, \; j\in \hat{J}_f^{k-1},$$

$$\mu_k(\nu)=0, \quad \mu_j(\nu)=(1-\nu)\mu_j^{k-1}, \; j\in \hat{J}_F^{k-1},$$

and (2.25b) by

$$\lambda_k(\nu)=0, \quad \lambda_j(\nu)=(1-\nu)\lambda_j^{k-1}, \; j\in \hat{J}_f^{k-1},$$

$$\mu_k(\nu)=\nu, \quad \mu_j(\nu)=(1-\nu)\mu_j^{k-1}, \; j\in \hat{J}_F^{k-1},$$

and obtain, similarly to (5.5.4), that

$$\sum_{j\in J_f^k} \lambda_j(\nu) g_f^j + \sum_{j\in J_F^k} \mu_j(\nu) g_F^j = (1-\nu)p^{k-1}+\nu g^k,$$

$$\sum_{j\in J_f^k}\lambda_j(\nu)\alpha_{f,j}^k+\sum_{j\in J_F^k}\mu_j(\nu)\alpha_{F,j}^k=(1-\nu)\alpha_p^k+\nu\alpha^k,$$

$$\lambda_j(\nu)\geq 0,\ j\in J_f^k,\ \mu_j(\nu)\geq 0,\ j\in J_F^k,\quad \sum_{j\in J_f^k}\lambda_j(\nu)+\sum_{j\in J_F^k}\mu_j(\nu)=1$$

for each $\nu\in[0,1]$, since $J_f^k=\hat{J}_f^{k-1}\cup\{k\}$ and $J_F^k=\hat{J}_F^{k-1}\cup\{k\}$ if $r_a^k=0$. Then use the fact that $\hat{w}^k$ is the optimal value of (5.1) to complete the proof, as before. (See also the proof of Lemma 3.5.3.) □

In Lemma 4.8 replace (4.38) and (4.30) by

$$\max\{|p^k|,\hat{\alpha}_p^k\}\leq\max\{\tfrac{1}{2}|g^k|^2+\alpha^k,(|g^k|^2+2\alpha^k)^{1/2}\},$$

$$\max\{|p^k|,|g^k|,\hat{\alpha}_p^k,1\}\leq C\quad\text{if}\quad|x^k-\bar{x}|\leq\bar{a},$$

without influencing the proof, since $\hat{w}^k=\frac{1}{2}|p^k|^2+\hat{\alpha}_p^k$.

Lemma 4.9 is replaced by

Lemma 5.3. Suppose that there exist a point $\bar{x}\in R^N$ and an infinite set $K\subset\{1,2,\ldots\}$ such that $x^k\xrightarrow{K}\bar{x}$ and $|x^{k+1}-x^k|\longrightarrow 0$ as $k\to\infty$, $k\in K$. Then

$$|\alpha_p^{k+1}-\tilde{\alpha}_p^k|\longrightarrow 0\quad\text{as}\quad k\to\infty,\quad k\in K.$$

Proof. As in the proof of Lemma 3.5.4, observe that

$$|\alpha_p^{k+1}-\tilde{\alpha}_p^k|\leq\max\left[\max\{|\alpha_{f,j}^{k+1}-\alpha_{f,j}^k|:j\in J_f^k\},\right.$$
$$\left.\max\{|\alpha_{F,j}^{k+1}-\alpha_{F,j}^k|:j\in J_F^k\}\right]$$

and use estimates of the form (5.37) and

$$|F_j^{k+1}-F_j^k|=|\langle g_F^j,x^{k+1}-x^k\rangle|\leq|g_F^j||x^{k+1}-x^k|$$

together with the boundedness of $\{g_F(y^j)\}_{j\in J^k}$ for $k\in K$. □

We conclude that Algorithm 5.1 is globally convergent in the sense of Theorem 4.11, Theorem 4.12 and Corollary 4.13.

6. Modifications of the Methods

In this section we describe several modifications of the methods discussed so far and analyze their convergence within the framework established in the preceding two sections.

We start by demonstrating global convergence of versions of the methods for convex problems presented in Chapter 5 that use general line search criteria allowing for arbitrarily small serious stepsizes.

To this end, suppose that problem (1.1) is convex and consider the following modified line search rules for Algorithm 5.3.1 and Algorithm 5.5.1.

Step 3 (Line search). Find two stepsizes t_L^k and t_R^k such that $0 \leq t_L^k \leq t_R^k$ and such that the two corresponding points defined by

$$x^{k+1}=x^k+t_L^k d^k \quad \text{and} \quad y^{k+1}=x^k+t_R^k d^k$$

satisfy

$$f(x^{k+1}) \leq f(x^k)+m_L t_L^k v^k, \tag{6.1a}$$

$$F(x^{k+1}) \leq 0, \tag{6.1b}$$

$$-\alpha(x^{k+1},y^{k+1})+ \langle g(y^{k+1}),d^k \rangle \geq m_R v^k \quad \text{if} \quad t_L^k < \bar{t} \tag{6.1c}$$

$$|y^{k+1}-x^{k+1}| \leq \bar{a}, \tag{6.1d}$$

$$t_R^k \leq \tilde{t}, \tag{6.1e}$$

where $\bar{a}$, m_L, m_R, $\bar{t}$ and $\tilde{t}$ are fixed, positive line search parameters satisfying $m_L < m_R < 1$ and $\bar{t} \leq 1 \leq \tilde{t}$, and

$$\alpha(x,y) = \begin{cases} f(x)-f(y)- \langle g_f(y),x-y \rangle & \text{if } y \in S, \\ -F(y)- \langle g_F(y),x-y \rangle & \text{if } y \notin S. \end{cases}$$

Recalling that in Algorithm 5.3.1 and Algorithm 5.5.2 line searches are performed with $v^k < 0$, we conclude that Line Search Procedure 3.2 can be used for finding stepsizes t_L^k and t_R^k satisfying (6.1).

We observe that, except for Step 3, Algorithm 5.3.1 is obtained from Algorithm 3.1 by deleting in the latter method Step 5 and Step 6, and setting $\gamma_f=\gamma_F=0$ and $r_a^k=0$ for all k. In other words, the constructions involving distance measures s_j^k and resetting strategies are not necessary in the convex case.

We shall now establish global convergence of the above-described modifications of the methods of Chapter 5.

Theorem 6.1. Suppose that problem (1.1) is convex and satisfies the Slater constraint qualification, i.e. f and F are convex and $F(\tilde{x})<0$ for some $\tilde{x}\in R^N$. Then Algorithm 5.3.1 with the modified line search rules (6.1) generates a sequence $\{x^k\}\subset S$ satisfying $f(x^k)\downarrow \inf\{f(x): x\in S\}$. Moreover, if f attains its infimum on S then $\{x^k\}$ converges to a solution of problem (1.1).

Proof. In view of the results of Section 5.4, it suffices to establish Lemma 4.10 for the modified algorithm. This can be done by exploiting the fact that the algorithm is a simplified version of Algorithm 3.1. Therefore, one readily checks that Lemma 4.6, analogues of Lemma 3.4.9 and Corollary 3.4.10, and Lemma 4.7 remain valid, and that Lemma 4.8 holds in virtue of the line search ewquirement (6.1d). Thus we only need to replace Lemma 4.9 by the following result: if $x^k \xrightarrow{K} \bar{x}$ and $|x^{k+1}-x^k|\xrightarrow{K} 0$ then $|\alpha_p^{k+1}-\tilde{\alpha}_p^k|\xrightarrow{K} 0$. To this end, observe that

$$\tilde{\alpha}_p^k = \nu_f^k\tilde{\alpha}_{f,p}^k+\nu_F^k\tilde{\alpha}_{F,p}^k = \nu_f^k[f(x^k)-\tilde{f}_p^k]-\nu_F^k\tilde{F}_p^k,$$

$$\alpha_p^{k+1} = \nu_f^k\alpha_{f,p}^{k+1}+\nu_F^k\alpha_{F,p}^{k+1} = \nu_f^k[f(x^{k+1})-f_p^{k+1}]-\nu_F^kF_p^{k+1}$$

by (5.4.3), so

$$\alpha^{k+1}-\tilde{\alpha}_p^k = \nu_f^k[f(x^{k+1})-f(x^k)]-\nu_f^k(f_p^{k+1}-\tilde{f}_p^k)-\nu_F^k(F_p^{k+1}-\tilde{F}_p^k)=$$

$$= \nu_f^k[f(x^{k+1})-f(x^k)]-\nu_f^k < p_f^k,x^{k+1}-x^k > -\nu_F^k < p_F^k,x^{k+1}-x^k > =$$

$$= \nu_f^k[f(x^{k+1})-f(x^k)]- < p^k,x^{k+1}-x^k > .$$

Hence

$$|\alpha_p^{k+1}-\tilde{\alpha}_p^k| \le |f(x^{k+1})-f(x^k)|+|p^k||x^{k+1}-x^k|,$$

since $\nu_f^k\in[0,1]$. Then, since $x^k \xrightarrow{K} \bar{x}$, $|x^{k+1}-x^k|\xrightarrow{K} 0$ and $\{p^k\}_{k\in K}$ is bounded in view of Lemma 4.8, we have $f(x^{k+1})-f(x^k)\xrightarrow{K} 0$ and $|p^k||x^{k+1}-x^k|\xrightarrow{K} 0$, so $|\alpha_p^{k+1}-\alpha_p^k|\xrightarrow{K} 0$, as required. This result enables us to establish Lemma 3.4.15 for the modified method, and then we may prove Lemma 4.10 by using parts (i)-(iii) of the proof of Lemma 3.4.16. □

We conclude from the above proof and the results of Section 5 that Theorem 6.1 is valid for Algorithm 5.5.1 with the modified line search criteria 6.1.

To sum up, we have shown that one may use the general line search criteria (6.1) in the methods for convex problems from Chapter 5 without impairing the global convergence results of Section 5.4 and Section 5.5.

Let us now consider versions of the methods that use the subgradient locality measures $\alpha_f(x^k,y^j)$ and $\alpha_F(x^k,y^j)$ instead of $\alpha_{f,j}^k$ and $\alpha_{F,j}^k$. First, suppose that in Algorithm 3.1 we solve at Step 1 the k-th search direction finding subproblem (2.34) (or its dual, which is of the form (3.1) with $\alpha_{f,j}^k$ and $\alpha_{F,j}^k$ replaced by $\alpha_f(x^k,y^j)$ and $\alpha_F(x^k,y^j)$, respectively), replace (3.5) by (2.35), and calculate a^k by

$$a^k = \max\{|x^k-y^j| : j \in J_f^k \cup J_F^k\}$$

if $\lambda_p^k=\mu_p^k=0$, and by

$$a^{k+1} = \max\{|x^{k+1}-y^j| : j \in J_f^{k+1} \cup J_F^{k+1}\}$$

at Step 6. In effect, this version is obtained by substituting s_j^k with $|x^k-y^j|$ everywhere in Algorithm 3.1. Making use of this observation and of the fact that

$$|x^{k+1}-y^j| \le |x^k-y^j|+|x^{k+1}-x^k|,$$

one can verify that the convergence analysis of Section 4 covers this version of Algorithm 3.1.

Reasoning as above, we conclude that if we replace s_j^k by $|x^k-y^j|$ everywhere in Algorithm 5.1 then the resulting method is globally convergent in the sense of Theorem 4.11, Theorem 4.12 and Corollary 4.13. Moreover, this method has (primal) search direction finding subproblems of the form (2.33), which, as it was shown in Section 2, reduce to the Mifflin (1982) subproblem (2.40) if the rules for choosing J_f^k and J_F^k satisfy the requirements (2.30). Therefore, this method may be regarded as an implementable and globally convergent version of the Mifflin (1982) algorithm. Further comparisons with the algorithm of Mifflin (1982) are given below.

For the sake of completeness of the theory, let us now consider a method that uses all the past subgradients for search direction finding at each iteration. The <u>method with subgradient accumulation</u> is obtained from Algorithm 5.1 by deleting Step 5 and Step 6 and setting

$$J_f^k = J_F^k=\{1,\dots,k\}$$

(or using (2.30) with $\hat{J}_f^k=J_f^k$ and $\hat{J}_F^k=J_F^k$ for all k). Thus the method has no resets and there is no need for selecting Lagrange multipliers of (2.14) to meet the requirement (5.3).

As far as convergence is concerned, we recall from Section 3.6 that methods with total subgradient accumulation, which have no subgradient deletion rules, require additional boundedness assumptions. In this context, consider the following assumption on sequences $\{x^k\}$ and $\{y^k\}$ generated by the above-described method with subgradient accumulation

$$\{x^k\} \text{ and } \{y^k\} \text{ are bounded.} \tag{6.2}$$

The above assumption is satisfied if

$$\text{the set } \bar{S}=\{x \in S : f(x) \le f(x^1)\} \text{ is bounded,} \tag{6.3}$$

since then $\{x^k\} \subset \bar{S}$ in view of the monotonocity of $\{f(x^k)\}$ and the feasibility of $\{x^k\}$, so $\{x^k\}$ is bounded, while $|y^k-x^k| \le \bar{a}/2$ for all k owing to the line search requirement (5.9e).

The results of Section 4 and Section 5 imply that the above-described method with subgradient accumulation is convergent in the sense of Theorem 4.11, Theorem 4.12 and Corollary 4.13 under the additional assumption (6.2). The same result holds if we replace s_j^k by $|x^k-y^j|$ everywhere in this version of Algorithm 5.1.

As observed in Section 3.6, it may be efficient to calculate subgradients not only at $\{y^k\}$ but also at $\{x^k\}$, and then use such additional subgradients for each search direction finding. This idea can be easily incorporated in all the methods discussed so far in this chapter. For instance, in Algorithm 3.1 we may let

$$\begin{aligned}
&y^{-j} = x^j \quad \text{for} \quad j=1,2,\ldots,\\
&g_f^j = g_f(y^j) \quad \text{and} \quad g_F^j = g_F(y^j) \quad \text{for} \quad j=\pm1,\pm2,\ldots,\\
&f_j(x)=f(y^j)+\langle g_f^j, x-y^j\rangle \quad \text{for} \quad j=\pm1,\pm2,\ldots,\\
&F_j(x)=F(y^j)+\langle g_F^j, x-y^j\rangle \quad \text{for} \quad j=\pm1,\pm2,\ldots,\\
&s_j^k = \begin{cases} |y^j-x^{|j|}|+\sum_{i=|j|}^{k-1}|x^{i+1}-x^i| & \text{if} \quad |j|<k,\\ |y^j-x^k| & \text{if} \quad j=\pm k,\end{cases}
\end{aligned} \tag{6.4}$$

and substitute (3.13) with the following

$$J_f^{k+1} = \hat{J}_f^k \cup \{k+1,-(k+1)\}, \tag{6.5a}$$

$$J_F^{k+1} = \hat{J}_F^k \cup \{k+1,-(k+1)\}. \tag{6.5b}$$

One may also use the following modification of (2.30)

$$J_f^{k+1} = \begin{cases} \hat{J}_f^k \cup \{k+1,-(k+1)\} & \text{if } y^{k+1} \in S, \\ \hat{J}_f^k & \text{if } y^{k+1} \notin S, \end{cases} \tag{6.6a}$$

$$J_F^{k+1} = \begin{cases} \hat{J}_F^k & \text{if } y^{k+1} \in S, \\ \hat{J}_F^k \cup \{k+1,-(k+1)\} & \text{if } y^{k+1} \notin S. \end{cases} \tag{6.6b}$$

Of course, if $x^{k+1}=y^{k+1}$, i.e. $y^{k+1}=y^{-(k+1)}$, then, for instance, (6.5a) should be replaced by $J_f^{k+1}=\hat{J}_f^k \cup \{k+1\}$.

It should be clear that the above-described use of additional subgradients does not influence the preceding convergence results, although it may lead to faster convergence in practice.

We want to add that one may use the Mifflin (1982) definition (2.36) of subgradient locality measures in all the methods described so far in this chapter. This will, for instance, involve replacing in Algorithm 3.1 relations (3.2) by (2.37), (3.8) by

$$\tilde{\alpha}_{f,p}^k=\max\{f(x^k)-\tilde{f}_p^k,\gamma_f(\tilde{s}_f^k)^2\},$$

$$\tilde{\alpha}_{F,p}^k=\max\{-\tilde{F}_p^k,\gamma_F(\tilde{s}_F^k)^2\},$$

and (3.12) by

$$g(y)=g_f(y) \quad \text{and} \quad \alpha(x,y)=\max\{f(x)-\overline{f}(x;y),\gamma_f|x-y|^2\} \quad \text{if} \quad y \in S,$$

$$g(y)=g_F(y) \quad \text{and} \quad \alpha(x,y)=\max\{-\overline{F}(x;y),\gamma_F|x-y|^2\} \quad \text{if} \quad y \notin S.$$

Such modifications do not impair the preceding convergence results.

We shall now show how to strengthen the existing results on convergence of the Mifflin (1982) algorithm. At the k-th iteration this algorithm finds $(d^k,\hat{v}^k)$ by solving subproblem (2.40), where $J^k=\{1,...,k\}$ and $\alpha(x^k,y^j)$ is defined by (2.36) and (2.39). In effect, we see that the algorithm uses the same direction finding subproblems as does the above-described method with subgradient accumulation, if in the latter method J_f^k and J_F^k are chosen by (2.30) with $\hat{J}_f^k=J_f^k$ and $\hat{J}_F^k=J_F^k$ for all k. The Mifflin (1982) line search requirements are the following

$$f(x^{k+1}) \le f(x^k)+m_L t_L^k \hat{v}^k,$$

$$F(x^{k+1}) \le 0,$$

$$-\alpha(x^{k+1},y^{k+1})+\langle g(y^{k+1}),d^k\rangle \ge m_R \hat{v}^k, \tag{6.7}$$

$$t_L^k = t_R^k = 0 \quad \text{if} \quad \langle g(x^k),d^k\rangle \ge m_R \hat{v}^k,$$

$$|y^{k+1}-x^{k+1}| \le \bar{a}.$$

It is easy to observe that any stepsizes t_L^k and t_R^k satisfying (6.7) automatically satisfy (5.9) if $\bar{t}=+\infty$. Moreover, Line Search Procedure 3.2 can be used for finding stepsizes t_L^k and t_R^k satisfying (6.7) if $\langle g_f(x^k),d^k \rangle \le m_R\hat{v}^k$; otherwise one can set $t_L^k=t_R^k=0$.

To sum up, we have shown that the Mifflin (1982) algorithm can be regarded as a version of the method with subgradient accumulation described above. Therefore, by the preceding results, the Mifflin algorithm is convergent in the sense of Theorem 4.11 and Theorem 4.12 under the additional assumption (6.2) (or the stronger assumption (6.3)). Our result subsumes a result of Mifflin (1982), who proved that his algorithm has at least one stationary accumulation point under the additional assumption (6.2).

7. Methods with Subgradient Deletion Rules

The algorithms described in the preceding section were obtained by incorporating in the methods for convex problems of Chapter 5 the techniques for dealing with nonconvexity through the use of subgradient locality measures introduced in Chapter 3. In the unconstrained case considered in Chapter 4, we showed that one can also take nonconvexity into account by using suitable subgradient delection rules for localizing the past subgradient information that determines the current polyhedral approximation to the objective function. Therefore, in this section we shall consider the use of subgradient deletion strategies for obtaining extensions of methods of Chapter 5 to the nonconvex case, which differ from the algorithms described so far.

We start by remarking that in practice the performance of methods with subgradient locality measures may be sensitive with respect to values of the distance measure parameters γ_f and γ_F; see Remark 4.2.1. For this reason, in Chapter 4 we studied methods that result from setting $\gamma_f=0$ even in the nonconvex case. To ensure convergence of such methods, we had to employ subgradient deletion rules based on estimating the degree of stationarity of the current iterate. Proceeding in the same sprit, we may set $\gamma_f=\gamma_F=0$ in Algorith 3.1 and use a simple resetting strategy of Section 4.2 to obtain the following method, which may be regarded as a combination of Algorithm 4.3.1 and Algorithm 5.3.1.

Algorithm 7.1.

Step 0 (Initialization). Select the starting point $x^1 \in S$ and a final accuracy tolerance $\varepsilon_s \geq 0$. Choose fixed positive line search parameters $m_L, m_R, \bar{a}, \bar{t}$ and $\bar{\theta}$ with $\bar{t} < 1$, $m_L < m_R < 1$ and $\bar{\theta} < 1$. Set $M_{g,f}$ and $M_{g,F}$ equal to the fixed maximum number of subgradients of f and F, respectively, that the algorithm may use for each search direction finding; $M_{g,f} \geq 2$ and $M_{g,F} \geq 2$. Choose a predicted shift in x at the first iteration $s^1 > 0$ and set $\theta^1 = \bar{\theta}$. Select a positive reset tolerance m_a and set reset indicators $r_a^1 = r_f^1 = r_F^1 = 1$. Set $y^1 = x^1$, $s_1^1 = 0$ and

$$J_f^1 = \{1\},\ g_f^1 = p_f^0 = g_f(y^1),\ f_1^1 = f_p^1 = f(y^1),$$

$$J_F^1 = \emptyset\ ,\ \ p_F^0 = 0 \in R^N\ ,\ F_p^1 = 0.$$

Set the counters k=1, l=0 and k(0)=1.

Step 1 (Direction finding). Find multipliers λ_j^k, $j \in J_f^k$, λ_p^k, μ_j^k, $j \in J_F^k$ and μ_p^k that solve the folowing k-th dual search direction finding subproblem

$$\underset{\lambda,\mu}{\text{minimize}}\ \tfrac{1}{2}\Big|\sum_{j \in J_f^k}\lambda_j g_f^j + \lambda_p p_f^{k-1} + \sum_{j \in J_F^k}\mu_j g_F^j + \mu_p p_F^{k-1}\Big|^2 +$$

$$+ \sum_{j \in J_f^k}\lambda_j \alpha_{f,j}^k + \lambda_p \alpha_{f,p}^k + \sum_{j \in J_F^k}\mu_j \alpha_{F,j}^k + \mu_p \alpha_{F,p}^k\ ,$$

$$\text{subject to}\quad \lambda_j \geq 0,\ j \in J_f^k,\ \lambda_p \geq 0,\ \mu_j \geq 0,\ j \in J_F^k,\ \mu_p \geq 0, \tag{7.1}$$

$$\sum_{j \in J_f^k}\lambda_j + \lambda_p + \sum_{j \in J_F^k}\mu_j + \mu_p = 1,$$

$$\lambda_p = 0 \quad \text{if} \quad r_f^k = 1,\ \mu_p = 0 \quad \text{if} \quad r_F^k = 1.$$

where

$$\alpha_{f,j}^k = |f(x^k) - f_j^k|,\ \alpha_{F,j}^k = |F_j^k|, \tag{7.2}$$

$$\alpha_{f,p}^k = |f(x^k) - f_p^k|,\ \alpha_{F,p}^k = |F_p^k|. \tag{7.3}$$

Compute ν_f^k and ν_F^k by (3.3), and $\tilde{\lambda}_j^k$, $j \in J_f^k$, $\tilde{\lambda}_p^k$, $\tilde{\mu}_j^k$, $j \in J_F^k$, and $\tilde{\mu}_p^k$ by (3.4). Calculate

$$(p_f^k, \tilde{f}_p^k) = \sum_{j \in J_f^k}\tilde{\lambda}_j^k(g_f^j, f_j^k) + \tilde{\lambda}_p^k(p_f^{k-1}, f_p^k), \tag{7.4a}$$

$$(p_F^k, \tilde{F}_p^k) = \sum_{j \in J_F^k}\tilde{\mu}_j^k(g_F^j, F_j^k) + \tilde{\mu}_p^k(p_F^{k-1}, F_p^k) \tag{7.4b}$$

$$p^k = \nu_f^k p_f^k + \nu_F^k p_F^k, \tag{7.5}$$

$$d^k = -p^k, \tag{7.6}$$

$$\tilde{\alpha}^k_{f,p} = |f(x^k) - \tilde{f}^k_p|, \tag{7.7a}$$

$$\tilde{\alpha}^k_{F,p} = |\tilde{F}^k_p|, \tag{7.7b}$$

$$v^k = -\{|p^k|^2 + \nu^k_f \tilde{\alpha}^k_{f,p} + \nu^k_F \tilde{\alpha}^k_{F,p}\}, \tag{7.8}$$

If $r^k_f = r^k_F = 1$ set

$$a^k = \max\{s^k_j : j \in J^k_f \cup J^k_F\}. \tag{7.9}$$

Step 2 (Stopping criterion). If $\max\{|p^k|, m_a a^k\} \le \varepsilon_s$ then terminate. Otherwise, go to Step 3.

Step 3 (Resetting test). If $|p^k| \le m_a a^k$ then go to Step 4; otherwise, go to Step 5.

Step 4 (Resetting). (i) If $r^k_a = 0$ then set $r^k_a = r^k_f = r^k_F = 1$, replace J^k_f and J^k_F by $\{j \in J^k_f : j \ge k - M_{g,f} + 2\}$ and $\{j \in J^k_F : j \ge k - M_{g,f} + 2\}$, respectively, and go to Step 1.

(ii) If $|J^k_f \cup J^k_F| > 1$ then delete the smallest number from J^k_f or J^k_F and go to Step 1.

(iii) Set $y^k = x^k$, $g^k_f = g_f(y^k)$, $f^k_k = f(y^k)$, $J^k_f = \{k\}$, $s^k_k = 0$ and go to Step 1.

Step 5 (Line search). By a line search procedure as discussed below, find two stepsizes t^k_L and t^k_R such that $0 < t^k_L < t^k_R$ and such that the two corresponding points defined by

$$x^{k+1} = x^k + t^k_L d^k \quad \text{and} \quad y^{k+1} = x^k + t^k_R d^k$$

satisfy $t^k_L \le 1$ and

$$f(x^{k+1}) \le f(x^k) + m_L t^k_L v^k, \tag{7.10a}$$

$$F(x^{k+1}) \le 0, \tag{7.10b}$$

$$t^k_R = t^k_L \quad \text{if} \quad t^k_L \ge \bar{t}, \tag{7.10c}$$

$$-\alpha(x^{k+1}, y^{k+1}) + \langle g(y^{k+1}), d^k \rangle \ge m_R v^k \quad \text{if} \quad t^k_L < \bar{t}, \tag{7.10d}$$

$$|y^{k+1} - x^{k+1}| \le \bar{a}, \tag{7.10e}$$

$$|y^{k+1} - x^{k+1}| \le \theta^k s^k \quad \text{if} \quad t^k_L = 0, \tag{7.10f}$$

$$|y^{k+1} - x^{k+1}| \le \bar{\theta} |x^{k+1} - x^k| \quad \text{if} \quad t^k_L > 0, \tag{7.10g}$$

where

$$g(y)=g_f(y) \quad \text{and} \quad \alpha(x,y)=\alpha_f(x,y) \quad \text{if} \quad y\in S, \tag{7.11a}$$

$$g(y)=g_F(y) \quad \text{and} \quad \alpha(x,y)=\alpha_F(x,y) \quad \text{if} \quad y\notin S, \tag{7.11b}$$

$$\alpha_f(x,y)=|f(x)-f(y)-\langle g_f(y),x-y\rangle|, \tag{7.12a}$$

$$\alpha_F(x,y)=|F(y)-\langle g_F(y),x-y\rangle|. \tag{7.12b}$$

Step 6. If $t_L^k=0$ set $s^{k+1}=s^k$ and $\theta^{k+1}=\bar{\theta}\theta^k$. Otherwise, i.e. if $t_L^k>0$, set $s^{k+1}=|x^{k+1}-x^k|$, $\theta^{k+1}=\bar{\theta}$, $k(l+1)=k+1$ and increase l by 1.

Step 7 (Subgradient updating). Select sets $\hat{J}_f^k$ and $\hat{J}_F^k$ satisfying

$$\hat{J}_f^k \subset J_f^k \quad \text{and} \quad |\hat{J}_f^k| \le M_{g,f}-2, \tag{7.13a}$$

$$\hat{J}_F^k \subset J_F^k \quad \text{and} \quad |\hat{J}_F^k| \le M_{g,f}-2, \tag{7.13b}$$

and set

$$J_f^{k+1}=\hat{J}_f^k \cup \{k+1\} \quad \text{and} \quad J_F^{k+1}=\hat{J}_F^k \quad \text{if} \quad y^{k+1}\in S, \tag{7.14a}$$

$$J_f^{k+1}=\hat{J}_f^k \quad \text{and} \quad J_F^{k+1}=\hat{J}_F^{k+1} \cup \{k+1\} \quad \text{if} \quad y^{k+1}\notin S. \tag{7.14b}$$

Set $g_f^{k+1}=g_f(y^{k+1})$ if $y^{k+1}\in S$, $g_F^{k+1}=g_F(y^{k+1})$ if $y^{k+1}\notin S$. Compute f_j^{k+1}, $j\in J_f^{k+1}$, f_p^{k+1}, F_j^{k+1}, $j\in J_F^{k+1}$, F_p^{k+1}, s_{k+1}^{k+1} and s_j^{k+1}, $j\in \hat{J}_f^k\cup\hat{J}_F^k$, by (3.14). Calculate

$$a^{k+1}=\max\{a^k+|x^{k+1}-x^k|,\, s_{k+1}^{k+1}\}. \tag{7.15}$$

Set $r_a^{k+1}=0$ and

$$r_f^{k+1}=\begin{cases} 1 & \text{if} \quad r_f^k=1 \quad \text{and} \quad \nu_f^k=0, \\ 0 & \text{if} \quad r_f^k=0 \quad \text{or} \quad \nu_f^k\neq 0, \end{cases} \tag{7.16a}$$

$$r_F^{k+1}=\begin{cases} 1 & \text{if} \quad r_F^k=1 \quad \text{and} \quad \nu_F^k=0, \\ 0 & \text{if} \quad r_F^k=0 \quad \text{or} \quad \nu_F^k\neq 0. \end{cases} \tag{7.16b}$$

Step 8. Increase k by 1 and go to Step 1.

A few remarks on the algorithm are in order.

By Lemma 5.2.3, subproblem (7.1) is the dual of the following k-th (primal) search direction finding subproblem

$$\begin{aligned}
&\underset{(d,\hat v)\in R^{N+1}}{\text{minimize}} \quad \tfrac{1}{2}|d|^2+\hat v,\\
&\text{subject to} \quad -\alpha^k_{f,j}+\langle g^j_f,d\rangle \le \hat v,\ j\in J^k_f,\\
&\qquad -\alpha^k_{f,p}+\langle p^{k-1}_f,d\rangle \le \hat v \quad \text{if} \quad r^k_f=0,\\
&\qquad -\alpha^k_{F,j}+\langle g^j_F,d\rangle \le \hat v,\ j\in J^k_F,\\
&\qquad -\alpha^k_{F,p}+\langle p^{k-1}_F,d\rangle \le \hat v \quad \text{if} \quad r^k_F=0.
\end{aligned} \tag{7.17}$$

Subproblems (7.1) and (7.17) are equivalent to subproblems (3.1) and (2.22), respectively, provided that $r^k_a=r^k_f=r^k_F$. However, in contrast with Algorithm 3.1, in Algorithm 7.1 we use two reset indicators r^k_f and r^k_F, which need not have values equal to that of r^k_a. The condition $r^k_f=1$ ($r^k_F=1$) implies that the (k-1)-st aggregate subgradient of f (of F) is ignored at k-th search direction finding. First, this is the case if a reset occurs at the k-th iteration, since then $r^k_a=r^k_f=r^k_F$ by the rules of Step 4 and Step 7. Secondly, we have $r^k_f=1$ ($r^k_F=1$) if after the latest reset, say at the $k_r(k)$-th iteration, no subgradient of f (of F) contributed to the search directions

$$d^j=-(\nu^j_f p^j_f+\nu^j_F p^j_F) \quad \text{for} \quad j=k_r(k),\dots,k-1$$

in the sense that $\nu^j_f=0$ ($\nu^j_F=0$) for $j=k_r(k),\dots,k-1$ (cf.(7.16)). This can occur, for instance, if $J^j_f=\emptyset$ ($J^j_F=\emptyset$) for $j=k_r(k),\dots,k-1$ (cf. 7.13)-(7.14)). Then there is nothing to be aggregated at such iterations with $J^j_f=0$ and $r^j_f=1$, ($J^j_F=\emptyset$ and $r^j_F=1$), so we must ignore the (k-1)-st aggregate subgradient of f (of F) by setting $r^k_f=1$ ($r^k_F=1$).

For example, suppose that the constraint function is inactive, i.e. $S=R^N$. Then, since $J^1_F=\emptyset$ and each y^k is feasible, (7.13) and (7.14) yield $J^k_F=\emptyset$ for all k, and so, because

$$\nu^k_F=\sum_{j\in J_F}\mu^k_j+\mu^k_p \quad \text{and} \quad \mu^k_p=0 \quad \text{if} \quad r^k_F=1$$

and $r^1_F=1$, we have $\nu^k_F=0$ and $r^k_F=1$ for all k by the rules of Step 4 (i) and Step 7. (Note that we have implicitly assumed that $\sum_{j\in J^k_F}\mu^k_j=0$ if $J^k_F=\emptyset$. The same convention of ignoring summations over empty sets is employed in Algorithm 7.1). Moreover, we have $\nu^k_f=1-\nu^k_F=1$, so that $r^k_f=0$ if $r^k_a=0$, for all k. Then it is easy to deduce that Algorithm 7.1 reduces to Algorithm 4.3.1, except for the rule of updating a^k via (7.9) if $r^k_f=1$, while Algorithm

4.3.1 uses (7.9) if $\lambda_p^k=0$.

Algorithm 7.1 is closely related to Algorithm 5.3.1 for convex constrained problems. First, both algorithms use the same subgradient aggregation rules, cf. (7.4) and (5.2.45). Secondly, their search direction finding subproblems coincide in the convex case if $r_f^k=r_F^k=0$. To see this, recall from Lemma 5.4.1 and Lemma 5.4.2 that in the convex case the subgradient aggregation rules ensure that $f(x^k)-f_j^k\geq 0$, $f(x^k)-f_p^k\geq 0$, $-F_j^k\geq 0$ and $-F_p^k\geq 0$ for all k, hence (7.2)-(7.3) becomes

$$\alpha_{f,j}^k=f(x^k)-f_j^k \quad\text{and}\quad \alpha_{f,p}^k=f(x^k)-f_p^k,$$

$$\alpha_{F,j}^k=-F_j^k \quad\text{and}\quad \alpha_{F,p}^k=-F_p^k,$$

corresponding to relations (5.4a,b). Therefore, if $r_f^k=r_F^k=0$, in this case subproblem (7.1) coincides with subproblem (5.4.12), which is equivalent to the k-th search direction finding subproblem (5.3.1) of Algorithm 5.3.1 by Lemma 5.4.10.

To discuss the resetting strategy of the algorithm, we shall need the following result on convex representations of aggregate subgradients, which is an analogue of Lemma 4.4.1 and Lemma 4.1. Let

$$J^k=J_f^k\cup J_F^k \quad \text{for all } k, \tag{7.18}$$

and observe that by (7.13)-(7.14) we have

$$J_f^k\cap J_F^k=\emptyset \quad \text{for all } k. \tag{7.19}$$

Lemma 7.2. Suppose $k\geq 1$ is such that Algorithm 7.1 did not stop before the k-th iteration, and let

$$k_f(k)=\max\{j : j\leq k \text{ and } r_f^j=1\}, \tag{7.20a}$$

$$\hat{J}_{f,r}^k=J_f^{k_f(k)}\cup\{j : k_f(k)<j\leq k \text{ and } y^j\in S\}, \tag{7.20b}$$

$$k_F(k)=\max\{j : j\leq k \text{ and } r_F^j=1\}, \tag{7.20c}$$

$$\hat{J}_{F,r}^k=J_F^{k_F(k)}\cup\{j : k_F(k)<j\leq k \text{ and } y^j\notin S\}, \tag{7.20d}$$

$$k_p(k)=\max\{j : j\leq k \text{ and } r_f^j=r_F^j=1\}, \tag{7.21a}$$

$$\hat{J}_p^k=J^{k_p(k)}\cup\{j : k_p(k)<j\leq k\}, \tag{7.21b}$$

$$k_r(k)=\max\{j : j\leq k \text{ and } r_a^j=1\}, \tag{7.21c}$$

$$\hat{J}_r^k=J^{k_r(k)}\cup\{j : k_r(k)<j\leq k\}, \tag{7.21d}$$

$$\tilde{J}^k = \{j : k_r(k)-M_g \le j \le k\}, \tag{7.22}$$

where $M_g=M_{g,f}+M_{g,F}$. Let $M=N+2$. Then

$$a^k = \max\{s_j^k : j \in \hat{J}_p^k\}, \tag{7.23}$$

$$\hat{J}_p^k = \hat{J}_r^k \subset \tilde{J}^k, \tag{7.24}$$

$$\hat{J}_{f,r}^k \neq \emptyset \quad \text{if} \quad \nu_f^k \neq 0, \tag{7.25a}$$

$$\hat{J}_{F,r}^k \neq \emptyset \quad \text{if} \quad \nu_F^k \neq 0. \tag{7.25b}$$

If $\hat{J}_{f,p}^k \neq \emptyset$ then there exist numbers $\hat{\lambda}_i^k$ and (N+2)-vectors $(y_f^{k,i}, f^{k,i}, s_f^{k,i})$, $i=1,\dots,M$, satisfying

$$(p_f^k, \tilde{f}_p^k) = \sum_{i=1}^{M} \hat{\lambda}_i^k (g(y_f^{k,i}), f^{k,i}),$$

$$\hat{\lambda}_i^k \ge 0, \; i=1,\dots,M, \; \sum_{i=1}^{M} \hat{\lambda}_i^k = 1,$$

$$(g(y_f^{k,i}), f^{k,i}, s_f^{k,i}) \in \{(g_f^j, f_j^k, s_j^k) : j \in \hat{J}_{f,p}^k\}, \; i=1,\dots,M, \tag{7,26}$$

$$|y_f^{k,i} - x^k| \le s_f^{k,i},$$

$$\max\{s_f^{k,i} : i=1,\dots,M\} \le a^k.$$

If $\hat{J}_{F,p}^k \neq \emptyset$ then there exist numbers $\hat{\mu}_i^k$ and (N+2)-vectors $(y_F^{k,i}, F^{k,i}, s_F^{k,i})$, $i=1,\dots,M$, satisfying

$$(p_F^k, \tilde{F}_p^k) = \sum_{i=1}^{M} \hat{\mu}_i^k (g(y_F^{k,i}), F^{k,i}),$$

$$\hat{\mu}_i^k \ge 0, \; i=1,\dots,M, \; \sum_{i=1}^{M} \hat{\mu}_i^k = 1,$$

$$(g(y_F^{k,i}), F^{k,i}, s_F^{k,i}) \in \{(g_F^j, f_j^k, s_j^k) : j \in \hat{J}_{F,p}^k\}, \; i=1,\dots,M, \tag{7.27}$$

$$|y_F^{k,i} - x^k| \le s_F^{k,i}, \; i=1,\dots,M,$$

$$\max\{s_F^{k,i} : i=1,\dots,M\} \le a^k.$$

Moreover, we have (4.3a) if f is convex, and (4.3b) if F is convex and $y^j \notin S$ for some $j \le k$.

<u>Proof</u>. Since either ν_f^j or ν_F^j must be positive, because they form a

convex combination, the algorithm's rules imply that $r_f^j=r_F^j=1$ only if $r_a^k=1$; otherwise either $r_f^j=0$ or $r_F^j=0$. Hence $k_p(k)=k_r(k)$ and $\hat{J}_p^k=\hat{J}_r^k$. Moreover, $k_f(k)\geq k_p(k)$ and $k_F(k)\geq k_p(k)$, so $\hat{J}_{f,p}^k\subset\hat{J}_{F,p}^k\subset\hat{J}_p^k$. Suppose that $\hat{J}_{f,p}^k=\emptyset$. Then $J_f^i=\emptyset$ and $r_f^i=1$ for $i=k_f(k),\ldots,k$, hence $J_f^k=\emptyset$, $r_f^k=1, \lambda_p^k=0$ and

$$\nu_f^k = \sum_{j\in J_f^k}\lambda_j^k + \lambda_p^k = 0.$$

This proves (7.25a). An analogous argument yields (7.25b). (7.23) can be established as in the proof of Lemma 3.4.1. (7.24) follows from the rules of Step 4 (i), (7.13)-(7.14) and the above-derived relation $\hat{J}_p^k=\hat{J}_r^k$ as in the proof of Lemma 4.4.1. The representations (7.26) and (7.27) follow from the subgradient aggregation rules (see the proofs of Lemma 3.4.1 and Lemma 3.4.3), (7.23) and the fact that $\hat{J}_{f,p}^k\cup\hat{J}_{F,p}^k\subset\hat{J}_p^k$. The proof of the assertion concerning the convex case is similar to the proof of Lemma 5.4.2, since the aggregate subgradients are always convex combinations of the past subgradients, even if $\hat{J}_{f,p}^k=\hat{J}_{F,p}^k=\emptyset$. □

The stopping criterion of Step 2 admits of the following interpretation. For any $x\in R^N$ and $\varepsilon\geq 0$, define the following outer approximation to $\hat{M}(x)$

$$\hat{M}(x;\varepsilon)=\mathrm{conv}\{\hat{M}(y): |y-x|\leq\varepsilon\}. \tag{7.28}$$

In view of (2.2), $\hat{M}(x;\varepsilon)$ may be regarded as a generalization of the Goldstein ε-subdifferential

$$\partial f(x;\varepsilon) = \mathrm{conv}\{\partial f(y) : |y-x| \leq \varepsilon\}$$

to the constrained case. Observe that at each iteration only one subgradient of the form

$$g^k=g(y^k)=\begin{cases} g_f(y^k)\in\hat{M}(y^k) & \text{if } y^k\in S,\\ g_F(y^k)\in\hat{M}(y^k) & \text{if } y^k\notin S,\end{cases} \tag{7.29}$$

is added to the set of subgradients that are aggregated at subsequent iterations. We deduce from Lemma 7.2 and (7.29) that

$$p_f^k \in \mathrm{conv}\{M(y^j) : |y^j-x^k|\leq a^k\} \quad \text{if} \quad \hat{J}_{f,p}^k\neq\emptyset, \tag{7.30a}$$

$$p_F^k \in \mathrm{conv}\{M(y^j) : |y^j-x^k|\leq a^k\} \quad \text{if} \quad \hat{J}_{F,p}^k\neq\emptyset. \tag{7.30b}$$

Therefore, since the algorithm's rules and (7.25) yield

$$p^k = \nu_f^k p_f^k + \nu_F^k p_F^k,$$

$$\nu_f^k \geq 0,\ \nu_F^k \geq 0,\ \nu_f^k + \nu_F^k = 1, \tag{7.31a}$$

$$\nu_f^k = 0 \quad \text{if} \quad \hat{J}_{f,p}^k = \emptyset, \tag{7.31b}$$

$$\nu_F^k = 0 \quad \text{if} \quad \hat{J}_{F,p}^k = \emptyset, \tag{7.31c}$$

we obtain from (7.30) and (7.28) the following analoque of (4.2.30)

$$p^k \in \hat{M}(x^k; a^k). \tag{7.32}$$

By (7.32), the algorithm stops at Step 2 when

$$p^k \in \hat{M}(x^k; \varepsilon_s/m_a),\ |p^k| \leq \varepsilon_s \quad \text{and} \quad x^k \in S, \tag{7.33}$$

i.e. when x^k is approximately stationary for f on S.

The resetting strategy of the algorithm, which is related to (7.32), is a direct extension of the strategy of Algorithm 4.3.1. We may add that it is possible to use other, more efficient strategies similar to those of Section 4.6. We shall return to this subject later on.

The line search rules of Step 5 are a direct juxtaposition of the rules of Algorithm 4.3.1 and Algorithm 3.1, cf. (4.3.7)-(4.3.10),(3.11)-(3.12) and (7.10)-(7.12). Therefore we may refer the reader to Section 3 and Section 4.3 for the motivation of such rules, and of the rules of Step 6.

The following extension of Line Search Procedure 4.3.2 can be used for finding stepsizes $t_L = t_L^k$ and $t_R = t_R^k$ satisfying the requirements of Step 5.

<u>Line Search Procedure 7.3.</u>

(i) Set $t_L = 0$ and $t = t_U = \min\{1, \bar{a}/|d^k|\}$.

(ii) If $f(x^k + td^k) \leq f(x^k) + m_L t v^k$ and $F(x^k + td^k) \leq 0$ set $t_L = t$; otherwise set $t_U = t$.

(iii) If $t_L \geq \bar{t}$ set $t_R = t_L$ and return.

(iv) If $-\alpha(x^k + t_L d^k, x^k + td^k) + \langle g(x^k + td^k), d^k \rangle \geq m_R v^k$ and either $t_L = 0$ and $t|d^k| \leq \theta^k s^k$ or $t - t_L \leq \bar{\theta} t_L$, then set $t_R = t$ and return.

(v) Set $t = t_L + \bar{\theta}(t_U - t_L)$ and go to (ii).

The following result can be established similarly to Lemma 3.3.

Lemma 7.4. If f and F are semismooth in the sense of (3.3.23) and (3.18) then Line Search Procedure 7.3 terminates with $t_L^k=t_L$ and $t_R^k=t_R$ satisfying (7.10).

The requirement (7.13) may be substituted by the following

$$\hat{J}_f^k \subset J_f^k \quad \text{and} \quad \hat{J}_F^k \subset J_F^k, \tag{7.34a}$$

$$|\hat{J}_f^k \cup \hat{J}_F^k| \le M_g-2, \tag{7.34b}$$

where $M_g \ge 2$ is a fixed, user-supplied upper bound on the number of stored subgradients. In view of (7.18) and (7.19), the simplest way of satisfying (7.34) is to delete some smallest numbers from $\hat{J}^k=J_f^k \cup J_F^k$ so as to obtain $|\hat{J}^k| \le M_g-2$ with $\hat{J}_f^k \cup \hat{J}_F^k=\hat{J}^k$. In fact, as far as convergence is concerned, the requirement (7.34a) can be substituted by the following more general rule

$$\hat{J}_f^k \subset \hat{J}_{f,p}^k \quad \text{and} \quad \hat{J}_F^k \subset \hat{J}_{F,p}^k,$$

i.e. any subgradient used since the latest reset can be stored, cf. (7.17a,b,c).

Observe that (7.10e), (7.14) and the rules of Step 4 yield the following analogue of (4.3.23) and (3.21b)

$$k \in J_f^k,\ g^k=g_f(y^k) \quad \text{and} \quad |y^k-x^k| \le \bar{a} \quad \text{if} \quad y^k \in S, \tag{7.35a}$$

$$k \in J_F^k,\ g^k=g_F(y^k) \quad \text{and} \quad |y^k-x^k| \le \bar{a} \quad \text{if} \quad y^k \notin S. \tag{7.35b}$$

Thus the latest subgradient is always used for the current search direction finding.

We shall now establish convergence of the algorithm. To save space we shall use suitable modifications of the results of Section 4.4 and Section 4.

We suppose that the final accuracy tolerance ε_s is set to zero and that each execution of Line Search Procedure 7.3 is finite (see Lemma 7.4, Remark 3.3.4 and Remark 3.4).

First, we observe that Lemma 7.2 can serve as a substitute for Lemma 4.1, Lemma 4.4.1, Lemma 4.4.2 and Lemma 4.4.4. Secondly, since Lemma 4.3 holds in view of (7.33), the assumption that $\varepsilon_s=0$ and the definition of stationary points, we may assume that the method generates as infinite sequence of points. Then (4.4.9)-(4.4.11) are easily verified, and we conclude that $\{f(x^k)\}$ is nonincreasing. Thirdly, we note that part (i) of Lemma 4.2 can be replaced by Lemma 4.4.5, and part (ii) by the following result.

Lemma 7.5. Suppose that a point $\bar{x} \in R^N$, N-vectors $\bar{p}_F, \bar{y}_F^i$ and $\bar{g}_F^i$, and numbers $\bar{F}_p, \bar{F}^i$ and $\bar{s}_F^i$, i=1,...,M, satisfy

$$(\bar{p}_F, \bar{F}_p) = \sum_{i=1}^{M} \bar{\mu}_i (\bar{g}_F^i, \bar{F}^i),$$

$$\bar{\mu}_i \geq 0, \ i=1,\ldots,M, \ \sum_{i=1}^{M} \hat{\mu}_i = 1,$$

$$\bar{g}_F^i \in \partial F(\bar{y}_F^i), \ i=1,\ldots,M,$$

$$\bar{F}^i = F(\bar{y}_F^i) + \langle \bar{g}_F^i, \bar{x} - \bar{y}_F^i \rangle, \ i=1,\ldots,M,$$

$$|\bar{y}_F^i - \bar{x}| \leq \bar{s}_F^i, \ i=1,\ldots,M$$

$$\max \{ \bar{s}_F^i : \bar{\mu}_i \neq 0 \} = 0,$$

$$F(\bar{y}_F^i) \geq 0, \ i=1,\ldots,M.$$

Then $\bar{p}_F \in \partial F(\bar{x})$ and $\bar{F}_p = F(\bar{x}) \geq 0$, so that $\bar{p}_F \in \hat{M}(\bar{x})$.

Proof. Set $\bar{s}_F = \sum_{i=1}^{M} \bar{\mu}_i \bar{s}_F^i$ and use part (i) of the proof of Lemma 4.2. □

Lemma 4.4.6 and Lemma 4.4 are replaced by the following

Lemma 7.6. Suppose that there exist a point $\bar{x} \in R^N$ and an infinite set $K \subset \{1,2,\ldots\}$ such that $x^k \xrightarrow{K} \bar{x}$ and $a^k \xrightarrow{K} 0$. Then there exist an infinite set $\bar{K} \subset K$, N-vectors $\bar{p}, \bar{p}_f$ and $\bar{p}_F$, and numbers $\bar{\nu}_f$ and $\bar{\nu}_F$ such that

$$p^k \xrightarrow{\bar{K}} \bar{p},$$

$$\bar{p} = \bar{\nu}_f \bar{p}_f + \bar{\nu}_F \bar{p}_F,$$

$$\bar{\nu}_f \geq 0, \ \bar{\nu}_F \geq 0, \ \bar{\nu}_f + \bar{\nu}_F = 1, \tag{7.36}$$

$$\bar{p}_f \in \partial f(\bar{x}), \ \bar{p}_F \in \partial F(\bar{x}),$$

$$F(\bar{x}) \geq 0 \quad \text{if} \quad \bar{\nu}_F \neq 0.$$

Moreover, $\bar{p} \in \hat{M}(\bar{x})$ and $\nu_f^k \tilde{\alpha}_{f,p}^k + \nu_F^k \tilde{\alpha}_{F,p}^k \xrightarrow{\bar{K}} 0$.

Proof. By (7.31), $\hat{J}_{f,p}^k \cup \hat{J}_{F,p}^k \neq \emptyset$ for all k, so at least one of the following two sets

$$K_f = \{k \in K : \hat{J}^k_{f,p} \neq \emptyset\} \quad \text{and} \quad K_F = \{k \in K : \hat{J}^k_{F,p} \neq \emptyset\}$$

is infinite. Suppose that K_F is finite. Then we have $\nu_F^k=0$ and (7.26) for all large $k \in K_f$, hence we may use (7.31a), (7.6) and (7.7a) to deduce, as in the proof of Lemma 3.4.6, (7.36) with $\bar{\nu}_f=1$, $\bar{\nu}_F=0$ and $\nu_f^k\tilde{\alpha}^k_{f,p} \xrightarrow{\bar{K}} 0$. A similar argument based on (7.27) and Lemma 7.5 yields (7.36) with $\bar{\nu}_f=0$, $\bar{\nu}_F=1$ and $\nu_F^k\tilde{\alpha}^k_{F,p} \xrightarrow{\bar{K}} 0$ if K_f is finite. In view of the preceding two results, and the fact that $K=K_f \cup K_F$, it remains to consider the case of an infinite set $\tilde{K}=K_f \cap K_F$. Then (7.26) and (7.27) hold for all $k \in \tilde{K}$, so the desired conclusion can be deduced from (7.31), (7.6)-(7.7), Lemma 4.4.5 and Lemma 7.5. (7.36) implies $\bar{p} \in M(\bar{x})$ in view of (2.2). □

Define the stationarity measure

$$w^k = \tfrac{1}{2}|p^k|^2 + \tilde{\alpha}^k_p, \tag{7.37a}$$

where

$$\tilde{\alpha}^k_p = \nu_f^k\tilde{\alpha}^k_{f,p} + \nu_F^k\tilde{\alpha}^k_{F,p}, \tag{7.37b}$$

at the kth iteration (at Step 5) of the algorithm, for all k. We have the following analogue of Lemma 4.4.7.

Lemma 7.7. (i) Suppose that for some point $\bar{x} \in R^N$ we have

$$\liminf_{k\to\infty} \max\{w^k, |\bar{x}-x^k|\}=0, \tag{7.38}$$

or equivalently

there exists an infinite set $K \subset \{1,2,\ldots\}$ such that

$$x^k \xrightarrow{K} \bar{x} \quad \text{and} \quad w^k \xrightarrow{K} 0. \tag{7.39}$$

Then $0 \in M(\bar{x})$ and $F(\bar{x}) \le 0$.

(ii) Relations (7.38) and (7.39) are equivalent to the following

$$\liminf_{k\to\infty} \max\{|p^k|, |\bar{x}-x^k|\} = 0.$$

Proof. Use the proof of Lemma 4.4.7, replacing the reference to Lemma 4.4.5 by the one to Lemma 7.6, and observe that any accumulation point of $\{x^k\} \subset S$ must be feasible, because $S=\{x \in R^N : F(x) \le 0\}$ is closed. □

Let $\hat{w}^k$ denote the optimal value of the k-th dual search direction finding subproblem (7.1), for all k. Then it is easy to verify that Lemma 4.6 holds for Algorithm 7.1. This result replaces Lemma 4.4.8. Also it is straightforward to check that Lemma 4.4.9 and Corollary 4.4.10 are true for Algorithm 7.1. Next, we may use (7.35) to establish Lemma 4.7 and Lemma 4.8, thus replacing Lemma 4.4.11 and Lemma 4.4.12.

One can prove Lemma 4.9 for Algorithm 7.1 as follows. If $x^k \xrightarrow{K} \bar{x}$ then $\{|p^k|\}_{k\in K}$ is bounded in view of Lemma 4.8, so $\{a^k\}_{k\in K}$ is bounded, because we have $a^k \le |p^k|/m_a$ at Step 5, for all k. Then one can use the representations (7.26) and (7.27) as in the proof of Lemma 7.6 to obtain the desired conclusion from relations of the form (4.33). This result substitutes Lemma 4.4.13.

It is easy to check that the proofs of Lemma 4.4.14 through Lemma 4.4.18 require no modifications. Thus we have obtained the following result.

Theorem 7.8. Algorithm 7.1 is globally convergent in the sense of Theorem 4.11, Theorem 4.12 and Corollary 4.13.

Let us pass to the method with subgradient selection. To save space, we give a shortened description.

Algorithm 7.9.

Step 0 (Initialization). Do Step 0 of Algorithm 7.1. Set $M_g \ge N+2$ equal to the fixed maximum number of subgradients of f and F that the algorithm may use for each search direction finding. Set $J^1=J_f^1 \cup J_F^1$.

Step 1 (Direction finding). Do Step 1 of Algorithm 5.1, setting $\gamma_f=\gamma_F=0$ in (5.2). Set $a^k=\max\{s_j^k : j \in \hat{J}_f^k \cup \hat{J}_F^k\}$.

Step 2 (Stopping criterion). Do Step 2 of Algorithm 7.1.

Step 3 (Resetting test). Do Step 3 of Algorithm 7.1.

Step 4 (Resetting). (i) Replace J^k by $\{j \in J^k : j \ge k-M_g+1\}$, and then J_f^k and J_F^k by $\{j \in J^k : y^j \in S\}$ and $\{j \in J^k : y^j \notin S\}$, respectively. Set $r_a^k=1$.

(ii) If $|J^k| > 1$ then delete the smallest number from J_f^k or J_F^k, set $J^k=J_f^k \cup J_F^k$ and go to Step 1.

(iii) Set $y^k=x^k$, $g_f^k=g_f(y^k)$, $f_k^k=f(y^k)$, $s_k^k=0$, $J^k=J_f^k=\{k\}$, $J_F^k=\emptyset$ and go to Step 1.

Step 5 (Line search). Do Step 5 of Algorithm 7.1, replacing v^k by $\hat{v}^k$ in (7.10).

Step 6. Do Step 6 of Algorithm 7.1.

Step 7 (Subgradient updating). Do Step 7 of Algorithm 7.1, ignoring (7.13) and (7.15)-(7.16). Set $J^{k+1}=J_f^{k+1}\cup J_F^{k+1}$.

Step 8. Increase k by 1 and go to Step 1.

The above method is a combination of Algorithm 4.5.1 and Algorithm 5.5.1. Note that the method's subgradient deletion rules are less complicated than those of Algorithm 7.1, since the method does not update the aggregate subgradients.

We may add that one can replace v^k by $\hat{v}^k$ in Line Search Procedure 7.3 for executing Step 5 of the method. Lemma 7.4 remains valid, since we have $\hat{v}^k<0$ at Step 5

We have the following convergence result.

Theorem 7.10. Algorithm 7.9 is globally convergent in the sense of Theorem 4.11, Theorem 4.12 and Corollary 4.13.

Proof. Replacing (7.20) and (7.21a,b) by

$$\hat{J}^k_{f,r}=\hat{J}^k_f \quad \text{and} \quad \hat{J}^k_{F,r}=\hat{J}^k_F \quad \text{for all } k,$$

$$\hat{J}^k_p=\hat{J}^k_r \quad \text{for all } k,$$

we obtain an analogue of Lemma 7.2. Then it is easy, albeit tedious, to obtain the desired conclusion by modifying the preceding results of this section in the spirit of Section 5. This task is left to the reader. □

Let us discuss modified resetting strategies for methods with subgradient deletion rules. In Section 4.6 one can find detailed motivation behind the use of such strategies in the unconstrained case. Most of those remarks apply also to the constrained case. Thus we want to decrease the frequency of resettings, since too frequent discarding of the aggregate subgradients leads to a loss of the accumulated past subgradient information, which can result in slow convergence.

The resetting strategy of Algorithm 7.1 and Algorithm 7.9 is similar to that of Algorithm 4.3.1. A reset occurs at the k-th iteration if $|p^k|\le m_a a^k$, i.e. when the length of the current search direction $|d^k|=|p^k|$ becomes much shorter than the value of the locality radius a^k, which estimates the radius of the ball around x^k from which the past subgradient information was aggregated to form d^k (see (7.32)). To reduce the number of resets, in Algorithm 4.6.1 we used aggregate distance measures $\tilde{s}^k_p$

and resetting tests of the form $|p^k| \le m_a \tilde{s}_p^k$, instead of $|p^k| \le m_a a^k$. An extension of this strategy to the constrained case is given in the following method.

Algorithm 7.11.

Step 0 (Initialization). Do Step 0 of Algorithm 7.1. Set $s_f^1 = s_F^1 = 0$.

Step 1 (Direction finding). Do Step 1 of Algorithm 7.1, setting

$$(p_f^k, \tilde{f}_p^k, \tilde{s}_f^k) = \sum_{j \in J_f^k} \tilde{\lambda}_j^k (g_f^j, f_j^k, s_j^k) + \tilde{\lambda}_p^k (p_f^{k-1}, f_p^k, s_f^k),$$
$$(p_F^k, \tilde{F}_p^k, \tilde{s}_F^k) = \sum_{j \in J_F^k} \tilde{\mu}_j^k (g_F^j, F_j^k, s_j^k) + \tilde{\mu}_p^k (p_F^{k-1}, F_p^k, s_F^k) \qquad (7.40)$$

instead of using (7.4). Set

$$\tilde{s}_p^k = \nu_f^k s_f^k + \nu_F^k \tilde{s}_F^k. \qquad (7.41)$$

Step 2 (Stopping criterion). If $\max\{|p^k|, m_a \tilde{s}_p^k\} \le \varepsilon_s$ then terminate. Otherwise, go to Step 3.

Step 3 (Resetting test). If $|p^k| \le m_a \tilde{s}_p^k$ then go to Step 4; otherwise, go to Step 5.

Step 4 (Resetting). Do Step 4 of Algorithm 7.1.

Step 5 (Line search). Do Step 5 of Algorithm 7.1.

Step 6. Do Step 6 of Algorithm 7.1.

Step 7 (Subgradient updating). Do Step 7 of Algorithm 7.1. Set

$$s_f^{k+1} = \tilde{s}_f^k + |x^{k+1} - x^k|,$$
$$s_F^{k+1} = \tilde{s}_F^k + |x^{k+1} - x^k|.$$

Step 8 (Distance resetting test). If $a^{k+1} \le \bar{a}$ then go to Step 10. Otherwise, set $r_a^{k+1} = r_f^{k+1} = r_F^{k+1} = 1$ and go to Step 9.

Step 9 (Distance resetting). Keep deleting from J_f^{k+1} and J_F^{k+1} indices with the smallest values until the reset value of a^{k+1} satisfies

$$a^{k+1} = \max\{s_j^{k+1} : j \in J_f^{k+1} \cup J_F^{k+1}\} \le \bar{a}/2.$$

Set $J^{k+1} = J_f^{k+1} \cup J_F^{k+1}$.

Step 10. Increase k by 1 and go to Step 1.

Observe that the above method uses the subgardient aggregation rules of Algorithm 3.1, cf. (3.5) and (7.40). Also the distance resetting strategy of Step 8 and Step 9, which ensures locally uniform boundedness of accumulated subgardients, is similar to the corresponding strategy of Algorithm 3.1. At the same time, the method uses the aggregate distance measure $\tilde{s}_p^k$ instead of the locality radius a^k in the stopping criterion and the resetting test of Algorithm 7.1. In effect, Algorithm 7.11 is related to Algorithm 7.1 and Algorithm 3.1 in the same way as was Algorithm 4.6.1 to Algorithm 4.3.1 and Algorithm 3.3.1. This observation facilitates establishing convergence of the method, since one may reason as in the proof of Theorem 4.6.2, where convergence of Algorithm 4.6.1 was deduced from the results concerning Algorithm 4.3.1 and Algorithm 3.3.1. We follow a similar path below.

To obtain a better insight into properties of the aggregate distance measures $\tilde{s}_p^k$, we present the following analogue of Lemma 4.1 and Lemma 7.2. Its simple proof is left to the reader.

Lemma 7.12. Lemma 7.2 holds for Algorithm 7.11 if we replace (7.23) by

$$\max\{s_j^k : j \in \hat{J}_p^k\} = a^k \le \bar{a} \tag{7.42}$$

and augment (7.26) and (7.27) by

$$(p_f^k, \tilde{f}_p^k, \tilde{s}_f^k) = \sum_{i=1}^{M} \hat{\lambda}_i^k (g(y_f^{k,i}), f^{k,i}, s_f^{k,i}) \tag{7.43}$$

and

$$(p_F^k, \tilde{F}_p^k, \tilde{s}_F^k) = \sum_{i=1}^{M} \hat{\mu}_i^k (g(y_F^{k,i}), F^{k,i}, s_F^{k,i}), \tag{7.44}$$

respectively.

We conclude from (7.41), (7.31) and Lemma 7.12 that $\tilde{s}_p^k$ is always a convex combination of the aggregate distance measures $\tilde{s}_f^k$ and $\tilde{s}_F^k$, which in turn indicate how far p_f^k and p_F^k are from $\hat{M}(x^k)$. Thus

$$(p^k, \tilde{s}_p^k) = \nu_f^k (p_f^k, \tilde{s}_f^k) + \nu_F^k (p_F^k, \tilde{s}_F^k),$$

$$\nu_f^k \ge 0, \ \nu_F^k \ge 0, \ \nu_f^k + \nu_F^k = 1,$$

hence the value of $\tilde{s}_p^k$ indicates how far p^k is from $\hat{M}(x^k)$. This justifies the stopping criterion of Step 2. In fact, by using Lemma 7.12, Lemma 4.2 and the proof of Lemma 4.3 one can show that if $\varepsilon_s = 0$ then the algorithm stops only at stationary points.

Supposing the method does not terminate, we have the following result.

Theorem 7.13. Algorithm 7.11 is globally convergent in the sense of Theorem 4.11, Theorem 4.12 and Corollary 4.13.

Proof. We give only an outline of the proof, which is similar to the proof of Theorem 6.2. First, we observe that we always have $|p^k| > m_a\tilde{s}_p^k$ at Step 5. Secondly, we deduce from (7.41), (7.31) and Lemma 7.12 that

$$\begin{aligned} &\tilde{s}_p^k = \tilde{s}_f^k \le a^k \quad \text{if} \quad \hat{J}_{F,p}^k = \emptyset, \\ &\tilde{s}_p^k = \tilde{s}_F^k \le a^k \quad \text{if} \quad \hat{J}_{f,p}^k = \emptyset, \\ &\tilde{s}_p^k \le \max\{\tilde{s}_f^k, \tilde{s}_F^k\} \le a^k \quad \text{if} \quad \hat{J}_{f,p}^k \cap \hat{J}_{F,p}^k \ne \emptyset, \\ &\hat{J}_{f,p}^k \cup \hat{J}_{F,p}^k \ne \emptyset, \end{aligned} \tag{7.45}$$

and that $\tilde{s}_p^k \xrightarrow{K} 0$ if and only if $\nu_f^k \tilde{s}_f^k \xrightarrow{K} 0$ and $\nu_F^k \tilde{s}_F^k \xrightarrow{K} 0$. Using the latter property and the representations of the aggregate subgradients of Lemma 7.12 as in the proof of Lemma 7.6, we obtain from Lemma 4.2 that $x^k \xrightarrow{K} \bar{x}$ and $\tilde{s}_p^k \xrightarrow{K} 0$ implies the existence of an infinite $\bar{K} \subset K$ such that $p^k \xrightarrow{\bar{K}} \bar{p} \in \hat{M}(\bar{x})$. This result replaces Lemma 7.6, and enables us to establish both Lemma 4.7 by using the fact that $\tilde{s}_p^k < |p^k|/m_a$ for all k, and an analogue of Lemma 4.4.16, in which the relations $b^k = \max\{|p^k|, m_a a^k\}$ and $0 \in \partial f(\bar{x})$ are replaced by $b^k = \max\{|p^k|, m_a\tilde{s}_p^k\}$ and $0 \in \hat{M}(\bar{x})$, From the analysis of Algorithm 7.1 we may derive analogues of Lemma 4.4.8 through Lemma 4.4.15. Then, since $\tilde{s}_p^k \le a^k$ by (7.45), one may establish Lemma 4.4.18 as in the proof of Theorem 6.2. □

We may add that one can modify Algorithm 7.9 in the spirit of Algorithm 7.11 without impairing the preceding global convergence results. Namely, in Algorithm 7.9 we may use the stopping criterion

$$\max\{|p^k|, m_a\tilde{s}_p^k\} \le \varepsilon_s$$

and the resetting test

$$|p^k| \le m_a\tilde{s}_p^k,$$

with $\tilde{s}_p^k$ generated by (7.40) with $\tilde{\lambda}_p^k = \tilde{\mu}_p^k = 0$, and replace Step 8 by Step 8, Step 9 and Step 10 of Algorithm 7.11. To establish Theorem 7.10 for the resulting method with subgradient selection, one may use the proof of Theorem 7.13.

The preceding algorithms of this section can be modified by using the resetting strategies of Wolfe and Mifflin described in Section 4.6.

We shall not dwell on this subject, since, as explained in Section 4.6, the resulting algorithms are convergent only in the sense that they have at least one stationary accumulation point under additional boundedness assumptions. More precisely, it is straightforward to establish analogues of Theorem 4.6.3 and Theorem 4.6.4 in the constrained case, formulating them with

$$S_f = \{x \in S : f(x) \leq f(x^1)\}.$$

We shall now present a convergent modification of Algorithm 7.11 with subgradient deletion rules that do not require a repetition of search direction finding whenever a subgradient is dropped. As discussed in Section 4.6, such rules decrease the work involved in additional quadratic programming calculations.

Algorithm 7.14.

Step 0 (Initialization). Do Step 0 of Algorithm 7.1. Choose a positive δ^1. Set $s_f^1 = s_F^1 = 0$.

Step 1 (Direction finding). Do Step 1 of Algorithm 7.1. Set

$$\tilde{s}_f^k = \sum_{j \in J_f^k} \tilde{\lambda}_j^k s_j^k + \tilde{\lambda}_p^k s_f^k \quad \text{and} \quad \tilde{s}_F^k = \sum_{j \in J_F^k} \tilde{\mu}_j^k s_j^k + \tilde{\mu}_p^k s_F^k,$$

$$\tilde{s}_p^k = \nu_f^k \tilde{s}_f^k + \nu_F^k \tilde{s}_F^k.$$

Step 2 (Stopping criterion). If $\max\{|p^k|, m_a \tilde{s}_p^k\} \leq \varepsilon_s$ then terminate. Otherwise, go to Step 3.

Step 3 (Resetting test). Set $\tilde{\delta}^k = \max\{|p^k|, m_a \tilde{s}_p^k\}$. If $\tilde{\delta}^k < \bar{\theta}\delta^k$ set $\delta^{k+1} = \tilde{\delta}^k$; otherwise set $\delta^{k+1} = \delta^k$. If $|p^k| \leq m_a \tilde{s}_p^k$ then go to Step 4; otherwise, go to Step 5.

Step 4 (Resetting). Replace J_f^k by $\{j \in J_f^k : s_j^k \leq \bar{\theta}\delta^{k+1}/m_a\}$ and J_F^k by $\{j \in J_F^k : s_j^k \leq \bar{\theta}\delta^{k+1}/m_a\}$. If $|J_f^k \cup J_F^k| < 1$ set $y^k = x^k$, $g_f^k = g_f(x^k)$, $f_k^k = f(x^k)$, $s_k^k = 0$, $J_f^k = \{k\}$ and $J_F^k = \emptyset$. Set $r_a^k = r_f^k = r_F^k = 1$ and go to Step 1.

Step 5 (Line search). Do Step 5 of Algorithm 7.1, replacing (7.10e-g) by

$$|y^{k+1} - x^{k+1}| \leq \bar{\theta}\delta^{k+1}/m_a. \tag{7.46}$$

Step 6 (Subgradient updating). Do Step 7 of Algorithm 7.1. Set

$$s_f^{k+1} = \tilde{s}_f^k + |x^{k+1} - x^k| \quad \text{and} \quad s_F^{k+1} = \tilde{s}_F^k + |x^{k+1} - x^k|.$$

Step 7 (Distance resetting test). If $a^{k+1} \le \bar{a}$ and $\delta^{k+1} = \delta^k$ go to Step 9; otherwise go to Step 8.

Step 8 (Distance resetting). Replace J_f^{k+1} by $\{j \in J_f^{k+1} : s_j^{k+1} \le \bar{\theta}\delta^{k+1}/m_a\}$ and J_F^k by $\{j \in J_F^{k+1} : s_j^{k+1} \le \bar{\theta}\delta^{k+1}/m_a\}$. If $a^{k+1} > \bar{\theta}\delta^{k+1}/m_a$ then set $r_a^{k+1} = r_f^{k+1} = r_F^{k+1} = 1$ and $a^{k+1} = \max\{s_j^{k+1} : j \in J_f^{k+1} \cup J_F^{k+1}\}$.

Step 9. Increase k by 1 and go to Step 1.

We note that Line Search Procedure 3.2 (with $\bar{a} = \bar{\theta}\delta^{k+1}/m_a$) can be used for executing Step 5 of the above method.

Algorithm 7.14 is globally convergent in the sense of Theorem 4.11, Theorem 4.12 and Corollary 4.13. To verify this claim, the reader may use the preceding results of this section and the arguments in the proof of Theorem 4.6.5.

We may add that the algorithms described so far may use additional subgradients of f calculated at points $\{x^k\}$; see Remark 4.6.7. This may enhance faster convergence.

8. Methods That Neglect Linearization Errors

In this section we shall consider simplified versions of the methods discussed in Section 7 that are obtained ny neglecting linearization errors at each search direction finding. The resulting dual search direction finding subproblems have special structure, which enables one to use the efficent and numerically stable Wolfe (1976) quadratic programming algorithm. In effect, such methods may require less work per iteration, but may converge more slowly than do the previously discussed algorithms (Lemarechal, 1982). We may add that similar methods were proposed in (Mifflin, 1977b; Polak, Mayne and Wardi, 1983).

Let us, therefore, consider the following modification of Algorithm 7.1. We set

$$\alpha_{f,j}^k = 0, \ j \in J_f^k, \ \alpha_p^k = 0, \ \alpha_{F,j}^k = 0, \ j \in J_F^k, \ \alpha_{F,p}^k = 0$$

in the k-th search direction finding subproblem (7.1). Then the k-th line search is performed with

$$v^k = -|p^k|^2,$$

$$\alpha(x,y) = 0 \quad \text{for all} \quad x \text{ and } y$$

instead of using the previous definitions (7.8) and (7.12). Then $v^k < 0$ in Step 5, so that Line Search Procedure 7.3 can be used as before, with Lemma 7.4 remaining true.

It is easy to verify that the above modification does not impair Theorem 4.11 and Corollary 4.13. However, Theorem 4.12 may not hold for the reasons discussed in Section 4.7.

To obtain stronger convergence properties in the convex case, one may use in the above-described version of Algorithm 7.1 the additional resetting test

$$|p^k|^2 \le m_\alpha \tilde{\alpha}_p^k,$$

i.e. the algorithm should go to Step 4 if either $|p^k| \le m_a a^k$ or $|p^k| \le$ $\le m_\alpha \tilde{\alpha}_p^k$, where

$$\tilde{\alpha}_p^k = \nu_f^k \tilde{\alpha}_{f,p}^k + \nu_F^k \tilde{\alpha}_{F,p}^k$$

and $m_\alpha > 0$ is a scaling parameter. We refer the reader to Section 4.7 for the motivation of such additional resetting tests. Moreover, one may use the proof of Theorem 4.1.7 and the results of Section 7 to show that the resulting method is globally convergent in the sense of Theorems 4.11 and 4.12 and Corollary 4.13. All these results are easily extended to the corresponding version with subgradient selection.

9. Phase I - Phase II Methods

The algorithms described in the preceding sections require a feasible starting point. In this section we shall discuss phase I - phase II methods which may use infeasible starting points. These methods are extensions of the algorithms of Section 5.7 to the nonconvex case, hence we refer the reader to Section 5.7 for their motivation and derivation.

Throughout this section we suppose that one can calculate $f(x)$ and $g_f(x) \in \partial f(x)$ at each $x \in R^N$. Also for simplicity we assume that F and g_F can be evaluated everywhere; see Remark 2.3 and Section 7 for a discussion of how to relax this assumption.

We recall that the phase I - phase II algorithms of Section 5.7 were obtained by modifying the definitions of linearization errors and the line seach rules of the feasible point methods of Section 5.2-5.6. Introducing similar modifications in Algorithm 3.1, we obtain the following phase I - phase II method.

Algorithm 9.1.

Step 0 (Initialization). Select a starting point $x^1 \in R^N$ and initialize the method according to the rules of Step 0 of Algorithm 3.1.

Step 1 (Direction finding). Find multipliers λ_j^k, $j \in J_f^k$, λ_p^k, μ_j^k, $j \in J_F^k$, and λ_p^k that

$$\underset{\lambda,\mu}{\text{minimize}}\ \tfrac{1}{2}\Big|\sum_{j\in J_f^k}\lambda_j g_f^j + \lambda_p p_f^{k-1} + \sum_{j\in J_F^k}\mu_j g_F^j + \mu_p p_F^{k-1}\Big|^2 +$$

$$+\sum_{j\in J_f^k}\lambda_j\big[\alpha_{f,j}^k + F(x^k)_+\big] + \lambda_p\big[\alpha_{f,p}^k + F(x^k)_+\big] + \sum_{j\in J_F^k}\mu_j\alpha_{F,j}^k + \mu_p\alpha_{F,p}^k$$

$$\text{subject to}\quad \lambda_j \ge 0,\ j\in J_f^k,\ \lambda_p \ge 0,\ \mu_j \ge 0,\ j\in J_F^k,\ \mu_p \ge 0, \tag{9.1}$$

$$\sum_{j\in J_f^k}\lambda_j + \lambda_p + \sum_{j\in J_F^k}\mu_j + \mu_p = 1,$$

$$\lambda_p = \mu_p = 0 \quad \text{if} \quad r_a^k = 1,$$

where

$$\alpha_{f,j}^k=\max\{|f(x^k)-f_j^k|,\gamma_f(s_j^k)^2\},\ \alpha_{F,j}^k=\max\{|F(x^k)_+-F_j^k|,\ \gamma_F(s_j^k)^2\},$$

$$\alpha_{f,p}^k=\max\{|f(x^k)-f_p^k|,\gamma_f(s_f^k)^2\},\ \alpha_{F,p}^k=\max\{|F(x^k)_+-F_p^k|,\ \gamma_F(s_F^k)^2\}. \tag{9.2}$$

Compute ν_f^k, ν_F^k, $(\tilde\lambda^k,\tilde\mu^k)$, a^k, $(p_f^k,\tilde f_p^k,\tilde s_f^k)$, $(p_F^k,\tilde F_p^k,\tilde s_F^k)$, p^k and d^k as in Step 1 of Algorithm 3.1. Set

$$\tilde\alpha_{f,p}^k=\max\{|f(x^k)-\tilde f_p^k|,\gamma_f(\tilde s_f^k)^2\},\ \tilde\alpha_{F,p}^k=\max\{|F(x^k)_+-\tilde F_p^k|,\gamma_F(\tilde s_F^k)^2\}, \tag{9.3a}$$

$$\tilde\alpha_p^k = \nu_f^k[\tilde\alpha_{f,p}^k+F(x^k)_+] + \nu_F^k\tilde\alpha_{F,p}^k, \tag{9.3b}$$

$$v^k = -\{|p^k|^2 + \tilde\alpha_p^k\}. \tag{9.4}$$

Step 2 (Stopping criterion). Set

$$w^k = \tfrac{1}{2}|p^k|^2 + \tilde\alpha_p^k. \tag{9.5}$$

If $w^k \le \varepsilon_s$ terminate; otherwise, continue.

Step 3 (Line search). If $F(x^k) \le 0$, find two stepsizes t_L^k and t_R^k satisfying the requirements of Step 3 of Algorithm 3.1. Othewise, i.e. if $F(x^k) > 0$, find two stepsizes t_L^k and t_R^k such that $0 \le t_L^k \le t_R^k$ and such that $x^{k+1}=x^k+t_L^k d^k$ and $y^{k+1}=x^k+t_R^k d^k$ satisfy

$$F(x^{k+1}) \leq F(x^k) + m_L t_L^k v^k, \tag{9.6a}$$

$$t_R^k = t_L^k \quad \text{if} \quad t_L^k \geq \bar{t}, \tag{9.6b}$$

$$-\alpha_F(x^{k+1}, y^{k+1}) + \langle g_F(y^{k+1}), d^k \rangle \geq m_R v^k \quad \text{if} \quad t_L^k < \bar{t}, \tag{9.6c}$$

$$|y^{k+1} - x^{k+1}| \leq \bar{a}/2, \tag{9.6d}$$

where

$$\alpha_F(x,y) = \max\{|F(x)_+ - \bar{F}(x;y)|, \ \gamma_F |x-y|^2\}. \tag{9.7}$$

<u>Step 4 (Subgradient updating)</u>. Do Step 4 of Algorithm 3.1.

<u>Step 5 (Distance resetting test)</u>. Do Step 5 of Algorithm 3.1.

<u>Step 6 (Distance resetting)</u>. Do Step 6 of Algorithm 3.1.

<u>Step 7</u>. Increase k by 1 and go to Step 1.

A few comments on the algorithm are in order.

The values of $\alpha_f(x,y)+F(x)_+$ and $\alpha_F(x,y)$ indicate how much the subgradients $g_f(y) \in \partial f(y)$ and $g_F(y) \in F(y)$ differ from being elements of $\hat{M}(x)$ (see (2.2)), respectively. In effect, the value of the stationarity measure w^k indicates how much x^k differs from being a stationary point for f on S. Note that the value of w^k may be small even if x^k is infeasible, since ν_f^k may be close to zero. In this case x^k is close to being a stationary point for F, since $\nu_F^k = 1 - \nu_f^k \approx 1$, $p^k = \nu_f^k p_f^k + \nu_F^k p_F^k \quad p_F^k$, and $w^k \approx \frac{1}{2}|p_F^k|^2 + \tilde{\alpha}_{F,p}^k$, i.e. 0 is close to $\partial F(x^k)$ and $x^k \notin S$. To exclude such cases, phase I - phase II methods typically require the Cottle constraint qualification, i.e. $0 \notin \partial F(x)$ for all $x \notin S$.

It is easy to see that at phase II, i.e. when $x^k \in S$, the method reduces to Algorithm 3.1 and maintains feasibility of successive iterates. Thus only the case of an infinitely long phase I ($x^k \notin S$ for all k) is of interest here.

When $F(x^k) > 0$, one may apply Line Search Procedure 3.3.1 to F in order to find stepsizes t_L^k and t_R^k satisfying (9.6). By Lemma 3.3.3, this procedure will terminate in a finite number of iterations if F has the semismoothness property (3.3.23). In fact, this procedure may be stopped whenever it finds any feasible point, since then phase II will begin at the next iteration of the method.

We shall now establish convergence of the method, assuming that $\varepsilon_S = 0$.

Lemma 9.2. If Algorithm 9.1 terminates at the k-th iteration then the point $\bar{x}=x^k$ satisfies $0\in\hat{M}(\bar{x})$. If additionally $F(x^k)\le 0$ or

$$0\notin \partial F(x) \quad \text{for all} \quad x\notin S, \tag{9.8}$$

then $\bar{x}\in S$ and $\bar{x}$ is stationary for f on S.

Proof. Use (9.3) in the proof of Lemma 4.3 for replacing (4.9) by

$$p_f^k\in\partial f(\bar{x}) \quad \text{and} \quad F(\bar{x})_+=0 \quad \text{if} \quad \nu_f^k\neq 0,$$

$$p_F^k\in\partial F(\bar{x}) \quad \text{and} \quad F(\bar{x})\ge 0 \quad \text{if} \quad \nu_F^k\neq 0$$

to deduce that $0\in\hat{M}(\bar{x})$. By (2.2), (9.8) implies $\bar{x}\in S$ if $0\in\hat{M}(\bar{x})$. □

In view of the above result, we shall assume from now on that the method calculates an infinite sequence $\{x^k\}$. Of course, phase II of the method is covered by the results of Section 4. Therefore we need only consider the case when the method stays at phase I.

Theorem 9.3. Suppose that Algorithm 9.1 generates an infinite sequence $\{x^k\}$ such that $F(x^k)>0$ for all k. Then every accumulation point $\bar{x}$ of $\{x^k\}$ satisfies $0\in\hat{M}(\bar{x})$. Moreover, $\bar{x}\in S$ and $\bar{x}$ is stationary for f on S if (9.8) holds.

Proof. To save space, we shall only indicate how to modify the results of Section 4 for Algorithm 9.1.

(i) Proceeding as in the proof of Lemma 9.2, use (9.3) in the proof of Lemma 4.5 to obtain the desired conclusion if (4.11) holds.

(ii) In view of (9.3)-(9.5), one may express $\hat{\alpha}_p^k$ in the formulation and the proof of Lemma 4.6 as follows

$$\hat{\alpha}_p^k = \nu_f^k[\hat{\alpha}_{f,p}^k + F(x^k)_+] + \nu_F^k\alpha_{F,p}^k. \tag{9.9}$$

(iii) By assumption, $F(x^k)>0$ for all k, hence, by the algorithm's rules, we have (4.24) with $g^k=g_F(y^k)$ and $\alpha^k=\alpha_F(x^k,y^k)$ if $t_L^{k-1}<\bar{t}$ (cf.(9.6c)). Therefore one may use (9.3) and (9.9) to establish Lemma 4.7 for Algorithm 9.1 with

$$\alpha_p^k=\nu_f^{k-1}[\alpha_{f,p}^k+F(x^k)_+]+\nu_F^{k-1}\alpha_{F,p}^k. \tag{9.10}$$

(iv) It is easy to establish Lemma 4.8 for Algorithm 9.1 by defining $\alpha(x,y)=\alpha_f(x,y)+F(x)_+$ if $y\in S$, $\alpha(x,y)=\alpha_F(x,y)$ if $y\notin S$, and setting $\alpha^k=\alpha(x^k,y^k)$ for all k.

(v) In the proof of Lemma 4.9 for Algorithm 9.1, replace (4.33b) by a relation similar to (4.33a) (with f substituted by F), and use (9.3b) and (9.10) together with the assumption that $F(x^k)_+=F(x^k)>0$ for all k to show that $|\alpha_p^{k+1}-\tilde{\alpha}_p^k|\xrightarrow{K} 0$.

(vi) Combining the above results as in Section 4, we see that Lemma 4.10 holds for Algorithm 9.1, so we have (4.10) and (4.11), and the desired conclusion follows from part (i) above. □

Reasoning as in Section 4, one may deduce from the above proof the following result.

<u>Corollary 9.4</u>. Suppose that $F(x^1)>0$, the set $\{x\in R^N: F(x)\leq F(x^1)\}$ is bounded and $\varepsilon_s>0$. Then Algorithm 9.1 will either terminate at phase I or switch to phase II at some iteration.

We conclude from the above results that if the method terminates at a significantly infeasible point x^k, then F is likely to have a stationary point $\tilde{x}$ with $0\in\partial F(\tilde{x})$ and $F(\tilde{x})>0$. This will happen if F has a positive minimum, i.e. no feasible point exists.

We end this section by remarking that if we neglected linearization errors, i.e. set $\alpha_{f,j}^k=\alpha_{f,p}^k=\alpha_{F,j}^k=\alpha_{F,p}^k=\alpha_p^k=0$ and $\alpha_f=\alpha_F=0$, then the method would become similar to a conceptual algorithm proposed in (Polak, Mayne and Wardi, 1983).

CHAPTER 7

Bundle Methods

1. Introduction

The methods for nonsmooth minimization discussed in the preceding chapters belong to the class of algorithms proposed by Lemarechal (1978a) and extended by Mifflin (1982). In Chapters 4 and 6 we showed that by neglecting linearization errors one obtains simplified versions of these methods which are in the class of algorithms introduced by Lemarechal (1975) and Wolfe (1975), and extended by Mifflin (1977b) and Polak, Mayne and Wardi (1983). This chapter is devoted to bundle methods which form the third remaining class of algorithms. These methods were proposed by Lemarechal (1976,1978b) in the unconstrained convex case, and extended by Lemarechal, Strodiot and Bihain (1981) to nonconvex problems with linear constraints.

A computational advantage of bundle methods over the algorithms discussed so far is that their search direction finding subproblems may be solved by an efficient quadratic programming algorithm of Mifflin (1978) which exploits the structure of these subproblems, while up till now no special-purpose quadratic programming algorithm has been developed for subproblems of Chapter 2. Moreover, preliminary computational experimence (Lemarechal, 1982; Strodiot, Nguyen and Heukemes, 1983) indicates that bundle methods are promising. However, so far no global convergence of such methods seems to have been established, and the analysis of (Lemarechal, Strodiot and Bihain, 1981; Strodiot, Nguyen and Heukemes, 1983) only shows that bundle methods can find an approximate solution in a finite number of iterations. Also extensions of bundle methods to nonlineary constrained problems have not been considered in the literature.

In this chapter we shall present new versions of bundle methods for convex and nonconvex problems, both unconstrained and constrained ones. Owing to the use of subgradient selection and aggregation techniques, the methods have flexible storage requirements and work per iteration which can be controlled by the user. Our rules for regulating the approximation tolerances of the methods, which are different from those in (Lemarechal, Strodiot and Bihain, 1981), enable us to establish global convergence of the methods under no additional assumptions on the problem functions. We also give line search procedures that are finite procedures for functions having the semismoothness properties

(3.3.23) and (6.3.18), which are weaker than those required in (Lemarechal, Strodiot and Bihain, 1981); see (Lemarechal, 1981). In effect, we establish theoretical results on these versions of bundle methods that are comparable to the ones obtained for other algorithms in the preceding chapters.

We start, in Section 2, by deriving bundle methods for convex unconstrained minimization. A method with subgradient aggregation is described in detail in Section 3, and its convergence is established in Section 4. Section 5 discusses a method with subgradient selection and its convergence. Useful modifications of the methods are described in Section 6. Then we extend the methods to the nonconvex unconstrained case in Section 7, to convex constrained problems in Section 8, and to the nonconvex constrained case in Section 9.

2. Derivation of the Methods

In this section we derive a bundle method for the unconstrained problem of minimizing a convex function $f : R^N \to R$ that is not necessarily differentiable. We suppose that we have a finite process for finding a subgradient $g_f(x) \in \partial f(x)$ of f at each given $x \in R^N$.

The algorithm to be described will generate sequences of points $\{x^k\} \subset R^N$, search directions $\{d^k\} \subset R^N$ and stepsizes $\{t_L^k\} \subset R_+$, related by $x^{k+1}=x^k+t_L^k d^k$ for $k=1,2,\ldots$, where x^1 is a given starting point. The sequence $\{x^k\}$ is intended to converge to the required solution. The method will also calculate trial points $y^{k+1}=x^k+t_R^k d^k$ for $k=1,2,\ldots$, and subgradients $g^k=g_f(y^k)$ for all k, where $y^1=x^1$ and the auxiliary stepsizes $t_R^k \geq t_L^k$ satisfy $t_R^k=t_L^k$ if $t_L^k > 0$, for all k.

Given a point $y \in R^N$, let

$$\overline{f}(x;y) = f(y) + \langle g_f(y), x-y \rangle \quad \text{for all } x$$

denote the corresponding linearization of f, and let

$$\alpha(x,y) = f(x) - \overline{f}(x;y)$$

denote the linearization error at any $x \in R^N$. At the k-th iteration, we shall have a nonempty set $J^k \subset \{1,\ldots,k\}$ and the linearizations $f_j(\cdot)=\overline{f}(\cdot;y^j)$, $j \in J^k$, given by the (N+1)-vectors (g^j, f_j^k) in the form

$$f_j(x) = f_j^k + \langle g^j, x-x^k \rangle \quad \text{for all } x,$$

where $f_j^k = \bar{f}(x^k; y^j)$ for $j \in J^k$. Let $\alpha_j^k = \alpha_j^k(x^k, y^j)$ for all $j \in J^k$. By convexity, $g^j \in \partial_{\alpha_j^k} f(x^k)$, i.e.

$$f(x) \geq f(x^k) + \langle g^j, x - x^k \rangle - \alpha_j^k \quad \text{for all } x,$$

and hence for any $e \geq 0$ the convex polyhedron

$$G^k(e) = \{ g \in R^N : g = \sum_{j \in J^k} \lambda_j g^j, \ \sum_{j \in J^k} \lambda_j \alpha_j^k \leq e, \quad \lambda_j \geq 0, \ j \in J^k, \ \sum_{j \in J^k} \lambda_j = 1 \} \tag{2.1}$$

is an inner approximation to the e-subdifferential of f at x^k

$$G^k(e) \subset \partial_e f(x^k) \quad \text{for all } e \geq 0,$$

that is, if $G^k(e)$ is nonempty, then

$$f(x) \geq f(x^k) + \max\{ \langle g, x - x^k \rangle : g \in G^k(e) \} - e \quad \text{for all } x.$$

Suppose that for some $e^k > 0$ $G^k(e^k)$ is nonempty and we want to find a direction $d \in R^N$ such that $f(x^k + d) < f(x^k) - e^k$. Letting $x = x^k + d$ and $e = e^k$, we see that d must satisfy

$$\langle g, d \rangle < 0 \quad \text{for all } g \text{ in } G^k(e),$$

i.e. we must find a hyperplane separating $G^k(e)$ from the origin. One way of finding such a hyperplane is to compute the element $p^k = \operatorname{Nr} G^k(e)$ of $G^k(e)$ that is nearest to the origin, since (see Lemma 1.2.12)

$$\langle g, p^k \rangle \geq |p^k|^2 \quad \text{for all } g \in G^k(e),$$

and hence if $d^k = -p^k$ is nonzero then

$$\langle g, d^k \rangle \leq -|p^k|^2 < 0 \quad \text{for all } g \in G^k(e). \tag{2.2}$$

(We may add that, since $\langle g, p^k / |p^k| \rangle$ is the length of the projection of g on the direction of p^k and $|p^k|$ is the distance of the hyperplane

$$H = \{ z \in R^N : \langle z, p^k \rangle = |p^k|^2 \}$$

from the origin, among the hyperplanes separating $G^k(e)$ and the null vector H is the furthest one from the origin.) Of course, there is no separation if $p^k = 0$, but then $0 = p^k \in G^k(e) \subset \partial_e f(x^k)$ and so $f(x) \geq$ $\geq f(x^k) - e$ for all x. In this case x^k minimizes f up to the accuracy of e^k, so the method may stop if the value of e^k is small enough. Otherwise one may decrease the value of e, compute new $p^k = NrG^k(e)$ etc. This process will either drive e to zero, indicating that x^k is optimal, or find a direction $d^k = -p^k$ satisfying (2.2).

We shall now give another motivation for the above construction. In Section 1.2 (see Lemma 1.2.13) we considered search direction finding subproblems of the form

$$\text{minimize } \hat{f}^k(x^k+d) + \tfrac{1}{2}|d|^2 \quad \text{over all } d \in R^N$$

with the approximation $\hat{f}^k$ to f around x^k given by

$$\hat{f}^k(x) = \max\{f(x^k) + \langle g, x-x^k \rangle : g \in \partial f(x^k)\} \quad \text{for all } x.$$

Since the use of $\hat{f}^k$ would require the knowledge of the full subdifferential $\partial f(x^k)$, in Chapter 2 we replaced $\hat{f}^k$ by the polyhedral approximation

$$\hat{f}_s^k(x) = \max\{f_j(x) : j \in J^k\} =$$
$$= \max\{f(x^k) + \langle g^j, x-x^k \rangle - \alpha_j^k : j \in J^k\}.$$

By neglecting the linearization errors α_j^k in the definition of $\hat{f}_s^k$, we obtain the simplified approximation

$$\hat{f}_{L,W}^k(x) = \max\{f(x^k) + \langle g^j, x-x^k \rangle : j \in J^k\} \quad \text{for all } x$$

used in the methods of Lemarechal (1975) and Wolfe (1975); see Section 4.7. Let us now consider the following approximation to f at x^k

$$\hat{f}_{B,s}^k(x) = \max\{f(x^k) + \langle g, x-x^k \rangle : g \in G^k(e^k)\} \text{ for all } x. \tag{2.3}$$

Observe that $\hat{f}_{B,s}^k$ reduces to $\hat{f}_{LW}^k$ whenever e^k is sufficiently large, i.e. $e^k \geq \max\{\alpha_j^k : j \in J^k\}$. On the other hand, if e^k is small enough then we may hope that $G^k(e^k)$, being a subset of $\partial_{e^k} f(x^k)$, is a good approximation of $\partial f(x^k)$. In this case $\hat{f}_{B,s}^k$ is close to the "conceptual" approximation $\hat{f}^k$. It is natural, therefore, to consider the following search direction finding subproblem

$$\text{minimize } \hat{f}_{B,s}^k(x^k+d) + \tfrac{1}{2}|d|^2 \quad \text{over all } d \in R^N. \tag{2.4}$$

Lemma 2.1. (i) Subproblem (2.4) has a unique solution d^k. (Recall that $G^k(e^k)$ is nonempty by assumption.)

(ii) Let λ_j^k, $j \in J^k$, denote a solution to the problem

$$\text{minimize } \frac{1}{2}\left|\sum_{j\in J^k}\lambda_j g^j\right|^2 ,$$

$$\text{subject to } \lambda_j \geq 0,\ j\in J^k,\quad \sum_{j\in J^k}\lambda_j = 1, \tag{2.5}$$

$$\sum_{j\in J^k}\lambda_j \alpha_j^k \leq e^k,$$

and let

$$p^k = \sum_{j\in J^k}\lambda_j^k g^j. \tag{2.6}$$

Then $p^k = \mathrm{Nr}\, G^k(e^k)$ and $d^k = -p^k$.

(iii) There exists a solution λ_j^k, $j\in J^k$, of subproblem (2.5) such that the set

$$\hat{J}^k = \{j\in J^k : \lambda_j^k \neq 0\} \tag{2.7a}$$

satisfies

$$|\hat{J}^k| \leq N+1. \tag{2.7b}$$

Such a solution can be obtained by solving the following linear programming problem by the simplex method

$$\text{minimize } \sum_{j\in J^k}\lambda_j \alpha_j^k ,$$

$$\begin{aligned} \text{subject to } & \sum_{j\in J^k}\lambda_j = 1, \\ & \sum_{j\in J^k}\lambda_j g^j = p^k, \\ & \lambda_j \geq 0,\ j\in J^k. \end{aligned} \tag{2.8}$$

Moreover, d^k solves the following reduced subproblem

$$\text{minimize } \hat{f}_{B,r}^k(x^k+d) + \tfrac{1}{2}|d|^2, \tag{2.9}$$

where

$$\hat{f}_{B,r}^k(x) = \max\{f(x^k) + \langle g, x-x^k\rangle : g\in \hat{G}^k\}$$

$$\hat{G}^k = \{g\in R^N : g = \sum_{j\in \hat{J}^k}\lambda_j g^j,\ \sum_{j\in \hat{J}^k}\lambda_j\alpha_j^k \leq e^k,$$

$$\lambda_j \geq 0,\ j\in \hat{J}^k,\ \sum_{j\in \hat{J}^k}\lambda_j = 1\}.$$

(iv) There exists a Lagrange multiplier $s^k \geq 0$ for the last constraint of (2.5) such that (2.5) is equivalent to the problem

$$\underset{\lambda}{\text{minimize}}\ \frac{1}{2}\Big|\sum_{j\in J^k}\lambda_j g^j\Big|^2 + s^k \sum_{j\in J^k}\lambda_j \alpha_j^k, \qquad (2.10)$$

$$\text{subject to} \quad \lambda_j \geq 0,\ j\in J^k, \quad \sum_{j\in J^k}\lambda_j = 1.$$

Proof. (i) Reasoning as in the proof of Lemma 2.2.1, we deduce that the strongly convex function $\phi^k(d)=\hat{f}^k_{B,s}(x^k+d)+\frac{1}{2}|d|^2$ has a unique minimizer d^k.

(ii) We have $p^k=\mathrm{Nr}G^k(e^k)$ by the definition of $\mathrm{Nr}\,G^k(e^k)$. Let $G=G^k(e^k)$. Making use of Caratheodory's theorem (Lemma 1.2.1), we deduce that there exist $N+1$ not necessarily different elements $\tilde{g}^i$ of G and numbers $\hat{\lambda}_i$, $i=1,\dots,N+1$, such that

$$p^k = \sum_{i=1}^{N+1}\hat{\lambda}_i\tilde{g}^i, \qquad (2.11a)$$

$$\hat{\lambda}_i \geq 0,\ i=1,\dots,N+1, \quad \sum_{i=1}^{N+1}\hat{\lambda}_i = 1. \qquad (2.11b)$$

Of course, $\hat{\lambda}_i$ solve the problem

$$\underset{\lambda}{\text{minimize}}\ \frac{1}{2}\Big|\sum_{i=1}^{N+1}\lambda_i\tilde{g}^i\Big|^2,$$

$$\text{subject to} \quad \lambda_i \geq 0,\ i=1,\dots,N+1, \quad \sum_{i=1}^{N+1}\lambda_i = 1,$$

hence Lemma 2.2.1 yields

$$[\langle \tilde{g}^i,\hat{d}\rangle - \hat{v}]\hat{\lambda}_i = 0 \quad \text{for} \quad i=1,\dots,N+1, \qquad (2.11c)$$

while the basic property of $\mathrm{Nr}\,G$ implies (see Lemma 1.2.12)

$$\langle \tilde{g},\hat{d}\rangle \leq \hat{v} \quad \text{for all} \quad \tilde{g}\in G, \qquad (2.11d)$$

where $\hat{d}=-p^k$ and $\hat{v}=-|p^k|^2$. From (2.11) and Lemma 1.2.5 we obtain

$$p^k \in \partial\hat{f}^k_{B,s}(x^k+\hat{d}),$$

hence $0\in\partial\hat{f}^k_{B,s}(x^k+\hat{d})+\hat{d}=\partial\phi^k(\hat{d})$ by Corollary 1.2.6. Thus $\hat{d}$ minimizes ϕ^k, so $-p^k=\hat{d}=d^k$ from part (i) above.

(iii) The simplex method will find an optimal basic solution of (2.8) with no more than $N+1$ positive components (Dantzig , 1963) such that it solves (2.5), since $\frac{1}{2}|p^k|^2$ is the optimal value of (2.5). Hence $-p^k=-Nr\,\hat{G}^k$ solves (2.9), by part (ii) above.

(iv) The existence of s^k follows from duality theory; see (Lemarechal, 1978b). □

Remark 2.2. The Mifflin (1978) quadratic programming algorithm will automatically find multipliers λ_j^k satisfying (2.7), and the multiplier s^k.

We conclude from the above lemma that subproblem (2.5) has many properties similar to those of the subproblems studied in Chapter 2, which were of the form (2.10) but with $s^k=1$. In particular, (2.9) is its reduced version. Therefore, according to the generalized cutting plane idea of Section 2.2, we may construct the (k+1)-st approximation to f by choosing J^{k+1} such that

$$J^{k+1} \supset \hat{J}^k \cup \{k+1\}$$

where $\hat{J}^k$ satisfies (2.7). This will define the method with subgradient selection, which uses at most N+3 past subgradients for search direction finding at any iteration.

We may now consider the method with subgradient aggregation. Suppose that at the beginning of the k-th iteration we have the (k-1)-st aggregate subgradient

$$(p^{k-1},f_p^k) \in \operatorname{conv}\{(g^j,f_j^k) : j=1,\dots,k-1\}.$$

Define the corresponding aggregate linearization

$$\tilde{f}^{k-1}(x) = f_p^k + \langle p^{k-1},x-x^k \rangle \quad \text{for all } x$$

and the linearization error at x^k

$$\alpha_p^k = f(x^k)-f_p^k.$$

Since for each x

$$f(x) \geq f(x^k)+\langle g^j,x-x^k \rangle -\alpha_j^k =$$

$$= f(x^k)+\langle g^j,x-x^k \rangle -[f(x^k)-f_j^k],$$

we have

$$f(x) \geq f(x^k) + \langle p^{k-1},x-x^k \rangle -\alpha_p^k \quad \text{for all } x.$$

Therefore for any $e \geq 0$ the convex polyhedron

$$G_a^k(e)=\{g\in R^N : g=\sum_{j\in J^k}\lambda_j g^j+\lambda_p p^{k-1},\ \sum_{j\in J^k}\lambda_j\alpha_j^k+\lambda_p\alpha_p^k\le e,$$
$$\lambda_j\ge 0,\ j\in J^k,\lambda_p\ge 0,\ \sum_{j\in J^k}\lambda_j+\lambda_p=1\} \tag{2.12}$$

satisfies $G_a^k(e)\subset\partial_e f(x^k)$. These relations are natural extensions of those of the method with subgradient selection described above. Hence if $G_a^k(e^k)\neq\emptyset$ then we may find the direction d^k by computing $p^k=$ Nr $G_a^k(e^k)$ and setting $d^k=-p^k$. This can be done by calculating multipliers λ_j^k, $j\in J^k$, and λ_p^k that

$$\begin{aligned}&\underset{\lambda}{\text{minimize}}\ \tfrac{1}{2}\Big|\sum_{j\in J^k}\lambda_j g^j+\lambda_p p^{k-1}\Big|^2,\\&\text{subject to}\ \lambda_j\ge 0,\ j\in J^k,\lambda_p\ge 0,\ \sum_{j\in J^k}\lambda_j+\lambda_p=1,\\&\qquad\sum_{j\in J^k}\lambda_j\alpha_j^k+\lambda_p\alpha_p^k\le e^k,\end{aligned} \tag{2.13}$$

and setting

$$p^k=\sum_{j\in J^k}\lambda_j^k g^j+\lambda_p^k p^{k-1}.$$

The corresponding primal search direction finding subproblem is to

$$\text{minimize}\ \hat{f}_{B,a}^k(x^k+d)+\tfrac{1}{2}|d|^2\quad\text{over all}\ d\in R^N,$$

where $\hat{f}_{B,a}^k$ is the k-th <u>aggregate approximation</u> to f

$$\hat{f}_{B,a}^k(x)=\max\{f(x^k)+\langle g,x-x^k\rangle : g\in G_a^k(e^k)\}\quad\text{for all } x. \tag{2.14}$$

Also one may calculate λ_j^k and λ_p^k by the Mifflin (1978) algorithm, although in this case λ_j^k need not necessarily satisfy (2.7). Moreover, we may use the subgradient aggregation rules of Chapter 2 to define the next aggregate linearization $\tilde{f}^k$ by computing

$$(p^k,\tilde{f}_p^k)=\sum_{j\in J^k}\lambda_j^k(g^j,f_j^k)+\lambda_p^k(p^{k-1},f_p^k) \tag{2.15}$$

and setting

$$\tilde{f}^k(x)=\tilde{f}_p^k+\langle p^k,x-x^k\rangle\quad\text{for all } x.$$

The next aggregate approximation $\hat{f}_{B,a}^{k+1}$ will use

$$J^{k+1}=\hat{J}^k\cup\{k+1\},$$

where $\hat{J}^k$ is any subset of J^k such that $G_a^{k+1}(e^{k+1})$ is nonempty. Observing that $G_a^k(e)$ is nonempty for any $e \geq 0$ if $\alpha_j^k=0$ for some $j \in J^k$ (set $\lambda_j=1$, $\lambda_p=0$ and $\lambda_i=0$ for $i \neq j$ in (2.12)), we conclude that it suffices to ensure that $\alpha_j^{k+1}=0$ for some j in J^{k+1}.

Thus we have motivated the search direction finding subproblems of the methods. These subproblems depend on the parameter e^k, which controls the accuracy with which the sets $G^k(e^k) \subset \partial_{e^k} f(x^k)$ and $G_a^k(e^k) \subset \partial_{e^k} f(x^k)$ approximate $\partial f(x^k)$. Rules for choosing the approximation tolerances e^k, as well as the associated line search criteria, will be discussed in Sections 3 and 6.

3. The Algorithm with Subgradient Aggregation

We may now describe in detail the method with subgradient aggregation for minimizing a convex function $f : R^N \longrightarrow R$. Several comments on its rules will be given below.

Algorithm 3.1.

Step 0 (Initialization). Select the starting point $x^1 \in R^N$, a stopping parameter $\varepsilon_s \geq 0$ and an approximation tolerance $e_a > 0$. Choose positive line search parameters m_L, m_R, m_e, $\bar{t}$ and $\tilde{t}$ satisfying $m_L < m_R < 1$, $m_e < 1$ and $\bar{t} \leq 1 \leq \tilde{t}$. Set $J^1=\{1\}$, $y^1=x^1$, $p^0=g^1=g_f(y^1)$, $f_p^1=f_1^1=f(y^1)$ and $e^1=e_a$. Set the counters $k=1$, $l=0$ and $k(0)=1$.

Step 1 (Direction finding). Find multipliers λ_j^k, $j \in J^k$, and λ_p^k that solve the k-th dual subproblem (2.13). Calculate the aggregate subgradient $(p^k, \tilde{f}_p^k)$ by (2.15). Set $d^k=-p^k$ and

$$v^k = -|p^k|^2.$$

Step 2 (Stopping criterion). Set

$$\tilde{\alpha}_p^k = f(x^k)-\tilde{f}_p^k. \tag{3.1}$$

If $\max\{|p^k|^2, \tilde{\alpha}_p^k\} \leq \varepsilon_s$, terminate; otherwise, continue.

Step 3 (Approximation tolerance decreasing). If $|p^k|^2 > \tilde{\alpha}_p^k$ then go to Step 4. Otherwise, i.e. if $|p^k|^2 \leq \tilde{\alpha}_p^k$, replace e^k by $m_e e^k$ and go to Step 1.

<u>Step 4 (Line search)</u>. By a line search procedure as given below, find two stepsizes t_L^k and t_R^k such that $0 \le t_L^k \le t_R^k \le \tilde{t}$ and $t_R^k = t_L^k$ if $t_L^k > 0$, and such that the two corresponding points defined by $x^{k+1} = x^k + t_L^k d^k$ and $y^{k+1} = x^k + t_R^k d^k$ satisfy

$$f(x^{k+1}) \le f(x^k) + m_L t_L^k v^k, \tag{3.2a}$$

$$t_L^k \ge \bar{t} \quad \text{or} \quad \alpha(x^k, x^{k+1}) > m_e e^k \quad \text{if} \quad t_L^k > 0, \tag{3.2b}$$

$$\alpha(x^k, y^{k+1}) \le m_e e^k \quad \text{if} \quad t_L^k = 0, \tag{3.2c}$$

$$< g_f(y^{k+1}), d^k > \ \ge m_R v^k \quad \text{if} \quad t_L^k = 0. \tag{3.2d}$$

<u>Step 5 (Approximation tolerance updating)</u>. If $t_L^k = 0$ (<u>null step</u>), set $e^{k+1} = e^k$; otherwise, i.e. if $t_L^k > 0$ (<u>serious step</u>), set $e^{k+1} = e_a$ and $k(l+1) = k+1$, and increase l by 1.

<u>Step 6 (Linearization updating)</u>. Choose a subset $\hat{J}^k$ of J^k containing $k(l)$ if $k(l) < k+1$, and set $J^{k+1} = \hat{J}^k \cup \{k+1\}$. Set $g^{k+1} = g_f(y^{k+1})$ and compute

$$\begin{aligned} f_{k+1}^{k+1} &= f(y^{k+1}) + < g^{k+1}, x^{k+1} - y^{k+1} >, \\ f_j^{k+1} &= f_j^k + < g^j, x^{k+1} - x^k > \quad \text{for} \quad j \in \hat{J}^k, \\ f_p^{k+1} &= \tilde{f}_p^k + < p^k, x^{k+1} - x^k >. \end{aligned} \tag{3.3}$$

<u>Step 7</u>. Increase k by 1 and go to Step 1.

A few remarks on the algorithm are in order.

The above subgradient aggregation rules are the same as those in Algorithm 2.3.1, since we always have

$$\lambda_j^k \ge 0, \ j \in J^k, \ \lambda_p^k \ge 0, \ \sum_{j \in J^k} \lambda_j^k + \lambda_p^k = 1. \tag{3.4}$$

Hence Lemmas 2.4.1 and 2.4.2 are valid for Algorithm 3.1. In particular we have

$$p^k \in \partial_\varepsilon f(x^k) \quad \text{for} \quad \varepsilon = \tilde{\alpha}_p^k, \tag{3.5}$$

$$f(x) \ge f(x^k) - |p^k||x - x^k| - \tilde{\alpha}_p^k \quad \text{for all } x, \tag{3.6}$$

cf. Remark 2.3.3. Therefore, if the algorithm terminates at the k-th iteration, then

$$f(x) \geq f(x^k) - \varepsilon_s^{1/2}(|x-x^k| + \varepsilon_s^{1/2}) \quad \text{for all } x, \tag{3.7}$$

This estimate justifies our stopping criterion and shows that x^k is optimal if $\varepsilon_s=0$.

Our rules for updating the approximation tolerances e^k stem from the following considerations. In view of (3.6), we aim at obtaining small values of both $|p^k|$ and $\tilde{\alpha}_p^k$ at some iteration. This will occur if both $|p^k|^2 \leq \tilde{\alpha}_p^k$ and the value of $\tilde{\alpha}_p^k$ is small. Thus a mechanism is needed for decreasing the value of $\tilde{\alpha}_p^k$ if $|p^k|^2 \leq \tilde{\alpha}_p^k$. Since

$$\tilde{\alpha}_p^k = f(x^k) - \tilde{f}_p^k = \sum_{j \in J^k} \lambda_j^k [f(x^k) - f_j^k] + \lambda_p^k [f(x^k) - f_p^k] =$$

$$= \sum_{j \in J^k} \lambda_j^k \alpha_j^k + \lambda_p^k \alpha_p^k \leq e^k \tag{3.8}$$

because λ_j^k and λ_p^k solve (2.13), we have the fundamental relation

$$\tilde{\alpha}_p^k \leq e^k. \tag{3.9}$$

Therefore, whenever $|p^k|^2 \leq \tilde{\alpha}_p^k$ occurs the algorithm decreases $e^k (m_e<1)$ and calculates new p^k and $\tilde{\alpha}_p^k$. Thus the upper bound on $\tilde{\alpha}_p^k$ is decreased, while the new $|p^k|$ cannot be smaller than the old one. Moreover, this reduction of e^k increases the accuracy of our approximation $\hat{f}_{B,a}^k$ of f around x^k, which is based on the set $G_a^k(e^k) \subset \partial_{e^k} f(x^k)$. We may add that, for simplicity, the algorithm uses the approximation tolerance $e^k = e_a$ after each serious step. Other, more efficient rules for updating e^k will be discussed in Section 6.

Our line search criteria ensure two basic prerequisites for convergence: a sufficient decrease of the objective value at a serious step, and a significant modification of the next approximation to f after a null step. We have, by (2.14) and (2.2),

$$\hat{f}_{B,a}^k(x^k + d^k) - f(x^k) = \max\{ \langle g, d^k \rangle : g \in G_a^k(e^k)\} = -|p^k|^2 =$$

$$= v^k \tag{3.10}$$

and $-v^k = |p^k|^2 > e^k > 0$ at line searches. Thus $v^k < 0$ may be regarded as an approximate directional derivative of f at x^k in the direction $d^k \neq 0$. Whenever a serious step is taken, the criteria (3.2a,b) make t_L^k sufficiently large so that x^{k+1} has a significantly smaller objective value than does x^k, since $\bar{t} > 0$, $m_e e^k > 0$ and $-m_L v^k > 0$. On the other hand, after a null step we have

$$g^{k+1} \in G_a^{k+1}(e^k) , \tag{3.11a}$$

$$< g^{k+1}, d^k > \ \geq -m_R|p^k|^2 > -|p^k|^2 . \tag{3.11b}$$

This follows from (3.2c,d) and the fact that $\alpha_{k+1}^{k+1}=\alpha(x^{k+1},y^{k+1})=\alpha(x^k, y^{k+1})$, $k+1 \in J^{k+1}$, $m_e e^k < e^k$, $v^k=-|p^k|^2$ and $m_R \in (0,1)$. Comparing (3.10) with (3.11) we see that d^{k+1} must differ from d^k after a null step, since then $e^{k+1}=e^k$. At the same time, (3.2c) implies

$$g^{k+1} \in \partial_\varepsilon f(x^{k+1}) \quad \text{for} \quad \varepsilon=\alpha(x^k,y^{k+1}) \leq m_e e^k .$$

This shows that when e^k decreases during a series of null steps then the algorithm collects only local subgradient information, i.e. g^{k+1} is close to $\partial f(x^{k+1})$.

The following line search procedure may be used for executing Step 4.

Line Search Procedure 3.2.

(a) Set $t_L=0$ and $t=t_U=1$. Choose m satisfying $m_L < m < m_R$, e.g. $m=(9m_L+m_R)/10$.

(b) If $f(x^k+td^k) \leq f(x^k)+ mtv^k$ set $t_L=t$; otherwise set $t_U=t$.

(c) If $f(x^k+td^k) \leq f(x^k)+m_L tv^k$ and either $t \geq \bar{t}$ or $\alpha(x^k,x^k+td^k) > m_e e^k$ set $t_L^k=t_R^k=t$ and return.

(d) If $\alpha(x^k,x^k+td^k) \leq m_e e^k$ and $t < \bar{t}$ and $< g_f(x^k+td^k), d^k > \ \geq m_R v^k$ set $t_R^k=t$, $t_L^k=0$ and return.

(e) Choose $t \in [t_L+0.1(t_U-t_L),\ t_U-0.1(t_U-t_L)]$ by some interpolation procedure and go to (ii).

We shall now establish convergence of the above procedure.

Lemma 3.3. Line Search Procedure 3.2 terminates in a finite number of iterations, finding stepsizes t_L^k and t_R^k satisfying (3.2).

Proof. Assume, for contradiction purposes, that the search does not terminate. Denote by t^i, $\tilde{t}_L^i$ and t_U^i the values of t, t_L and t_U after the i-th execution of step (b) of the procedure, so that $t^i \in \{\tilde{t}_L^i, t_U^i\}$, for all i. Since $\tilde{t}_L^i \leq \tilde{t}_L^{i+1} \leq t_U^{i+1} \leq t_U^i$ and $t_U^{i+1}-\tilde{t}_L^{i+1} \leq 0.9(t_U^i-\tilde{t}_L^i)$ for all i, there exists $\hat{t} \geq 0$ such that $\tilde{t}_L^i \uparrow \hat{t}$ and $t_U^i \downarrow \hat{t}$. Let $x=x^k$,

$d=d^k$, $v=v^k$ and

$$TL = \{t \geq 0 : f(x+td) \leq f(x)+mtv\}.$$

Since $\{\tilde{t}_L^i\} \subset TL$, $\tilde{t}_L^i \uparrow \hat{t}$ and f is continuous, we have $\hat{t} \in LT$, i.e.

$$f(x+\hat{t}d) \leq f(x)+m\hat{t}v. \tag{3.12a}$$

Since $t_U^i \downarrow \hat{t}$, $\hat{t} \in LT$ and $t_U^i \notin LT$ if $t_U^i=t^i$, the set $I=\{i:t^i=t_U^i\}$ is infinite and $t^i > \hat{t}$ and

$$f(x+t^id) > f(x)+mt^iv \quad \text{for all} \quad i \in I. \tag{3.12b}$$

We shall consider the following two cases.

(i) Suppose that $\hat{t} > 0$. Since, by (3.12a),

$$f(x+\hat{t}d) \leq f(x)+m_L\hat{t}v-\varepsilon$$

with $\varepsilon=-(m-m_L)\hat{t}v > 0$ ($m > m_L$ and $v < 0$), and $t^i \xrightarrow{I} \hat{t}$, we have

$$f(x+t^id) \leq f(x)+m_Lt^iv \quad \text{for large} \quad i \in I$$

from the continuity of f. Therefore at step (c) we must have $\alpha(x,x+t^id) \leq m_e e^k$ and $t^i < \bar{t}$ for all large $i \in I$, and hence at step (d)

$$\langle \bar{g}^i,d \rangle < m_Rv \quad \text{for large} \quad i \in I, \tag{3.12c}$$

where $\bar{g}^i=g_f(x+t^id)$ for all i.

(ii) Suppose that $\hat{t}=0$. Then $t^i \xrightarrow{I} 0$ and $\alpha(x,x+t^id) \xrightarrow{I} 0$, since

$$\alpha(x,x+t^id) \leq f(x)-f(x+t^id)+t^i|\bar{g}^i||d|,$$

f is continuous and $\{\bar{g}^i\}$ is bounded (from the local boundedness of ∂f; see Lemma 1.2.2). Hence $\alpha(x,x+t^id) < m_e e^k$ at step (d) for large $i \in I$, because $m_e e^k > 0$, so we have (3.12c) at step (d).

Making use of (3.12) and the fact that $0 < m < m_R$, one may obtain the desired conclusion as in the proof of Lemma 3.3.3, since f, being a convex function, has the semismootheness property (3.3.23) (see Remark 3.3.4). □

We refer the reader to Remark 3.3.5 for a discussion of interpolation formulae which may be used at step (e) of Line Search Procedure 3.2.

Practical rules for deciding which past subgradients should be dropped at Step 6 should weigh speed of convergence against storage and work per iteration; see Remark 2.2.4. We may add that one may use additional subgradients for search direction finding when the objective function is a max function, see Remark 2.2.5. We also note that the subgradient $g_f(x^k)$ is always used for search direction finding at the k-th iteration, i.e. we have

$$k(l)\in J^k,\ g^{k(l)}=g_f(x^k) \text{ and } \alpha^k_{k(l)}=0 \quad \text{if} \quad k(l)\le k<k(l+1). \qquad (3.13)$$

This property, which will be established in the next section, ensures that the k-th dual subproblem (2.13) is feasible, since $\alpha_j^k=0$ for some $j\in J^k$.

4. Convergence.

In this section we shall show that each sequence $\{x^k\}$ generated by Algorithm 3.1 minimizes f, i.e. $f(x^k)\downarrow \inf\{f(x):x\in R^N\}$, and that $\{x^k\}$ converges to a minimum point of f whenever f attains its infimum. Naturally, convergence results assume that the stopping parameter ε_s is set to zero. To save space, our analysis will rely on the results of Chapters 2 and 3.

We start ny recalling the following result of the preceding section.

Lemma 4.1. If Algorithm 3.1 terminates at the k-th iteration, then x^k is a minimum point of f.

From now on we assume that the algorithm does not stop.

Note that Step 1 may be executed more than once at each iteration, since each decrease of the approximation tolerance e^k involves a repetition of the search direction finding. This is not reflected in our notation, since it should be clear from the context that the various relations, such as (3.4)-(3.9), hold upon each completion of Step 1 at any iteration.

We shall now analyze the case when the algorithm produces a finite sequence $\{x^k\}$.

Lemma 4.2. Suppose that at the k-th iteration Algorithm 3.1 cycles infinitely between Steps 1 and 3. Then $0\in\partial f(\bar{x})$ for $\bar{x}=x^k$.

Proof. If at the k-th iteration there are infinitely many returns from Step 3 to Step 1, then $|p^k|$ and $\tilde{\alpha}_p^k$ tend to zero, since we have $|p^k|^2\le\tilde{\alpha}_p^k\le e^k$ at Step 3, and e^k is replaced by $m_e e^k$ with $m_e\in(0,1)$.

Using this in (3.6), we obtain $f(x) \geq f(x^k)$ for all x, so $0 \in \partial f(x^k)$. □

In view of the above lemma, we may assume in what follows that the algorithm generates an infinite sequence $\{x^k\}$, executing Step 1 finitely many times at each iteration.

The following result is crucial for convergence.

Lemma 4.3. Suppose that for some $\bar{x} \in R^N$ we have

$$\liminf_{k\to\infty} \max\{|p^k|, \tilde{\alpha}_p^k, |\bar{x}-x^k|\} = 0, \tag{4.1}$$

or equivalently

$$\text{there exists an infinite set } K \subset \{1,2,\ldots\} \text{ such that } x^k \xrightarrow{K} \bar{x}, \quad |p^k| \xrightarrow{K} 0 \text{ and } \tilde{\alpha}_p^k \xrightarrow{K} 0. \tag{4.2}$$

Then $0 \in \partial f(\bar{x})$.

Proof. (4.1) and (4.2) are equivalent, since $\tilde{\alpha}_p^k \geq 0$ for all k. If (4.2) holds, we may let $k \in K$ tend to infinity in (3.6) and use the continuity of f (f is locally Lipschitzian as a convex function on R^N) to obtain $f(x) \geq f(\bar{x})$ for all x, i.e. $0 \in \partial f(\bar{x})$. □

Consider the following condition for some fixed $\bar{x} \in R^N$

$$\text{there exists an infinite set } K \subset \{1,2,\ldots\} \text{ such that } x^k \xrightarrow{K} \bar{x}. \tag{4.3}$$

In view of the above lemma, our aim is to show that $\max\{|p^k|, e^k\} \xrightarrow{K} 0$.

We start by collecting a few useful results. By the rules of Step 5, we have

$$x^k = x^{k(l)} \quad \text{for } k=k(l), k(l)+1, \ldots, k(l+1)-1,$$

where we set $k(l+1)=\infty$ if the number l of serious steps stays bounded, i.e. if $x^k=x^{k(l)}$ for some fixed l and all $k \geq k(l)$. At Step 4 we always have

$$|p^k|^2 > e^k > 0, \tag{4.4}$$

hence $d^k=-p^k \neq 0$ and $v^k=-|p^k|^2 < 0$. Therefore the line search requirement (3.2a) with $m_L > 0$ ensures that the sequence $\{f(x^k)\}$ is nonincreasing and

$$f(x^{k+1}) < f(x^k) \quad \text{if} \quad x^{k+1} \neq x^k.$$

These line search properties yield the following auxiliary result.

Lemma 4.4. (i) Suppose that the sequence $\{f(x^k)\}$ is bounded from below. Then

$$\sum_{k=1}^{\infty} \{t_L^k|p^k|^2+t_L^k\tilde{\alpha}_p^k\} < \infty. \tag{4.5}$$

(ii) If (4.3) holds then (4.5) is satisfied and

$$f(x^k)\downarrow f(\bar{x}) \text{ as } k\to\infty, \tag{4.6}$$

$$t_L^k|p^k|^2 \longrightarrow 0 \text{ as } k\to\infty. \tag{4.7}$$

Proof. (i) By the line search criterion (3.2a),

$$f(x^1)-f(x^{k+1})=f(x^1)-f(x^2)+\ldots+f(x^k)-f(x^{k+1}) \geq$$

$$\geq -m_L \sum_{i=1}^{k} t_L^i v^i = m_L \sum_{i=1}^{k} t_L^i |p^i|^2.$$

Since $m_L > 0$ and $0 \leq \tilde{\alpha}_p^k \leq |p^k|^2$ at line searches and $\{f(x^k)\}$ is nonincreasing, the above inequality yields (4.5) if $\{f(x^k)\}$ is bounded from below.

(iii) If (4.3) holds then (4.6) follows from the continuity of f and the monotonicity of $\{f(x^k)\}$. Hence $f(x^k) \geq f(\bar{x})$ for all k, so (4.5) holds and we have (4.7), as desired. □

We shall now show that the properties of the dual search direction finding subproblems ensure locally uniform reductions of $|p^k|$ after null steps.

Lemma 4.5. Let $\bar{x} \in R^N$, $\bar{a} > 0$ and $B=\{y \in R^N : |y-\bar{x}| \leq \bar{a}\}$ be given. Then there exists C independent of k such that

$$\max\{|p^k|,|g^{k+1}|,1\} \leq C \quad \text{if} \quad x^k \in B. \tag{4.8}$$

Moreover, if $x^k \in B$, $t_L^{k-1}=0$ and $e^k=e^{k-1}$ for some $k > 1$ then

$$\tfrac{1}{2}|p^k|^2 \leq \phi_C(\tfrac{1}{2}|p^{k-1}|^2), \tag{4.9}$$

where the function ϕ_C is defined (for the fixed value of the line search parameter $m_R \in (0,1)$) by

$$\phi_C(t) = t-(1-m_R)^2t^2/(8C^2). \tag{4.10}$$

Proof. (i) Observe that, by (2.15), $|p^k|^2/2$ is the optimal value of the k-th dual subproblem (2.13), for any k.

(ii) Suppose that $k(l) \le k < k(l+1)$, so that $x^k = x^{k(l)}$. Observe that $t_R^{k-1} = t_L^{k-1}$, and hence $y^k = x^k = x^{k(l)}$ and $g^k = g_f(x^{k(l)})$ if $k = k(l)$. Combining this with the fact that $k(l) \in J^k$ by the rules of Steps 5 and 6, and that $\alpha_{k(l)}^k = \alpha(x^k, x^{k(l)}) = \alpha(x^{k(l)}, x^{k(l)}) = 0$, we obtain

$$k(l) \in J^k, \; g^{k(l)} = g_f(x^k) \;\text{ and }\; \alpha_{k(l)}^k = 0.$$

Hence the multipliers

$$\lambda_{k(l)} = 1, \; \lambda_j = 0 \;\text{ for }\; j \in J^k \setminus \{k(l)\}, \; \lambda_p = 0$$

are feasible for the k-th subproblem (2.13), so its optimal value

$$\tfrac{1}{2}|p^k|^2 \le \tfrac{1}{2}|g^{k(l)}|^2 + \alpha_{k(l)}^k \le \tfrac{1}{2}|g_f(x^k)|^2. \tag{4.11}$$

(iii) Observing that $|d^k| = |p^k|$ and $|y^{k+1} - \bar{x}| \le |y^{k+1} - x^k| + |x^k - \bar{x}| \le t_R^k |d^k| + \bar{a} \le \tilde{t}|d^k| + \bar{a}$ and $g^{k+1} = g_f(y^{k+1})$ if $x^k \in B$, we deduce from (4.11) and the local boundedness of g_f the existence of a constant C which is larger than $|p^k|, |y^{k+1} - \bar{x}|$ and $|g^{k+1}|$ whenever $x^k \in B$. This yields (4.8).

(iv) Suppose that $x^k \in B$, $t_L^{k-1} = 0$ and $e^k = e^{k-1}$. Then $x^{k-1} = x^k \in B$, so

$$\max\{|p^{k-1}|, |g^k|, 1\} \le C \tag{4.12}$$

from (4.8). Since $e^k = e^{k-1}$, $m_e < 1$ and $x^k = x^{k-1}$, (3.9) and (3.2c) yield $\alpha_p^k = \tilde{\alpha}_p^{k-1} \le e^k$ and $\alpha_k^k = \alpha(x^k, y^k) \le m_e e^k \le e^k$, while $k \in J^k$ by the rules of Step 6, so the multipliers

$$\lambda_k(\nu) = \nu, \; \lambda_j(\nu) = 0 \;\text{ for }\; j \in J^k \setminus \{k\}, \; \lambda_p(\nu) = 1 - \nu$$

are feasible in (2.13) for each $\nu \in [0,1]$. Therefore the optimal value of (2.13) satisfies

$$\tfrac{1}{2}|p^k|^2 \le \min\{|(1-\nu)p^{k-1} + \nu g^k|^2 : \nu \in [0,1]\}. \tag{4.13}$$

Since $t_L^{k-1} = 0$, (3.2d) yields

$$\langle g^k, d^{k-1} \rangle \ge m_R v^{k-1}. \tag{4.14}$$

Using (4.12), (4.14) and the fact that $m_R \in (0,1)$, $d^{k-1} = -p^{k-1}$ and $v^{k-1} = -|p^{k-1}|^2$, we deduce from Lemma 2.4.10 that the right side of inequality (4.13) is no larger than $\phi_C(|p^{k-1}|^2/2)$, so (4.9) holds, as required. □

We are now ready to analyze the case of a finite number of serious steps of the method.

Lemma 4.6. Suppose that $x^k = x^{k(l)} = \bar{x}$ for some fixed l and all $k \ge k(l)$. Then (4.1) holds and $0 \in \partial f(\bar{x})$.

Proof. Suppose that $x^k = \bar{x}$ for all $k \ge k(l)$ and let $K = \{k : |p^k|^2 \le \tilde{\alpha}_p^k\}$. We shall consider two cases.

(i) Suppose that K is infinite. Then, since $0 \le e^{k+1} \le e^k$ for $k \ge k(l)$, and $e^{k+1} \le m_e e^k$ if $k \in K$ with the fixed $m_e \in (0,1)$, we have $e^k \xrightarrow{K} 0$. Since $|p^k|^2 \le \tilde{\alpha}_p^k \le e^k$ for all $k \in K$, we obtain (4.2), and hence (4.1) and $0 \in \partial f(\bar{x})$ from Lemma 4.3.

(ii) Suppose that K is finite, i.e. $|p^k|^2 > \tilde{\alpha}_p^k$ for all large k. Since $x^k = \bar{x}$ and $t_L^{k-1} = 0$ for $k > k(l)$, by the rules of Step 5 we have $e^k = e > 0$ for all $k > k(l)$. But then Lemma 4.5 yields $|p^k| \downarrow 0$, so, since $\tilde{\alpha}_p^k < |p^k|^2$ for large k, both $|p^k|$ and $\tilde{\alpha}_p^k$ tend to zero, and Lemma 4.3 yields $0 \in \partial f(\bar{x})$. □

It remains to consider the case of an infinite number of serious steps. To this end, let $K_l = \{k : k(l) \le k < k(l+1)\}$ and let b^l denote the minimum value taken on by $\max\{|p^k|, \tilde{\alpha}_p^k\}$ in Step 2 at iterations $k \in K_l$, for all l. Note that b^l is well-defined if $l \to \infty$, since then there can by only finitely many executions of Step 1 at any iteration.

Lemma 4.7. Suppose that there exist a point $\bar{x} \in R^N$ and an infinite set $L \subset \{1,2,\ldots\}$ such that $x^{k(l)} \to \bar{x}$ as $l \to \infty$, $l \in L$. Then $\liminf_{l \in L} b^l = 0$ and $0 \in \partial f(\bar{x})$.

Proof. Suppose that $x^{k(l)} \xrightarrow{L} \bar{x}$. We shall consider two cases.

(i) Suppose that $b^l \xrightarrow{\bar{L}} 0$ for some infinite set $\bar{L} \in L$. Since $x^k = x^{k(l)}$ for all $k \in K_l$ and $x^{k(l)} \to \bar{x}$ as $l \to \infty$, $l \in \bar{L}$, from the definition of b^l we deduce (4.2). Hence $0 \in \partial f(\bar{x})$ by Lemma 4.3.

(ii) Suppose that $\{b^l\}_{l \in L}$ is bounded away from zero. Then there exists $\varepsilon > 0$ such that on each entrance to Step 3 we have

$$\max\{|p^k|^2, \tilde{\alpha}_p^k\} \ge \varepsilon \quad \text{for all } k \in K_l \text{ and large } l \in L. \tag{4.15}$$

By the algorithm's rules, for any l and k such that $k \in K_l$ and $k+1 \in K_l$

we have $e^{k(l)}=e_a>0$, $e^{k+1}\le m_e e^k$ if $|p^k|\le\tilde{\alpha}_p^k\le e^k$ at Step 3, and $e^{k+1}=e^k$ otherwise. Therefore, if e^k approached zero for some $k\in K_l$ and large $l\in L$, then so would $|p^k|^2$ and $\tilde{\alpha}_p^k$, which would contradict (4.15). Thus $e^k\ge\varepsilon_e>0$ for some ε_e and all $k\in K_l$ and large $l\in L$. In particular,

$$e^k\ge\varepsilon_e>0 \quad \text{for all large } k\in K, \tag{4.16}$$

where $K=\{k(l+1)-1 : l\in L\}$. Also, since $|p^k|^2>\tilde{\alpha}_p^k$ at Step 4, (4.15) yields

$$|p^k|^2\ge\varepsilon \quad \text{for all large } k\in K. \tag{4.17}$$

Since $\bar{x}$ is an accumulation point of $\{x^k\}$, Lemma 4.4 yields $t_L^k|p^k|^2\to 0$ as $k\to\infty$. Combining this with (4.17) and the fact that

$$t_L^k|p^k|^2=|t_L^k d^k||p^k|=|x^{k+1}-x^k||p^k| \quad \text{for all } k,$$

we obtain $t_L^k\xrightarrow{K}0$ and $|x^{k+1}-x^k|\xrightarrow{K}0$. But $t_L^k>0$ for all $k\in K$, so we deduce from (3.2b), the fact that $\bar{t}>0$ is fixed, and (4.16) that

$$\alpha(x^k,x^{k+1})>m_e\varepsilon_e>0 \quad \text{for all large } k\in K. \tag{4.18}$$

Since $x^k\xrightarrow{K}\bar{x}$ and $|x^{k+1}-x^k|\xrightarrow{K}0$, we have

$$\alpha(x^k,x^{k+1})=f(x^k)-f(x^{k+1})-\langle g_f(x^{k+1}),x^k-x^{k+1}\rangle\xrightarrow{K}0 \tag{4.19}$$

from the continuity of f and the local boundedness of g_f. This contradicts (4.18). Therefore, $\{b^l\}_{l\in L}$ cannot be bounded away from zero, and case (i) above yields the desired conclusion. □

Combining Lemmas 4.6 and 4.7 we obtain

<u>Theorem 4.8</u>. Every accumulation point $\bar{x}$ of an infinite sequence $\{x^k\}$ generated by Algorithm 3.1 satisfies $0\in\partial f(\bar{x})$.

Our next result states that the global convergence properties of the method are the same as those of the algorithms considered in Chapter 2.

<u>Theorem 4.9</u>. Every infinite sequence $\{x^k\}$ calculated by Algorithm 3.1 minimizes f, i.e. $f(x^k)\downarrow\inf\{f(x) : x\in R^N\}$ as $k\to\infty$. Moreover, $\{x^k\}$ converges to a minimum point of f whenever f attains its infimum.

<u>Proof</u>. In virtue of Theorem 4.9 and the fact that we have (3.5), $t_L^k\le\tilde{t}$

for all k, and (4.5) if $\{f(x^k)\}$ is bounded from below, the proofs of Lemma 2.4.14, Theorem 2.4.15 and Theorem 2.4.16 are valid for Algorithm 3.1. □

The next results provide further substantiation of our stopping criterion.

Corollary 4.10. If f has a minimum point and the stopping parameter ε_s is positive, then Algorithm 3.1 terminates in a finite number of iterations.

Proof. If the assertion were false then Lemma 2.4.14, which holds for Algorithm 3.1 owing to (3.5a) and Lemma 4.4(i), would imply that $\{x^k\}$ is bounded and has some accumulation point $\bar{x}$ if $\{x^k\}$ is infinite, while the proof of Lemma 4.2 shows that the method must stop if $\{x^k\}$ is finite and $\varepsilon_s > 0$. Then Lemmas 4.6 and 4.7 would yield that $\max\{|p^k|, \tilde{\alpha}_p^k\} \le \varepsilon_s$ for some k, and hence the method would stop, a contradiction. □

Corollary 4.11. If the level set $S_f=\{x \in R^N : f(x) \le f(x^1)\}$ is bounded and the stopping parameter ε_s is positive, then Algorithm 3.1 terminates in a finite number of iterations.

Proof. Since $\{x^k\} \subset S_f$ is bounded and $\varepsilon_s > 0$, we may use either the proof of Lemma 4.2 or Lemmas 4.6 and 4.7 to show that $\max\{|p^k|, \tilde{\alpha}_p^k\} \le \varepsilon_s$ for some k. □

5. The Algorithm with Subgradient Selection.

In this section we shall state and analyze in detail the method with subgradient selection introduced in Section 2.

Algorithm 5.1.

Step 0 (Initialization). Do Step 0 of Algorithm 3.1.

Step 1 (Direction finding). Find multipliers λ_j^k, $j \in J^k$, which solve the k-th dual subproblem (2.5) and are such that the corresponding set $\hat{J}^k = \{j \in J^k : \lambda_j^k \ne 0\}$ has at most N+1 elements. Set

$$(p^k, \tilde{f}_p^k) = \sum_{j \in J^k} \lambda_j^k (g^j, f_j^k), \tag{5.1}$$

$d^k = -p^k$ and $v^k = -|p^k|^2$.

Step 2 (Stopping criterion). Do Step 2 of Algorithm 3.1.

Step 3 (Approximation tolerance decreasing). Do Step 3 of Algorithm 3.1.

Step 4 (Line search). Do Step 4 of Algorithm 3.1.

Step 5 (Approximation tolerance updating). Do Step 5 of Algorithm 3.1.

Step 6 (Linearization updating). Set

$$J^{k+1}=\hat{J}^k\cup\{k+1,\ k(l)\}. \tag{5.2}$$

Set $g^{k+1}=g_f(y^{k+1})$ and calculate f_j^{k+1}, $j\in J^{k+1}$, by (3.3).

Step 7. Increase k by 1 and go to Step 1.

A few comments on the method are in order.

The subgradient selection and aggregation rules of the method are the same as those of Algorithm 2.5.1, since we always have

$$\lambda_j^k\geq 0,\ j\in\hat{J}^k,\quad \sum_{j\in\hat{J}^k}\lambda_j^k=1 \tag{5.3}$$

from Lemma 2.1 and the construction of $\hat{J}^k$. Hence, by the results of Section 2.5, Lemmas 2.4.1 and 2.4.2 are true for Algorithm 5.1 and we have (3.5). In view of (5.1) and (5.3), we may set $\lambda_p^k=0$ in (3.7) to obtain

$$\tilde{\alpha}_p^k\leq e^k$$

as in Algorithm 3.1.

The remarks in Section 3 on the line search criteria of Algorithm 3.1 apply to Algorithm 5.1 if one replaces $\hat{f}_{B,a}^k$ and G_a^k in (3.8)-(3.9) by $\hat{f}_{B,s}^k$ and G^k, respectively; see (2.3) and (2.13). Of course, Line Search Procedure 3.2 may be used for executing Step 4 of Algorithm 5.1.

The requirement (5.2) results in relation (3.11), which ensures that the k-th dual subproblem (2.5) is feasible, i.e. the algorithm is well-defined. We refer the reader to Remarks 2.2.5 and 2.5.3 on the possible use of additional subgradients for search direction finding.

As far as convergence is concerned, it is easy to verify that all the results of Section 4 hold for Algorithm 5.1. In fact, only part (iv) of the proof of Lemma 4.5 needs a minor modification. To this end, define the multipliers

$$\lambda_k(\nu)=\nu,\lambda_j(\nu)=(1-\nu)\lambda_j^{k-1}\text{ for } j\in\hat{J}^{k-1},\lambda_j(\nu)=0\text{ for } j\in J^k\setminus(\hat{J}^{k-1}\cup\{k\})$$

and use (2.5.4), which follows from (5.1)-(5.3), to deduce that (4.13) holds, as before.

To sum up, Algorithm 5.1 is a globally convergent method in the sense of Theorem 4.9 and Corollaries 4.10 and 4.11.

6. Modified Line Search Rules and Approximation Tolerance Updating Strategies.

In this section we shall describe modified rules for updating the approximation tolerances $\{e^k\}$ of Algorithm 3.1. These rules are more efficient than the original ones. We shall also discuss modified line search requirements that are similar to those proposed by Lemarechal, Strodiot and Bihain (1981).

The motivation for our modified approximation tolerance updating strategies is practical and stems from the following observation. Algorithm 3.1 resets e^k to the fixed value of $e_a > 0$ after each serious step, and then each decreasing of e^k involves solving an additional "idle" quadratic programming subproblem. This strategy has two drawbacks. First, too small a value of e^k may result in many function and subgradient evaluations at initial iterations, when the line search procedure may need many contractions for finding y^{k+1} sufficiently close to x^k so that $\alpha(x^k, y^{k+1}) \le m_e e^k$ at a null step. Secondly, when the algorithm is close to a solution then, in general, d^k is a descent direction only if e^k is small enough, so that $G_a^k(e^k)$ is close to $\partial f(x^k)$. Hence later iterations may require many solutions of quadratic programming subproblems only to reduce the values of e^k.

The first drawback may be eliminated by allowing the method to choose after a serious step any value of e^k not smaller than e_a. For instance, following Lemarechal, Strodiot and Bihain (1981), one may set

$$e^{k+1} = \max\{e_a, -t_L^k v^k\} \tag{6.1a}$$

or

$$e^{k+1} = \max\{e_a, f(x^k) - f(x^{k+1})\} \tag{6.1b}$$

at Step 5 of Algorithm 3.1 if $t_L^k > 0$. This will enable the method to use e^k larger then e_a at initial iterations. At the same time, one may easily verify that this modification does not impair the convergence results of Section 4.

The above modification does not eliminate the need for the fixed threshold $\varepsilon_a > 0$, hence it has the second drawback mentioned above. For this reason, consider the following modification of Algorithm 3.1.

Algorithm 6.1.

Step 0 (Initialization). Do Step 0 of Algorithm 3.1. Set $\delta^1=e_a$ and $e^1=m_e\delta^1$.

Step 1 (Direction finding). Do Step 1 of Algorithm 3.1.

Step 2 (Stopping criterion). Do Step 2 of Algorithm 3.1.

Step 3 (Approximation tolerance decreasing). Set

$$\tilde{\delta}^k = \max\{|p^k|^2,\ \tilde{\alpha}_p^k\}. \tag{6.2}$$

If $|p^k|^2 > \tilde{\alpha}_p^k$ then go to Step 4. Otherwise, i.e. if $|p^k|^2 \le \tilde{\alpha}_p^k$, set $\delta^k=\tilde{\delta}^k$, $e^k=m_e\delta^k$ and go to Step 1.

Step 4 (Line search). Do Step 4 of Algorithm 3.1.

Step 5 (Approximation tolerance updating). Set $\delta^{k+1}=\delta^k$ and $e^{k+1}=m_e\delta^{k+1}$.

Step 6 (Linearization updating). Do Step 6 of Algorithm 3.1.

Step 7. Increase k by 1 and go to Step 1.

Remark 6.2. If is easy to see that Algorithm 6.1 calculates monotonically nonincreasing sequences of positive numbers $\{\delta^k\}$ and $\{e^k\}$ satisying $e^k=m_e\delta^k$ for all k, if it does not stop. To this end, observe that since $\tilde{\alpha}_p^k \le e^k=m_e\delta^k$ from (3.9), δ^k is set equal to $\tilde{\delta}^k$ only if $|p^k|^2 \le \tilde{\alpha}_p^k$ and

$$\tilde{\delta}^k = \max\{|p^k|^2, \tilde{\alpha}_p^k\} = \tilde{\alpha}_p^k \le e^k=m_e\delta^k.$$

In this case δ^k is reduced by the factor $m_e \in (0,1)$; otherwise, δ^k is unchanged. It follows that either δ^k and e^k eventually stay constant and there are no returns from Step 3 to Step 1, or they converge to zero and the method returns from Step 3 to Step 1 infinitely often with $\max\{|p^k|^2,\tilde{\alpha}_p^k\} \le \delta^k$.

We conclude from the above remark that Algorithm 6.1 automatically reduces e^k when it approaches an optimal point, which is indicated by a small value of $\max\{|p^k|^2,\tilde{\alpha}_p^k\}$.

Let us now analyze convergence of the method. Making use of Remark 6.2, one may easily establish Lemma 4.2 for Algorithm 6.1. Therefore, we may suppose in what follows that the method generates an infinite sequence $\{x^k\}$. Next, we observe that the proofs of Lemmas 4.3-4.6 need not be modified. Lemma 4.7 is replaced by the following result.

Lemma 6.3. Suppose that Algorithm 6.1 generates an infinite sequence $\{x^k\}$ such that for some $\hat{x} \in R^N$ one has $f(\hat{x}) \le f(x^k)$ for all k. Then there exists $\bar{x} \in R^N$ such that $0 \in \partial f(\bar{x})$ and $x^k \to \bar{x}$ as $k \to \infty$. Moreover, $\liminf_{k\to\infty} \tilde{\delta}^k = 0$.

Proof. Suppose that $f(x^k) \ge f(\hat{x})$ for all k. We may use Lemma 4.4(i), (3.9) and the fact that $t_L^k \le \tilde{t}$ for all k to deduce that Lemma 2.4.14 holds for Algorithm 6.1. Hence $\{x^k\}$ is bounded and has an accumulation point x satisfying $f(\bar{x}) \le f(x^k)$ for all k by Lemma 4.4(ii). Using the proof of Theorem 2.4.15, we deduce that in fact $x^k \to \bar{x}$ as $k \to \infty$. In view of Remark 6.2, we shall consider two cases.

(i) Suppose that e^k stays constant for all large k. Then the desired conclusion follows from Lemma 4.6 and the proof of Lemma 4.7, which is valid for constant e^k.

(ii) Suppose that e^k tends to zero. Then $\delta^k \downarrow 0$ and for infinitely many k-s we have $\max\{|p^k|^2, \tilde{\alpha}_p^k\} \le \delta^k$, while $x^k \to \bar{x}$, so (4.2) holds and Lemma 4.3 yields the desired conclusion. □

We conclude from Lemmas 6.3 and 4.4(ii) that Theorem 4.8 holds for Algorithm 6.1. Then it is straigtforward to verify that Theorem 4.9 and Corollaries 4.10 and 4.11 are true for Algorithm 6.1, since their proofs did not depend on the choice of $\{e^k\}$.

To sum up, we have proved that Algorithm 6.1 is globally convergent in the sense of Theorem 4.9 and Corollaries 4.10 and 4.11.

We shall now consider modified line search rules that are similar to those in (Lemarechal, Strodiot and Bihain, 1981). To this end, recall from Lemma 2.1(iv) that the k-th dual subproblem (2.13) is equivalent to the following

$$\begin{aligned} &\underset{\lambda}{\text{minimize}}\ \tfrac{1}{2}\Big|\sum_{j\in J^k}\lambda_j g^j+\lambda_p p^{k-1}\Big|^2+\sum_{j\in J^k}\lambda_j s^k\alpha_j^k+\lambda_p s^k\tilde{\alpha}_p^k, \\ &\text{subject to}\ \ \lambda_j\ge 0,\ j\in J^k,\ \lambda_p\ge 0,\ \sum_{j\in J^k}\lambda_j+\lambda_p=1, \end{aligned} \tag{6.3}$$

where $s^k \ge 0$ is the Lagrange multiplier for the last constraint of (2.13), satisfying

$$s^k\Big(\sum_{j\in J^k}\lambda_j^k\alpha_j^k+\lambda_p^k\alpha_p^k-e^k\Big)=0.$$

By (3.8),

$$\tilde{\alpha}_p^k = \sum_{j \in J^k} \lambda_j^k \alpha_j^k + \lambda_p^k \alpha_p^k,$$

hence we have $s^k \tilde{\alpha}_p^k = s^k e^k$. Combining the preceding relations with (2.15) and invoking Lemma 2.2.1, we see that (6.3) is the dual to the problem

$$\begin{aligned} &\underset{(d,v) \in R^{N+1}}{\text{minimize}} \ \tfrac{1}{2}|d|^2 + v, \\ &\text{subject to} \ -s^k \alpha_j^k + \langle g^j, d \rangle \le v, \ j \in J^k, \\ &\qquad\qquad -s^k \alpha_p^k + \langle p^{k-1}, d \rangle \le v, \end{aligned} \tag{6.4}$$

which has a unique solution $(d^k, \tilde{v}^k)$ with $d^k = -p^k$ (p^k is given by (2. 15)) and

$$\tilde{v}^k = -\{|p^k|^2 + s^k \tilde{\alpha}_p^k\} = -\{|p^k|^2 + s^k e^k\}. \tag{6.5}$$

Note that for $s^k = 1$ (6.3)-(6.5) reduce to the corresponding relations of Algorithm 2.3.1 (see also (3.2.30)-(3.2.31)), while if $s^k = 0$ then $\tilde{v}^k$ reduces to the variable $v^k = -|p^k|^2$ used by Algorithm 3.1; see (3.10). Therefore $\tilde{v}^k$ may be regarded as an approximate directional derivative of f at x^k in the direction d^k.

Let us, therefore, consider the use of $\tilde{v}^k$ in place of v^k in Algorithm 3.1. This amounts to replacing v^k by $\tilde{v}^k$ in the line search criteria (3.2). Since $s^k \ge 0$, $e^k > 0$ and $v^k < 0$ at line searches, we have $\tilde{v}^k \le v^k < 0$, i.e. $\tilde{v}^k < 0$, so the modified criteria may be interpreted similarly to the original ones, and Line Search Procedure 3.2 will meet these requirements.

We shall now establish convergence of the resulting method. One may easily verify that among the results of Section 4 only part (iv) of the proof of Lemma 4.5 is invalidated if we have $\tilde{v}^k < v^k$ for some k. But Lemma 4.5 was instrumental only in part (ii) of the proof of Lemma 4.6. Therefore, we need only prove the following result.

Lemma 6.4. Consider the version of Algorithm 3.1 that uses $\tilde{v}^k$ given by (6.5) instead of v^k, for all k, and assume that $m_e + m_R \le 1$. Suppose that $x^k = \bar{x}$, $t_L^k = 0$ and $e^k = e > 0$ for some fixed $\bar{x}$, e and all large k. Then $|p^k| \downarrow 0$ as $k \to \infty$.

Proof. (i) With no loss of generality, suppose that $x^k = \bar{x}$, $t_L^k = 0$ and $e^k = e > 0$ for all k. Then (3.2c,d) with $\tilde{v}^k$ replaced by v^k yield

$\alpha_{k+1}^{k+1} \le m_e e$ and

$$< g^{k+1},p^k > \le m_R|p^k|^2+m_R e s^k \tag{6.6a}$$

for all k. Moreover, since $(d^{k+1},\tilde{v}^{k+1})$ solves the (k+1)-st subproblem (6.4), while $k+1 \in J^{k+1}$, we have $-s^{k+1}\alpha_{k+1}^{k+1} + < g^{k+1},d^{k+1} > \le \tilde{v}^{k+1}$, so

$$\begin{aligned} < g^{k+1},p^{k+1} > &\ge |p^{k+1}|^2+s^{k+1}e-s^{k+1}\alpha_{k+1}^{k+1} \ge \\ &\ge |p^{k+1}|^2+(1-m_e)e\, s^{k+1} , \end{aligned} \tag{6.6b}$$

since $\alpha_{k+1}^{k+1} \le m_e e$, for all k. Subtracting (6.6a) from (6.6b) and rearraging terms yields

$$s^{k+1} \le s^k m_R/(1-m_e)-c^k/(1-m_e)e,$$

where

$$c^k = |p^{k+1}|^2-m_R|p^k|^2- < g^{k+1},p^{k+1}-p^k > , \tag{6.7}$$

hence the fact that $m_R \le 1-m_e$ implies

$$s^{k+1} \le s^k - c^k/(1-m_e)e \quad \text{for all k.} \tag{6.8}$$

(ii) From the proof of Lemma 4.5 we deduce that $\{g^k\}$ is bounded and that $p^{k+1}=Nr[p^{k+1},p^k]$, for all k. Therefore $\{|p^k|\}$ is monotonically nonincreasing, and (6.6b) and the positivity of $(1-m_e)e$ imply the existence of a constant $\bar{s}$ such that $s^k \le \bar{s}$ for all k.

(iii) Since $\{p^k\}$ is bounded, there exists $\bar{p} \in R^N$ and an infinite set $K \subset \{1,2,...\}$ such that $p^k \xrightarrow{K} \bar{p}$. Assume, for contradiction purposes, that $\bar{p} \ne 0$, since $|p^k| \downarrow 0$ otherwise. We shall show that $p^{k+1} \xrightarrow{K} \bar{p}$. To this end, suppose that $p^{k+1} \xrightarrow{\tilde{K}} \tilde{p}$ for some $\tilde{p}$ and an infinite set $\tilde{K} \subset K$. Since $p^{k+1}=Nr[p^{k+1},p^k]$, Lemma 1.2.12 implies $< p^{k+1},p^k> \ge |p^{k+1}|^2$. Passing to the limit with $k \in \tilde{K}$ and using the monotonicity of $\{|p^k|\}$ yields $< \tilde{p},\bar{p} > \ge |\tilde{p}||\bar{p}| > 0$, so $\tilde{p}=\bar{p} \ne 0$ from elementary properties of the inner product. This shows that p^k and p^{k+1} have a common limit $\bar{p} \ne 0$ as $k\to\infty$, $k \in K$. By (6.7) and the boundedness of $\{g^k\}$, we have $c^k \xrightarrow{K} \bar{c}$ with positive $\bar{c}=(1-m_R)|\bar{p}|$, since $m_R < 1$.

(iv) Let k_c satisfy $k_c \bar{c}/2 > \bar{s}(1-m_e)e$. From part (iii) above we deduce the existence of k_p such that $c^k \ge \bar{c}/2$ for $k=k_p,k_p+1,\dots,k_p+k_c$.

But $s^{k_p} \le \bar{s}$, so (6.8) yields $s^{k_p+k_c+1} \le \bar{s} - k_c \bar{c}/2(1-m_e)e < 0$, contradicting the nonnegativity of $\{s^k\}$. Hence $\bar{p}=0$ and the proof is complete. $\square$

We conclude that the use of $\{\tilde{v}^k\}$ in Algorithm 3.1 does not impair the global convergence results of Section 4, provided that $m_e + m_R \le 1$.

We may add that all the above-described modifications may be used in Algorithm 5.1, without impairing the results of Section 5.

7. Extension to Nonconvex Unconstrained Problems.

In this section we shall present boundle methods for the problem of minimizing a locally Lipschitzian function $f: R^N \to R$ that is not necessarily convex or differentiable. As before, we suppose that for any $x \in R^N$ one can find in finite time a subgradient $g_f(x) \in \partial f(x)$ of f at x.

Our extension of the bundle method from Section 3 to the nonconvex case will use the techniques for dealing with nonconvexity developed in Chapter 3. The relevant features of this approach are the following.

First, we observe that for convex f the k-th dual search direction finding subproblem (2.5) uses automatic weighing of the past subgradients g^j by their linearization errors $\alpha_j^k = f(x^k) - f_j^k \ge 0$ in the sense that g^j contributes to d^k with a relatively small weight $\lambda_j^k > 0$ if the value of α_j^k is large. This important property is ensured by the last constraint of (2.5) and the nonnegativity of α_j^k.

Secondly, we recall that in the convex case linearization errors may serve as subgradient locality measures in the sense that

$$g^j \in \partial_\varepsilon f(x^k) \quad \text{for} \quad \varepsilon = \alpha_j^k \ge 0,$$

so that $g^j = g_f(y^j) \in \partial f(y^j)$ is close to $\partial f(x^k)$ if α_j^k is small, even if y^j is far from x^k. This property need not, of course, hold if f is nonconvex. Therefore, in Chapter 3 we used subgradient locality measures of the form

$$\alpha(x,y) = \max\{|f(x) - \bar{f}(x;y)|, \gamma|x-y|^2\} \tag{7.1}$$

for indicating how much the subgradient $g_f(y)$ differs from being a member of $\partial f(x)$, where $f(x)-\overline{f}(x;y)$ is the error with which the linearization of f calculated at y

$$\overline{f}(x;y) = f(y) + \langle g_f(y), x-y \rangle \tag{7.2}$$

approximates f at x, and γ is a positive parameter, which may be set to zero when f is convex. The save storage, we approximated $\alpha(x^k,y^j)$ by

$$\alpha_j^k = \max\{|f(x^k)-f_j^k|, \gamma(s_j^k)^2\}, \tag{7.3}$$

where $f_j^k=\overline{f}(x^k;y^j)$ and

$$s_j^k = |y^j-x^j| + \sum_{i=j}^{k-1} |x^{i+1}-x^i|$$

is an overestimate of $|x^k-y^j|$.

Thirdly, in Chapter 3 we used a simple resetting strategy involving, at the k-th iteration, the locality radius a^k that estimated the radius of the ball around x^k from which past subgradients g^j had been collected to form the k-th aggregate subgradient. This strategy ensured locally uniform boundedness of such subgradients, which was important for global convergence under no additional boundedness assumptions.

In view of the above remarks, it is natural to extend Algorithm 3.1 to the nonconvex case by using subgradient locality measures of the form (7.2)-(7.3) and the resetting strategy of Chapter 3. The resulting method is given below.

Algorithm 7.1.

Step 0 (Initialization). Select the starting point $x^1 \in R^N$, a stopping parameter $\varepsilon_s \geq 0$ and an approximation tolerance $e_a > 0$. Choose positive line search parameters m_L, m_R, m_e, $\overline{t}$ and $\tilde{t}$ satisfying $m_L < m_R < 1$, $m_e < 1$ and $\overline{t} \leq 1 \leq \tilde{t}$. Select a distance measure parameter $\gamma > 0$ ($\gamma=0$ if f is convex) and a resetting tolerance $\overline{a} > 0$. Set $J^1=\{1\}$, $y^1=x^1$, $p^0=g^1=g_f(y^1)$, $f_p^1=f_1^1=f(y^1)$, $s_p^1=s_1^1=0$ and $e^1=e_a$. Set $a^1=0$ and the reset indicator $r_a^1=1$. Set the counters k=1, l=0 and k(0)=1.

Step 1 (Direction finding). Find multipliers λ_j^k, $j \in J^k$, and λ_p^k that solve the k-th dual subproblem

$$\underset{\lambda}{\text{minimize}}\ \tfrac{1}{2}\Big|\sum_{j \in J^k}\lambda_j g^j + \lambda_p p^{k-1}\Big|^2,$$

$$\text{subject to} \quad \lambda_j \geq 0,\ j \in J^k,\ \lambda_p \geq 0,\ \sum_{j \in J^k} \lambda_j + \lambda_p = 1,\ \lambda_p = 0 \text{ if } r_a^k = 1, \tag{7.4}$$
$$\sum_{j \in J^k} \lambda_j \alpha_j^k + \lambda_p \alpha_p^k \leq e^k,$$

where α_j^k are given by (7.3) and

$$\alpha_p^k = \max\{|f(x^k) - f_p^k|,\ \gamma(s_p^k)^2\}. \tag{7.5}$$

Set

$$(p^k, \tilde{f}_p^k, \tilde{s}_p^k) = \sum_{j \in J^k} \lambda_j^k (g^j, f_j^k, s_j^k) + \lambda_p^k (p^{k-1}, f_p^k, s_p^k), \tag{7.6}$$

$d^k = -p^k$ and $v^k = -|p^k|^2$. If $\lambda_p^k = 0$ set $a^k = \max\{s_j^k : j \in J^k\}$.

Step 2 (Stopping criterion). Set

$$\tilde{\alpha}_p^k = \max\{|f(x^k) - \tilde{f}_p^k|,\ \gamma(\tilde{s}_p^k)^2\}. \tag{7.7}$$

If $\max\{|p^k|^2, \tilde{\alpha}_p^k\} \leq \varepsilon_s$, terminate; otherwise, continue.

Step 3 (Approximation tolerance decreasing). If $|p^k|^2 > \tilde{\alpha}_p^k$ then go to Step 4; otherwise, i.e. if $|p^k|^2 \leq \tilde{\alpha}_p^k$, replace e^k by $m_e e^k$ and go to Step 1.

Step 4 (Line search). By a line search procedure (e.g. Line Search Procedure 3.2), find two stepsizes t_L^k and t_R^k such that $0 \leq t_L^k \leq t_R^k \leq \tilde{t}$ and $t_R^k = t_L^k$ if $t_L^k > 0$, and such that the two corresponding points defined by $x^{k+1} = x^k + t_L^k d^k$ and $y^{k+1} = x^k + t_R^k d^k$ satisfy

$$f(x^{k+1}) \leq f(x^k) + m_L t_L^k v^k, \tag{7.8a}$$

$$t_L^k \geq \bar{t} \quad \text{or} \quad \alpha(x^k, x^{k+1}) > m_e e^k \quad \text{if} \quad t_L^k > 0, \tag{7.8b}$$

$$\alpha(x^k, y^{k+1}) \leq m_e e^k \quad \text{if} \quad t_L^k = 0, \tag{7.8c}$$

$$\langle g_f(y^{k+1}), d^k \rangle \geq m_R v^k \quad \text{if} \quad t_L^k = 0, \tag{7.8d}$$

where $\alpha(x,y)$ is defined by (7.1).

Step 5 (Approximation tolerance updating). If $t_L^k = 0$ (null step), set $e^{k+1} = e^k$; otherwise, i.e. if $t_L^k > 0$ (serious step), choose $e^{k+1} \geq e_a$ (e.g. by (6.1)), set $k(l+1) = k+1$, and increase l by 1.

Step 6 (Linearization updating). Choose $\hat{J}^k \subset J^k$ such that the set $J^{k+1} = \hat{J}^k \cup \{k+1\}$ contains $k(l)$, i.e. $k(l) \in \hat{J}^k$ if $k(l) < k+1$. Set $g^{k+1} = g_f(y^{k+1})$ and compute

$$f_{k+1}^{k+1} = f(y^{k+1}) + \langle g^{k+1}, x^{k+1} - y^{k+1} \rangle,$$

$$f_j^{k+1} = f_j^k + \langle g^j, x^{k+1} - x^k \rangle \quad \text{for } j \in \hat{J}^k, \tag{7.9}$$

$$f_p^{k+1} = \tilde{f}_p^k + \langle p^k, x^{k+1} - x^k \rangle,$$

$$s_{k+1}^{k+1} = |y^{k+1} - x^{k+1}|,$$

$$s_j^{k+1} = s_j^k + |x^{k+1} - x^k| \quad \text{for } j \in \hat{J}^k, \tag{7.10}$$

$$s_p^{k+1} = \tilde{s}_p^k + |x^{k+1} - x^k|.$$

Step 7 (Distance resetting test). Set $a^{k+1} = \max\{a^k + |x^{k+1} - x^k|, s_{k+1}^{k+1}\}$. If $a^{k+1} \le \bar{a}$ or $t_L^k = 0$ then set $r_a^{k+1} = 0$ and go to Step 9; otherwise, set $r_a^{k+1} = 1$ and go to Step 8.

Step 8 (Distance resetting). Delete from J^{k+1} all indices j with $s_j^{k+1} > \bar{a}/2$, and set $a^{k+1} = \max\{s_j^{k+1} : j \in J^{k+1}\}$.

Step 9. Increase k by 1 and go to Step 1.

A few comments on the method are in order.

The subgradient selection and aggregation rules of the method are borrowed from Algorithm 3.3.1, hence the properties of the aggregate subgradients $(p^k, \tilde{f}_p^k, \tilde{s}_p^k)$ may be deduced from Lemmas 3.4.1-3.4.3. Moreover, since the stationarity measure

$$w^k = \tfrac{1}{2}|p^k|^2 + \tilde{\alpha}_p^k$$

satisfies

$$w^k \le 2\max\{|p^k|^2, \tilde{\alpha}_p^k\} \quad \text{and} \quad \max\{|p^k|^2, \tilde{\alpha}_p^k\} \le 2w^k, \tag{7.11}$$

the stopping criterion of the method may be interpreted similarly to that of Algorithm 3.3.1; see Section 3.3. Also we have (3.5) and (3.7) if f happens to be convex.

The method updates the approximation tolerances e^k according to the modified rules of Section 6. Note that we still have $\tilde{\alpha}_p^k \le e^k$ for all k, since Lemma 3.4.8 yields a suitable extension of relation (3.8). We do not use the seemingly more efficient strategy of Algorithm 6.1, because it may impair global convergence in the nonconvex case. We may add that in practice one should set e_a equal to a small positive number, e.g. $e_a=\varepsilon_s=10^{-6}$.

Line Search Procedure 3.2 may be used for executing Step 4 of the method. It is easy to verify that this procedure will terminate in a finite number of iterations if f has the semismoothness property (3.3.23); see the proof of Lemma 3.3.

The subgradient deletion rules of Step 7 and 8 differ slightly from those of Algorithm 3.3.1 in that no distance reset can occur after a null step. At the same time, we still have (3.13), since $s_{k(l)}^k = |x^k - y^{k(l)}| = |x^k - x^{k(l)}| = 0$, while in Step 8 $s_{k+1}^{k+1} = |y^{k+1} - x^{k+1}| = 0$, since $t_L^k = t_R^k > 0$, so $k+1$ cannot be deleted from J^{k+1}.

Let us now analyze convergence of the method. To this end, we shall need the following analogue of Lemma 4.3.

Lemma 7.2. Suppose that Algorithm 7.1 generates an infinite sequence $\{x^k\}$ such that $x^k \xrightarrow{K} \bar{x}$, $|p^k| \xrightarrow{K} 0$ and $\tilde{\alpha}_p^k \xrightarrow{K} 0$ for some $\bar{x}$ and an infinite set $K \subset \{1,2,\dots\}$. Then $0 \in \partial f(\bar{x})$.

Proof. Inspecting the proofs of Lemmas 3.4.1, 3.4.2, 3.4.6 and 3.4.7, we see that $0 \in \partial f(\bar{x})$ if $\{a^k\}_{k \in K}$ is bounded. Therefore we need only show that $\{a^k\}_{k \in K}$ is bounded. Let $k(l) \le k < k(l+1)$, so that $x^j = x^{k(l)}$ if $k(l) \le j \le k$. From the proof of Lemma 3.4.1 we deduce that the s_j^k-s which form $a^k = \max\{s_j^k : j \in \hat{J}_p^k\}$ may be divided into two disjoint groups. The first group comprises $s_j^k \le \bar{a}$ with $j \le k(l)$, since the rules of Steps 7 and 8 ensure $a^{k(l)} \le \bar{a}$ after a serious step, while s_j^k stay constant between serious steps. Since the second group contains $s_j^k = |y^j - x^j| = t_R^j |d^j| \le \tilde{t}|p^j|$ with $x^j = x^k$ and $k(l) < j \le k$, while $x^k \xrightarrow{K} \bar{x}$, we deduce from the proof of Lemma 4.5 that such s_j^k-s are uniformly bounded for $k \in K$, because so are the corresponding p^j-s. Hence $\{a^k\}_{k \in K}$ is bounded, as desired. □

Combining the above proof with the proof of Lemma 4.2, we deduce that Lemma 4.2 holds for Algorithm 7.1. Similarly we obtain that if the method stops at the k-th iteration and $\varepsilon_s=0$, then $0 \in \partial f(x^k)$. Next, one easily checks that the proofs of Lemmas 4.4-4.7 need not be modified, so Theorem 4.8 and Corollary 4.11 are true for Algorithm 7.1. Moreover, if f is convex then $f(x^k)-\tilde{f}_p^k \le \tilde{\alpha}_p^k$ for all k (see Lemma 3.4.2), hence in the convex case Theorem 4.9 and Corollary 4.10 hold for Algorithm 7.1. We conclude that Algorithm 7.1 is a globally convergent method.

We may add that one may modify the line search criteria of the method by replacing v^k in (7.8) by $\tilde{v}^k$ defined by (6.5). This modification will not impair the preceding global convergence results, since the proof of Lemma 6.4 remains valid.

8. Bundle Methods for Convex Constrained Problems.

In this section we shall present bundle methods for solving the following convex minimization problem

$$\text{minimize } f(x), \text{ subject to } F(x) \le 0, \tag{8.1}$$

where the functions $f: R^N \to R$ and $F: R^N \to R$ are convex, but not necessarily differentiable. We assume that the Slater constraint qualification is fulfilled, i.e. $F(\tilde{x}) < 0$ for some $\tilde{x} \in R^N$, so that the feasible set

$$S = \{x \in R^N : F(x) \le 0\}$$

has a nonempty interior. Moreover, we suppose that we have a finite process for calculating $f(x)$ and a subgradient $g_f(x) \in \partial f(x)$ of f at each $x \in S$, and $F(x)$ and a subgradient $g_F(x) \in \partial F(x)$ of F at each $x \in R^N \setminus S$. For simplicity of exposition, we shall initially assume that one can compute $f(x), g_f(x)$, $F(x)$ and $g_F(x)$ at any x.

Our extension of the bundle methods from Sections 3 and 5 to the constrained case will use the approach of Chapter 5. To this end, we recall from Lemma 5.2.1 that in terms of the improvement function

$$H(y;x) = \max\{f(y)-f(x), F(y)\} \tag{8.2}$$

and its subdifferential $\partial H(y;x)$ at y (for a fixed x)

$$\partial H(y;x) = \begin{cases} \partial f(y) & \text{if } f(y)-f(x) > F(y), \\ \text{conv}\{\partial f(y)\ \partial F(y)\} & \text{if } f(y)-f(x) = F(y), \\ \partial F(y) & \text{if } f(y)-f(x) < F(y), \end{cases} \tag{8.3}$$

a necessary and sufficient condition for a point $\bar{x} \in R^N$ to minimize f on S is given by each of the following equivalent relations

$$\min\{H(y;\bar{x}) : y \in R^N\} = H(\bar{x};\bar{x}) = 0, \tag{8.4}$$

$$0 \in \partial H(\bar{x};\bar{x}).$$

Thus testing if a point $x \in S$ is optimal may be done by trying to find a direction of descent for the convex function $H(\cdot,x)$ at x. If such a direction exists, then moving from x along this direction will yield a better point; otherwise, x is optimal.

The above remarks suggest the following extension of the bundle methods. Suppose that at the k-th iteration we have a point $x^k \in S$. We would like to find a direction of descent for the convex function

$$H^k(x) = \max\{f(x)-f(x^k),F(x)\}$$

at x^k. Treating H^k as our temporary objective function to be minimized, we may proceed as in Section 2 by finding a direction $d^k = -\mathrm{Nr}G^k(e)$, where $G^k(e) \subset \partial_e H^k(x^k)$ is a certain approximation to $\partial H^k(x^k)$ and $e \geq 0$ is the approximation tolerance. For constructing $G^k(e)$ we may use past subgradient $g_f^j = g_f(y^j)$ and $g_F^j = g_F(y^j)$ calculated at trial points y^j, $j=1,\ldots,k$, and the corresponding linearizations

$$f_j(x) = f(y^j) + \langle g_f^j, x-y^j \rangle ,$$

$$F_j(x) = F(y^j) + \langle g_F^j, x-y^j \rangle$$

of f and F, respectively. We recall from Lemma 5.7.2 that g_f^j and g_F^j may be regarded as ε-subgradients of H^k at x^k, since in terms of the linearization errors

$$\alpha_{f,j}^k = f(x^k) - f_j^k \quad \text{and} \quad \alpha_{F,j}^k = -F_j^k, \tag{8.6}$$

where $f_j^k = f_j(x^k)$ and $F_j^k = F_j(x^k)$, we have $(x^k \in S)$

$$g_f^j \in \partial_\varepsilon H^k(x^k) \quad \text{for} \quad \varepsilon = \alpha_{f,j}^k,$$

$$g_F^j \in \partial_\varepsilon H^k(x^k) \quad \text{for} \quad \varepsilon = \alpha_{F,j}^k.$$

It follows, as in Section 2, that for any $e \geq 0$ the convex polyhedron

$$G^k(e)=\{g\in R^N: g=\sum_{j\in J_f^k}\lambda_j g_f^j+\sum_{j\in J_F^k}\mu_j g_F^j,\ \sum_{j\in J_f^k}\lambda_j\alpha_{f,j}^k+\sum_{j\in J_F^k}\mu_j\alpha_{F,j}^k\le e,$$

$$\lambda_j\ge 0, j\in J_f^k, \mu_j\ge 0, j\in J_F^k, \sum_{j\in J_f^k}\lambda_j+\sum_{j\in J_F^k}\mu_j=1\},\quad (8.7)$$

where $J_f^k\cup J_F^k\subset\{1,\dots,k\}$, satisfies $G^k(e)\subset\partial_e H^k(x^k)$. Hence finding d^k by computing $p^k=\mathrm{Nr}G^k(e^k)$ and setting $d^k=-p^k$ corresponds to the construction used by the bundle method with subgradient selection from Section 5.

To derive a bundle method with subgradient aggregation, suppose that at the k-th iteration we have two (k-1)-st aggregate subgradients

$$(p_f^{k-1},f_p^k)\in\mathrm{conv}\{(g_f^j,f_j^k): j=1,\dots,k-1\},$$
$$(p_F^{k-1},F_p^k)\in\mathrm{conv}\{(g_F^j,F_j^k): j=1,\dots,k-1\}$$

of f and F, respectively. Once again Lemma 5.7.2 yields that

$$p_f^{k-1}\in\partial_\varepsilon H^k(x^k)\quad\text{for}\quad \varepsilon=\alpha_{f,p}^k,$$
$$p_F^{k-1}\in\partial_\varepsilon H^k(x^k)\quad\text{for}\quad \varepsilon=\alpha_{F,p}^k,$$

with the linearization errors

$$\alpha_{f,p}^k=f(x^k)-f_p^k\quad\text{and}\quad \alpha_{F,p}^k=-F_p^k.\quad (8.8)$$

Hence for $e\ge 0$ the polyhedron

$$G_a^k(e)=\{g\in R^N: g=\sum_{j\in J^k}\lambda_j g^j+\lambda_p p^{k-1}+\sum_{j\in J_F^k}\mu_j g_F^j+\mu_p p_F^{k-1},$$

$$\sum_{j\in J_f^k}\lambda_j\alpha_{f,j}^k+\lambda_p\alpha_{f,p}^k+\sum_{j\in J_F^k}\mu_j\alpha_{F,j}^k+\mu_p\alpha_{F,p}^k\le e,$$

$$\lambda_j\ge 0, j\in J_f^k, \lambda_p\ge 0, \mu_j\ge 0, j\in J_F^k, \mu_p\ge 0,$$

$$\sum_{j\in J_F^k}\lambda_j+\lambda_p+\sum_{j\in J_F^k}\mu_j+\mu_p=1\}\quad (8.9)$$

satisfies $G_a^k(e)\subset\partial_e H^k$. Therefore finding d^k by calculating $p^k=\mathrm{Nr}G_a^k(e^k)$ and setting $d^k=-p^k$ corresponds to the construction used by the bundle method with subgradient aggregation from Section 3.

Moreover, $v^k=-|p^k|^2$ may be regarded as an approximate directional derivative of H^k at x^k in the direction d^k; cf. (3.10).

Having derived the search direction finding subproblems, we may state the bundle method with subgradient aggregation for solving problem (8.1).

Algorithm 8.1.

Step 0 (Initialization). Select a starting point $x^1 \in S$, a stopping parameter $\varepsilon_s \geq 0$ and an approximation tolerance $\varepsilon_a > 0$. Choose positive line search parameters m_L, m_R, m_e, $\bar{t}$ and $\tilde{t}$ satisfying $m_L < m_R < 1$, $m_e < 1$ and $\bar{t} \leq 1 \leq \tilde{t}$. Set $J_f^1=J_F^1=\{1\}$, $y^1=x^1$, $p_f^0=g_f^1=g_f(y^1)$, $p_F^0=g_F^1=g_F(y^1)$, $f_p^1=f_1^1=f(y^1)$, $F_p^1=F_1^1=F(y^1)$, and $e^1=e_a$. Set the counters $k=1$, $l=0$ and $k(0)=1$.

Step 1 (Direction finding). Find multipliers λ_j^k, $j \in J_f^k$, λ_p^k, μ_j^k, $j \in J_F^k$, and λ_p^k that solve the k-th dual subproblem

$$\underset{\lambda,\mu}{\text{minimize}}\ \tfrac{1}{2}\Big|\sum_{j \in J_F^k}\lambda_j g_f^j+\lambda_p p_f^{k-1}+\sum_{j \in J_F^k}\mu_j g_F^j+\mu_p p_F^{k-1}\Big|^2,$$

$$\begin{aligned} \text{subject to}\quad & \lambda_j \geq 0,\ j \in J_f^k,\ \lambda_p \geq 0,\ \mu_j \geq 0,\ j \in J_F^k,\ \mu_p \geq 0, \\ & \sum_{j \in J_f^k}\lambda_j + \lambda_p + \sum_{j \in J_F^k}\mu_j + \mu_p = 1, \\ & \sum_{j \in J_f^k}\lambda_j\alpha_{f,j}^k+\lambda_p\alpha_{f,p}^k+\sum_{j \in J_F^k}\mu_j\alpha_{F,j}^k+\mu_p\alpha_{F,p}^k \leq e^k. \end{aligned} \tag{8.10}$$

Compute the scaled multipliers

$$\nu_f^k = \sum_{j \in J_f^k}\lambda_j^k + \lambda_p^k \quad \text{and} \quad \nu_F^k = \sum_{j \in J_F^k}\mu_j^k + \mu_p^k,$$

$$\begin{aligned} &\tilde{\lambda}_j^k=\lambda_j^k/\nu_f^k \ \text{ for } j \in J_f^k \ \text{ and } \ \tilde{\lambda}_p^k=\lambda_p^k/\nu_f^k \ \text{ if } \nu_f^k \neq 0, \\ &\tilde{\lambda}_k^k=1,\ \tilde{\lambda}_j^k=0 \ \text{ for } j \in J_f^k \setminus \{k\} \ \text{ and } \ \tilde{\lambda}_p^k=0 \ \text{ if } \nu_f^k=0, \\ &\tilde{\mu}_j^k=\mu_j^k/\nu_F^k \ \text{ for } j \in J_F^k \ \text{ and } \ \tilde{\mu}_p^k=\mu_p^k/\nu_F^k \ \text{ if } \nu_F^k \neq 0, \\ &\tilde{\mu}_k^k=1,\ \tilde{\mu}_j^k=0 \ \text{ for } j \in J_F^k \setminus \{k\} \ \text{ and } \ \tilde{\mu}_p^k=0 \ \text{ if } \nu_F^k=0. \end{aligned} \tag{8.11}$$

Calculate the aggregate subgradients

$$(p_f^k,\tilde f_p^k) = \sum_{j\in J_f^k}\tilde\lambda_j^k(g_f^j,f_j^k)+\tilde\lambda_p^k(p_f^{k-1},f_p^k),$$
$$(p_F^k,\tilde F_p^k) = \sum_{j\in J_F^k}\tilde\mu_j^k(g_F^j,F_j^k)+\tilde\mu_p^k(p_F^{k-1},F_p^k), \tag{8.12}$$

$$p^k=\nu_f^k p_f^k + \nu_F^k p_F^k, \tag{8.13}$$

and set $d^k=-p^k$ and $v^k=-|p^k|^2$.

<u>Step 2 (Stopping criterion)</u>. Set

$$\tilde\alpha_{f,p}^k = f(x^k)-\tilde f_p^k \quad\text{and}\quad \tilde\alpha_{F,p}^k = -\tilde F_p^k, \tag{8.14}$$

$$\tilde\alpha_p^k = \nu_f^k\tilde\alpha_{f,p}^k + \nu_F^k\tilde\alpha_{F,p}^k. \qquad 8.15$$

If $\max\{|p^k|^2,\tilde\alpha_p^k\}\le\varepsilon_s$, terminate; otherwise, continue.

<u>Step 3 (Approximation tolerance decreasing)</u>. If $|p^k|^2 > \tilde\alpha_p^k$ then go to Step 4. Otherwise, i.e. if $|p^k|^2\le\tilde\alpha_p^k$, replace e^k by $m_e e^k$ and go to Step 1.

<u>Step 4 (Line search)</u>. By a line search procedure as given below, find three not necessarily different stepsizes t_L^k, t_R^k and t_B^k such that $0\le t_L^k\le t_R^k\le\tilde t$ and $t_L^k\le t_B^k\le 10t_L^k$, and $t_R^k=t_L^k$ if $t_L^k>0$, and such that the three corresponding points defined by $x^{k+1}=x^k+t_L^k d^k$, $y^{k+1}=x^k+t_R^k d^k$ and $\tilde y^{k+1}=x^k+t_B^k d^k$ satisfy

$$f(x^{k+1})\le f(x^k)+m_L t_L^k v^k, \tag{8.16a}$$

$$F(x^{k+1})\le 0, \tag{8.16b}$$

$$t_L^k\ge\bar t \quad\text{or}\quad \max\{\alpha(x^k,x^{k+1}),\alpha(x^k,\tilde y^{k+1})\}>m_e e^k \quad\text{if}\quad t_L^k>0, \tag{8.16c}$$

$$\alpha(x^k,y^{k+1})\le m_e e^k \quad\text{if}\quad t_L^k=0, \tag{8.16d}$$

$$\langle g(y^{k+1}),d^k\rangle \ge m_R v^k \quad\text{if}\quad t_L^k=0, \tag{8.16e}$$

where

$$g(y)=g_f(y) \quad\text{and}\quad \alpha(x,y)=f(x)-f(y)-\langle g_f(y),x-y\rangle \quad\text{if}\quad y\in S,$$
$$g(y)=g_F(y) \quad\text{and}\quad \alpha(x,y)=-F(y)-\langle g_F(y),x-y\rangle \quad\text{if}\quad y\notin S. \tag{8.17}$$

Step 5 (Approximation tolerance updating). If $t_L^k=0$ (null step), set $e^{k+1}=e^k$; otherwise, i.e. if $t_L^k>0$ (serious step), choose $e^{k+1}\geq e_a$ (e.g. by (6.1)), set $k(l+1)=k+1$, and increase l by 1.

Step 6 (Linearization updating). Choose $\hat{J}_f^k\subset J_f^k$ and $\hat{J}_F^k\subset J_F^k$ such that $k(l)\in\hat{J}_f^k$ if $k(l)<k+1$, and set $J_f^{k+1}=\hat{J}^k\cup\{k+1\}$ and $J_F^{k+1}=\hat{J}_F^k\cup\{k+1\}$. Set $g_f^{k+1}=g_f(y^{k+1})$, $g_F^{k+1}=g_F(y^{k+1})$ and calculate f_j^{k+1}, $j\in J_f^{k+1}$, f_p^{k+1}, F_j^{k+1}, $j\in J_F^{k+1}$, and F_p^{k+1} by (6.3.14).

Step 7. Increase k by 1 and go to Step 1.

A few remarks on the algorithm are in order.

The subgradient aggregation rules of the method are taken from Algorithm 5.3.1. Hence the properties of the aggregate subgradients are expressed by Lemmas 5.4.1, 5.4.2 and 5.7.2. In particular, we have

$$p^k\in\partial_\varepsilon H(x^k,x^k)\quad\text{for}\quad\varepsilon=\tilde{\alpha}_p^k. \tag{8.18}$$

Therefore, if the algorithm terminates at the k-th iteration then $p^k\in\partial_{\varepsilon_s}H(x^k;x^k)$ and $|p^k|^2\leq\varepsilon_s$. This estimate justifies our stopping criterion and shows that x^k is stationary for f on S if $\varepsilon_s=0$, since stationary points $\bar{x}$ satisfy $0\in\partial H(\bar{x};\bar{x})$ and are optimal for problem (8.1).

The method updates the approximation tolerances e^k according to modified rules of Section 6. Note that we always have $\tilde{\alpha}_p^k\leq e^k$, since the proof of Lemma 6.4.6 yields a suitable extension of relation (3.8). It is worth adding that the method may also use the efficient strategy of Algorithm 6.1 for regulating e^k, as will be shown below.

The line search criteria (8.16) extend (3.2) to the constrained case in a way that ensures monotonicity $(f(x^{k+1})\leq f(x^k))$, feasibility $(\{x^k\}\subset S)$, sufficiently large t_L^k at serious steps, and a significant enrichment of $G^k(e^k)$ after a null step (one may derive (3.11) from (8.16d,e) and (8.17) by defining

$$\begin{aligned} g^k&=g_f(y^k)\quad\text{and}\quad\alpha^k=\alpha_{f,k}^k\quad\text{if}\quad y^k\in S,\\ g^k&=g_F(y^k)\quad\text{and}\quad\alpha^k=\alpha_{F,k}^k\quad\text{if}\quad y^k\notin S\end{aligned} \tag{8.19}$$

and noting that

$$g^k=g(y^k)\quad\text{and}\quad\alpha^k=\alpha(x^{k-1},y^k)\quad\text{if}\quad t_L^{k-1}=0, \tag{8.20}$$

for all $k > 1$). The nontrivial aspect of this extension consists in allowing for a serious step when $\alpha(x^k,\tilde{y}^{k+1}) > m_e e^k$, which indicates, by the properties of the function α and the fact that $m_e e^k$ is positive, that $|\tilde{y}^{k+1}-x^k|$, and hence also $|x^{k+1}-x^k|$, are sufficiently large, so that significant progress occurs. This follows from the fact that by construction

$$|x^{k+1}-x^k| \le |\tilde{y}^{k+1}-x^k| \le 10|x^{k+1}-x^k|. \tag{8.21}$$

The reason for our introduction of the additional stepsize t_B^k will become clear in the analysis of convergence of the following procedure for finding stepsizes $t_L=t_L^k$, $t_R=t_R^k$ and $t_B=t_B^k$ to meet the requirements of Step 4 with $x=x^k$, $d=d^k$, $e=e^k$ and $v=v^k$.

Line Search Procedure 8.2.

(a) Set $t_L=0$ and $t=t_U=1$.

(b) If $F(x+td) \le 0$ and $f(x+td) \le f(x)+m_L tv$ set $t_L=t$; otherwise set $t_U=t$.

(c) If $t_L > 0$ and either $t_L \ge \bar{t}$ or $\max\{\alpha(x,x+t_L d), \alpha(x,x+t_U d)\} > m_e e$ set $t_R=t_L$ and $t_B=t_U$, and return.

(d) If $\alpha(x,x+td) < m_e e$ and $\langle g(x+td),d\rangle \ge m_R v$ set $t_R=t$ and $t_L=t_B=0$, and return.

(e) Choose $t \in [t_L+0.1(t_U-t_L),\ t_U-0.1(t_U-t_L)]$ by some interpolation procedure and go to (b).

Lemma 8.3. Line Search Procedure 8.2 terminates in a finite number of iterations with stepsizes $t_L^k=t_L$, $t_R^k=t_R$ and $t_B^k=t_B$ satisfying the requirements of Step 4 of Algorithm 8.1.

Proof. We shall use a combination of the proofs of Lemmas 3.3 and 6.3.3. Assume, for contradiction purposes, that the search does not terminate. Let

$$TL=\{t \ge 0 : f(x+td) \le f(x)+m_L tv \quad \text{and} \quad F(x+td) \le 0\}.$$

Denote by t^i, $\tilde{t}_L^i$ and t_U^i the values of t, t_L and t_U after the i-th execution of step (b) of the procedure, and let $I=\{i : t^i=t_U^i\}$, $I_f=\{i : t^i=t_U^i$ and $F(x+t^i d) \le 0\}$ and $I_F=\{i : t^i=t_U^i$ and $F(x+t^i d) > 0\}$, so that $I=I_f \cup I_F$. We deduce, as in the proofs of Lemmas 3.3 and 6.3.3, the existence of $\hat{t}\geq 0$ such that $\tilde{t}_L^i \uparrow t$, $t_U^i \downarrow \hat{t}$ and $\hat{t}\in LT$, and that the set I is infinite. We shall consider the following two cases.

(i) Suppose that $\hat{t}>0$. Then, since $\tilde{t}_L^i \uparrow \hat{t}$, we have $\tilde{t}_L^i>0$ for large i, and, since $t^i\in\{\tilde{t}_L^i, t_U^i\}$ for all i, the rules of step (c) imply that step (d) is entered with $\alpha(x,x+t^i d)<m_e e$ for large i. Therefore in step (d)

$$\langle g(x+t^i d),d\rangle < m_R v \quad \text{for all large i.} \tag{8.22}$$

(ii) Suppose that $\hat{t}=0$. Then $\tilde{t}_L^i=0$ and $t^i=t_U^i$ for all i, so $t^i\downarrow 0$ and, by (8.17) and the local boundedness of g_f and g_F, we have $\alpha(x,x+t^i d)\to 0$. Hence $\alpha(x,x+t^i d)<m_e e$ at step (d) for large i, because $m_e e>0$, so we again obtain (8.22).

Making use of the above results, one may proceed as in the proof of Lemma 6.3.3 to derive a contradiction between (8.22) and the fact that f and F, being convex functions, have the semismoothness properties (3.3.23) and (6.3.18) (see Remarks 3.3.4 and 6.3.4). Therefore the search terminates. It is easy to show that (8.16) holds at termination. Also the first positive $\tilde{t}_L^i$ must satisfy $\tilde{t}_L^i=t^i>0.1t_U^i$ by the choice of t at step (e), and then we have $t_U^{i+1}\leq t_U^i\leq 10\tilde{t}_L^i\leq 10\tilde{t}_L^{i+1}$. This shows that $t_B\leq 10t_L$ at termination. □

The rules for choosing J_f^{k+1} at Step 6 yield the following analogue of (3.13)

$$k(l)\in J_f^k,\ g_f^{k(l)}=g_f(x^k) \quad \text{and} \quad \alpha_{f,k(l)}^k=0 \quad \text{if} \quad k(l)\leq k<k(l+1), \tag{8.23}$$

which ensures that the constraints of the k-th subproblem (8.10) are consistent $(G_a^k(e^k)\neq\emptyset)$. We may add that Remarks 5.3.4 and 6.2.3 and Section 6.7 indicate how to modify the choice of J_f^{k+1} and J_F^{k+1} and impose the additional constraint $\mu_p=0$ in subproblem (8.10) at certain iterations in order to treat the case when one cannot compute $f(x)$ and $g_f(x)$ at $x\notin S$, and $F(x)$ and $g_F(x)$ at $x\in S$. Also the subsequent proofs may be easily modified to cover such modifications. We shall leave this task to the reader.

We shall now establish convergence of the method by modifying the analysis of Section 4 with the help of the results of Section 5.4. To this end, define the stationarity measure

$$w^k=\tfrac{1}{2}|p^k|^2+\tilde{\alpha}_p^k \tag{8.24a}$$

and observe that we always have

$$w^k \le 2\max\{|p^k|^2+\tilde{\alpha}_p^k\} \quad \text{and} \quad \max\{|p^k|^2,\tilde{\alpha}_p^k\} \le 2w^k. \tag{8.24b}$$

We showed above that if the method terminates at the k-th iteration, and $\varepsilon_s=0$, then x^k solves problem (8.1). Therefore, we shall assume from now on that the method does not stop. We shall now show that Lemmas 4.2-4.7 hold for Algorithm 8.1 if one replaces in their formulations and proofs the relation $0 \in \partial f(\bar{x})$ by $0 \in \partial H(\bar{x};\bar{x})$.

First, we observe that Lemmas 4.2 and 4.3 may be established for Algorithm 8.1 by combining their original proofs with the proof of Lemma 5.4.7 and using (8.24). In view of the line search rules (8.1a,b), Lemma 4.4 is valid for Algorithm 8.1 with the additional assertion that $F(\bar{x}) \le 0$, which follows from the continuity of F and the feasibility of $\{x^k\}$. In the proof of Lemma 4.5, use (8.23) and the proof of Lemma 6.4.8 in parts (ii)-(iii), and (8.19)-(8.20) for deriving (4.13) as in the proof of Lemma 5.4.11. The proof of Lemma 4.6 need not be modified. In the proof of Lemma 4.7, replace $\alpha(x^k,x^{k+1})$ in (4.18)-(4.19) by $\max\{\alpha(x^k,x^{k+1}),\ \alpha(x^k,\tilde{y}^{k+1})\}$, observing that this replacement is valid in virtue of (8.16c), (8.21) and the elementary property of α

$$\alpha(x,y)\to 0 \quad \text{if} \quad x,y \to \bar{x} \in S,$$

which is a consequence of (8.17) and the local boundedness of g_f and g_F. In effect, we obtain the following analogue of Theorem 4.8.

<u>Theorem 8.4</u>. Every accumulation point of a sequence $\{x^k\}$ generated by Algorithm 8.1 is stationary for f on S.

Combining the above theorem with the above-described extension of Lemma 4.4 and the fact that we have (8.18) and $t_L^k \le \tilde{t}$ for all k, one may use the corresponding proofs of Sections 4 and 5.4 to establish the following results.

<u>Theorem 8.5</u>. Every sequence $\{x^k\}$ calculated by Algorithm 8.1 minimizes f on S: $\{x^k\} \subset S$ and $f(x^k)\downarrow\inf\{f(x):x\in S\}$. Moreover, $\{x^k\}$ converges to a solution of problem (8.1) whenever problem (8.1) has any solutions.

<u>Corollary 8.6</u>. If problem (8.1) has a solution and the stopping parameter ε_s is positive, then Algorithm 8.1 terminates in a finite number of iterations.

<u>Corollary 8.7</u>. If the level set $\{x \in S : f(x) \le f(x^1)\}$ is bounded and the

stopping parameter ε_s is positive, then Algorithm 8.1 terminates in a finite number of iterations.

Let us now consider the method with subgradient selection. This method is obtained from Algorithm 8.1 by using the additional constraints $\lambda_p=0$ and $\mu_p=0$ in subproblem (8.10) and demanding that the sets

$$\hat{J}_f^k=\{j\in J_f^k:\lambda_j^k\neq 0\} \quad\text{and}\quad \hat{J}_F^k=\{j\in J_F^k:\mu_j^k\neq 0\} \tag{8.25a}$$

should satisfy

$$|\hat{J}_f^k\cup\hat{J}_F^k|\leq N+1. \tag{8.25b}$$

The required multipliers λ_j^k and μ_j^k may be found by the Mifflin (1978) algorithm; see Remark 2.2.

One may easily verify that the method with subgradient selection is globally convergent in the sense of Theorem 8.5 and Corollaries 8.6-8.7. To this end, it suffices to modify the preceding convergence analysis of the method with subgradient aggregation by using (5.5.2)-(5.5.4) in the proof of an analogue of Lemma 4.5.

Algorithm 8.1 may be modified by incorporating the approximation tolerance updating strategy of Algorithm 6.1. This will not impair the preceding convergence results, since an analogue of Lemma 6.3 (in which one uses the additional assumption that $\hat{x}\in S$ and then asserts that $0\in\partial H(\bar{x};\bar{x})$ and $\bar{x}\in S$) may be established by combining the proof of Lemma 6.3 with the proofs of Lemma 5.4.14 and Theorem 5.4.15.

The line search rules of Algorithm 8.1 may be modified by replacing v^k with $\tilde{v}^k$ given by (6.5), where s^k is the Lagrange multiplier for the last constraint of (8.10) (see Lemma 2.1(iv)), and imposing the additional requirement $m_e+m_R\leq 1$ on the choice of the line search parameters. Since subproblem (8.10) is of the form (2.5) and we have (8.19)-(8.20) with $k\in J_f^k$ if $y^k\in S$ and $k\in J_F^k$ if $y^k\notin S$, the proof of Lemma 6.4 remains valid for this modification of Algorithm 8.1. Therefore, this modification of Algorithm 8.1 retains all the convergence properties of the original method.

We may add that the two above-described modifications of Algorithm 8.1 may be incorporated in the method with subgradient selection with no essential changes in its convergence analysis.

9. Extensions to Nonconvex Constrained Problems.

In this section we shall present extensions of the bundle methods of the preceding section for solving the following problem

$$\text{minimize } f(x), \quad \text{subject to } F(x) \le 0, \tag{9.1}$$

where the functions $f: R^N \to R$ and $F: R^N \to R$ are locally Lipschitzian but not necessarily convex or differentiable.

Our extension of the bundle methods for convex constrained minimization to the nonconvex case will use the techniques for dealing with nonconvexity developed in Sections 6.2-6.5. Thus we only need to modify those constructions of Section 8 which depended on the convexity of the problem functions.

First, we recall that in the nonconvex case it is convenient to use the following outer approximation to the subdifferential $\partial H(x;x)$ of the improvement function $H(\cdot;x)$ at x

$$\hat{M}(x) = \begin{cases} \partial f(x) & \text{if } F(x) < 0, \\ \operatorname{conv}\{\partial f(x) \cup \partial F(x)\} & \text{if } F(x) = 0, \\ \partial F(x) & \text{if } F(x) > 0. \end{cases} \tag{9.2}$$

Then a point $\bar{x} \in R^N$ is stationary for f on S if it satisfies the necessary optimality condition $0 \in \hat{M}(\bar{x})$ and $\bar{x} \in S$. Also for any $x \in S$ and $y \in R^N$ the subgradient locality measures

$$\alpha_f(x,y)=\max\{|f(x)-\bar{f}(x;y)|,\ \gamma_f|x-y|^2\},$$

$$\alpha_F(x,y)=\max\{|\bar{F}(x;y)|,\ \gamma_F|x-y|^2\},$$

defined in terms of the linearizations

$$\bar{f}(x;y)=f(y)+\langle g_f(y),x-y\rangle \quad \text{and} \quad \bar{F}(x;y)=F(y)+\langle g_F(y),x-y\rangle,$$

indicate how much the subgradients $g_f(y) \in \partial f(y)$ and $g_F(y) \in \partial F(y)$ differ from being elements of $\hat{M}(x)$, respectively, where γ_f (γ_F) is a positive parameter which may be set to zero if f (F) is convex. More compactly, we may define the subgradient mapping $g(\cdot)$ and its locality measure $\alpha(x,\cdot)$ at any $x \in S$ by

$$\begin{aligned} &g(y)=g_f(y) \quad \text{and} \quad \alpha(x,y)=\alpha_f(x,y) \quad \text{if } y \in S, \\ &g(y)=g_F(y) \quad \text{and} \quad \alpha(x,y)=\alpha_F(x,y) \quad \text{if } y \notin S. \end{aligned} \tag{9.3}$$

Using the above concepts and incorporating the subgradient aggregation techniques of Algorithm 6.3.1 into Algorithm 8.1, we obtain the following method.

Algorithm 9.1.

Step 0 (Initialization). Select a starting point $x^1 \in S$, a stopping parameter $\varepsilon_s \geq 0$ and an approximation tolerance $\varepsilon_a \geq 0$. Choose positive line search parameters m_L, m_R, m_e, $\bar{t}$ and $\tilde{t}$ satisfying $m_L < m_R < 1$, $m_e < 1$ and $\bar{t} \leq 1 \leq \tilde{t}$. Select distance measure parameters $\gamma_f > 0$ and $\gamma_F > 0$ ($\gamma_f = 0$ if f is convex; $\gamma_F = 0$ if F is convex), and a positive resetting tolerance $\bar{a}$. Set $y^1 = x^1$, $J_f^1 = J_F^1 = \{1\}$, $p_f^0 = g_f^1 = g_f(y^1)$, $p_F^0 = g_F^0 = g_F(y^1)$, $f_p^1 = f_1^1 = f(y^1)$, $F_p^0 = F_1^1 = F(y^1)$, $s_f^1 = s_F^1 = s_1^1 = 0$ and $e^1 = e_a$. Set $a^1 = 0$ and $r_a^1 = 1$. Set the counters $k=1$, $l=0$ and $k(0)=1$.

Step 1 (Direction finding). Find multipliers λ_j^k, $j \in J_f^k$, λ_p^k, μ_j^k, $j \in J_F^k$, and μ_p^k that

$$
\begin{aligned}
&\underset{\lambda,\mu}{\text{minimize}}\ \tfrac{1}{2}\Big|\sum_{j \in J_f^k} \lambda_j g_f^j + \lambda_p p_f^{k-1} + \sum_{j \in J_F^k} \mu_j g_F^j + \mu_p p_F^{k-1}\Big|^2, \\
&\text{subject to } \lambda_j \geq 0,\ j \in J_f^k,\ \lambda_p \geq 0,\ \mu_j \geq 0,\ j \in J_F^k,\ \mu_p \geq 0, \\
&\qquad \sum_{j \in J_f^k} \lambda_j + \lambda_p + \sum_{j \in J_F^k} \mu_j + \mu_p = 1, \qquad (9.4) \\
&\qquad \lambda_p = \mu_p = 0 \quad \text{if} \quad r_a^k = 1, \\
&\qquad \sum_{j \in J_f^k} \lambda_j \alpha_{f,j}^k + \lambda_p \alpha_{f,p}^k + \sum_{j \in J_F^k} \mu_j \alpha_{F,j}^k + \mu_p \alpha_{F,p}^k \leq e^k,
\end{aligned}
$$

where the subgradient locality measures are defined by

$$
\begin{aligned}
&\alpha_{f,j}^k = \max\{|f(x^k) - f_j^k|,\ \gamma_f (s_j^k)^2\} \quad \text{and} \quad \alpha_{F,j}^k = \max\{|F_j^k|, \gamma_F (s_j^k)^2\}, \\
&\alpha_{f,p}^k = \max\{|f(x^k) - f_p^k|,\ \gamma_f (s_f^k)^2\} \quad \text{and} \quad \alpha_{F,p}^k = \max\{|F_p^k|, \gamma_F (s_F^k)^2\}.
\end{aligned} \qquad (9.5)
$$

Calculate the scaled multipliers ν_f^k, ν_F^k, $\tilde{\lambda}_j^k$, $j \in J_f^k$, $\tilde{\lambda}_p^k$, $\tilde{\mu}_j^k$, $j \in J_F^k$, and $\tilde{\mu}_p^k$ by (8.11). Compute the aggregate subgradients

$$
\begin{aligned}
&(p_f^k, \tilde{f}_p^k, \tilde{s}_f^k) = \sum_{j \in J_f^k} \tilde{\lambda}_j^k (g_f^j, f_j^k, s_j^k) + \tilde{\lambda}_p^k (p_f^{k-1}, f_p^k, s_f^k), \\
&(p_F^k, \tilde{F}_p^k, \tilde{s}_F^k) = \sum_{j \in J_F^k} \tilde{\mu}_j^k (g_F^j, F_j^k, s_j^k) + \tilde{\mu}_p^k (p_F^{k-1}, F_p^k, s_F^k),
\end{aligned} \qquad (9.6)
$$

$$p^k = \nu_f^k p_f^k + \nu_F^k p_F^k$$

and the corresponding locality measures

$$\tilde{\alpha}_{f,p}^k = \max\{|f(x^k)-\tilde{f}_p^k|, \gamma_f(\tilde{s}_f^k)^2\} \text{ and } \tilde{\alpha}_{F,p}^k = \max\{|F_p^k|, \gamma_F(s_F^k)^2\},$$
$$\tilde{\alpha}_p^k = \nu_f^k \tilde{\alpha}_{f,p}^k + \nu_F^k \tilde{\alpha}_{F,p}^k. \tag{9.7}$$

Set $d^k = -p^k$ and $v^k = -|p^k|^2$. If $\lambda_p^k = \mu_p^k = 0$ set $a^k = \max\{s_j^k : j \in J_f^k \cup J_F^k\}$.

<u>Step 2 (Stopping criterion)</u>. If $\max\{|p^k|^2, \tilde{\alpha}_p^k\} \le \varepsilon_s$, terminate; otherwise continue.

<u>Step 3 (Approximation tolerance decreasing)</u>. If $|p^k|^2 > \tilde{\alpha}_p^k$, go to Step 4; otherwise, replace e^k by $m_e e^k$ and go to Step 1.

<u>Step 4 (Line search)</u>. By a line search procedure (e.g. Line Search Procedure 8.2), find three not necessarily different stepsizes t_L^k, t_R^k and t_B^k satisfying the requirements of Step 4 of Algorithm 8.1 with g and α defined by (9.3).

<u>Step 5 (Approximation tolerance updating)</u>. If $t_L^k = 0$, set $e^{k+1} = e^k$; otherwise, choose $e^{k+1} \ge e_a$ (e.g. by (6.1)), set $k(l+1) = k+1$, and increase l by 1.

<u>Step 6 (Linearization updating)</u>. Choose $\hat{J}_f^k \subset J_f^k$ and $\hat{J}_F^k \subset J_F^k$ such that $k(l) \in \hat{J}_f^k$ if $k(l) < k+1$, and set $J_f^{k+1} = \hat{J}_f^k \cup \{k+1\}$ and $J_F^{k+1} = \hat{J}_F^k \cup \{k+1\}$. Set $g_f^{k+1} = g_f(y^{k+1})$ and $g_F^{k+1} = g_F(y^{k+1})$. Calculate f_j^{k+1}, $j \in J_f^{k+1}$, f_p^{k+1}, F_j^{k+1}, $j \in J_F^{k+1}$, F_p^{k+1}, s_j^k, $j \in J_f^{k+1} \cup J_F^{k+1}$, s_f^{k+1} and s_F^{k+1} by (6.3.14).

<u>Step 7 (Distance resetting test)</u>. Set $a^{k+1} = \max\{a^k + |x^{k+1} - x^k|, s_{k+1}^{k+1}\}$. If $a^{k+1} \le \bar{a}$ or $t_L^k = 0$ then set $r_a^{k+1} = 0$ and go to Step 9; otherwise, set $r_a^{k+1} = 1$ and go to Step 8.

<u>Step 8 (Distance resetting)</u>. Delete from J_f^{k+1} and J_F^{k+1} all indices j with $s_j^{k+1} > \bar{a}/2$, and set $a^{k+1} = \max\{s_j^{k+1} : j \in J_f^{k+1} \cup J_F^{k+1}\}$.

<u>Step 9</u>. Increase k by 1 and go to Step 1.

A few comments on the algorithm are in order.

Since the method uses the subgradient aggregation rules of Algorithm 6.3.1, the properties of the aggregate subgradients (9.6) may be deduced from Lemma 6.4.1. Moreover, the stationarity measure

$$w^k = \frac{1}{2}|p^k|^2+\tilde{\alpha}_p^k \tag{9.8a}$$

satisfies

$$w^k \le 2\max\{|p^k|^2,\tilde{\alpha}_p^k\} \quad \text{and} \quad \max\{|p^k|^2,\tilde{\alpha}_p^k\} \le 2w^k, \tag{9.8b}$$

hence the stopping criterion of the method may be interpreted similarly to that of Algorithm 6.3.1; see Section 6.3.

The method updates the approximation tolerances e^k according to the modified rules of Section 6. It is easy to see that we always have $\tilde{\alpha}_p^k \le e^k$ as in Algorithm 3.1, since the proof of Lemma 6.4.6 provides a suitable extension of relation (3.8). We may add that, as in Section 7, we do not use the approximation tolerance updating strategy of Algorithm 6.1, because it may impair global convergence in the nonconvex case.

Line Search Procedure 8.2 may be used for executing Step 4 of the method. Using the proof of Lemma 8.3, one readily verifies that this procedure will terminate in a finite number of iterations if f and F have the semismoothness properties (3.3.23) and (6.3.18), respectively.

The subgradient deletion rules of Step 7 and 8 are borrowed from Algorithm 7.1, i.e. they differ slightly from those of Algorithm 6.3.1 in that no distance reset can occur after a null step. Thus the latest subgradients can never be deleted, i.e. $k+1 \in J_f^{k+1} \cap J_F^{k+1}$ at Step 9, since $s_{k+1}^{k+1}=|y^{k+1}-x^{k+1}|=0$ if $t_L^k > 0$. Moreover, as in Algorithm 7.1, we have relation (8.23), with ensures that the constraints of the k-th subproblem (9.4) are consistent (the set $G_a^k(e^k)$ defined by (8.9) is nonempty). We may add that Remarks 5.3.4 and 6.2.3 and Section 6.7 indicate how to modify the choice of J_f^{k+1} and J_F^{k+1} and impose the additional constraint $\mu_p=0$ in subproblem (9.4) at certain iterations in order to treat the case when one cannot, or does not want to, compute $f(x)$ and $g_f(x)$ at $x \notin S$, and $F(x)$ and $g_F(x)$ at $x \in S$.

We shall now establish convergence of the method, assuming that $\varepsilon_s=0$.

<u>Theorem 9.2</u>. Suppose that Algorithm 9.1 generates a sequence $\{x^k\}$. Then:

(i) If $\{x^k\}$ is finite, then its last element x^k is stationary for f on S.

(ii) If $\{x^k\}$ is infinite, then every accumulation point $\bar{x}$ of $\{x^k\}$ is stationary for f on S.

(iii) If f and F are convex and $F(\tilde{x})<0$ for some $\tilde{x}$, then $\{x^k\}$ minimizes f on S, i.e. $\{x^k\}\subset S$ and $f(x^k)\downarrow\inf\{f(x):x\in S\}$. Moreover, $\{x^k\}$ converges to a minimum point of f on S whenever f attains its infimum on S.

Proof. The proof may be constructed by modifying the analysis of Section 4 with the aid of the results of Sections 7,8 and 6.4. Since a formal proof would involve lengthy repetitions of the preceding results, we only give an outline of the required analysis.

(i) Suppose that $\bar{x}=x^k$ is the last point generated by the method. If the algorithm stops at the k-th iteration, then $\max\{|p^k|^2,\tilde{\alpha}_p^k\}\le\varepsilon_s=0$, so $w^k=0$ by (9.8), and the proof of Lemma 6.4.3 yields the stationarity of x^k. The case when the method cycles infinitely between Steps 1 and 3 may be easily analyzed by using the proof of Lemma 4.2 and Lemma 6.4.5, which is valid for Algorithm 9.1, as will be shown below.

(ii) Suppose that $\{x^k\}$ is infinite and has an accumulation point $\bar{x}$. Reasoning as in the proof of Lemma 7.2, one may deduce from (8.23) and the proofs of Lemmas 6.4.1 and 6.4.5 that Lemma 6.4.5 is true for Algorithm 9.1. This yields, by (9.8), an analogue of Lemma 4.3. The proofs of analogues of Lemmas 4.4-4.7 are similar to the ones discussed in Section 8 in connection with Algorithm 8.1. Combining all these results, one shows that $\bar{x}$ is stationary for f on S.

(iii) Suppose that problem (9.1) is convex and satisfies the Slater constraint qualification. Then the properties of Algorithm 9.1 are essentially those of Algorithm 8.1, e.g. we have (8.18) and $t_L^k\le\tilde{t}$ for all k, so the desired conclusion may be deduced from part (ii) of the theorem and the results of Section 8. □

From the above proof we deduce easily that Corollary 8.7 is true for Algorithm 9.1.

We may add that Algorithm 9.1 reduces to a method with subgradient selection if one uses the additional constraint $\lambda_p=\mu_p=0$ in subproblem (9.4) and calculates multipliers satisfying (8.25). Also its line search rules may be modified by replacing v^k with $\tilde{v}^k$ as in Section 9. Following the analysis of Section 8, one may show that such modifications do not impair the preceding convergence results.

CHAPTER 8

Numerical Examples

1. Introduction

In this chapter we give numerical results for several optimization problems. Our intention is to give the reader some feeling for the speed of convergence he/she can expect when solving problems with several variables by the methods discussed in the preceding chapters. Also we think that it is too early to compare the efficiency of the various existing algorithms. For these reasons, we only describe results obtained with a very unsophisticated implementation of the algorithms with subgradient deletion rules from Chapters 4 and 6.

The algorithms were programmed in FORTRAN on a PDP 11/70 computer with the relative accuracy of 10^{-17} in double precision (seventeen-digits precision). Since the number N of variables was relatively small, the algorithms used subgradient selection and at most $M_g=N+4$ subgradients for each search direction finding (see Sections 4.5 and 6.7). The following standard values of line search and locality parameters were used: $m_L=0.1$, $m_R=0.5$, $\bar{a}=10^3$, $\bar{t}=0.01$, $\theta=0.5$, $s^1=1$, $m_a=10^{-3}$. The stopping criterion

$$|d^k| > m_a a^k \quad \text{and} \quad w^k \le \varepsilon_s$$

was employed with various values of the final accuracy tolerance ε_s.

In the next section k denotes the final (or current) iteration number, Lf is the total number of the objective function/subgradient evaluations, whereas LF is the number of the constraint function/ /subgradient evaluations. We also give the solution $\bar{x}$ to the problem in question whenever it is known.

2. Numerical Results

2.1. Shor's Problem

$$f(x) = \max\{f_i(x): i=1,\dots,10\}, \quad x \in R^5,$$

$$f_i(x) = b_i \sum_{j=1}^{5} (x_j - a_{ij})^2, \quad i=1,\dots,10,$$

$$(b_i) = (1,5,10,2,4,3,1.7,2.5,6,3.5),$$

$$(a_{ij})^T = \begin{bmatrix} 0 & 2 & 1 & 1 & 3 & 0 & 1 & 1 & 0 & 1 \\ 0 & 1 & 2 & 4 & 2 & 2 & 1 & 0 & 0 & 1 \\ 0 & 1 & 1 & 1 & 1 & 1 & 1 & 1 & 2 & 2 \\ 0 & 1 & 1 & 2 & 0 & 0 & 1 & 2 & 1 & 0 \\ 0 & 3 & 2 & 2 & 1 & 1 & 1 & 1 & 0 & 0 \end{bmatrix},$$

$$\bar{x} = (1.12434, 0.97945, 1.47770, 0.92023, 1.12429),$$

$$f(\bar{x}) = 22.60016,$$

$$x^1 = (0,0,0,0,1), \; f(x^1) = 80.$$

The results for various values of ε_s are given in Table 2.1. We also have

$$x^{49} = (1.12433, 0.97943, 1.47749, 0.92027, 1.12425).$$

Table 2.1

ε_s	k	$f(x^k)$	Lf
10^{-4}	34	22.60021	64
10^{-5}	41	22.60017	76
10^{-6}	49	22.60016	90

2.2. Lemarechal´s Problem MAXQUAD

$$f(x) = \max\{\langle A^l x, x\rangle - \langle b^l, x\rangle : l=1,\ldots,5\}, \quad x \in R^{10},$$

$$A^l_{ij} = A^l_{ji} = \exp(i/j)\cos(i-j)\sin(l), \quad l \neq j,$$

$$A^l_{ii} = i\sin(i)/10 + \sum_{j\neq i} A^l_{ij},$$

$$b_i^l = \exp(i/l)\sin(i-l),$$

$$\bar{x} = (-0.1263,-0.0346,-0.0067,0.2668,0.0673,\\ 0.2786, 0.0744, 0.1387,0.0839,0.0385),$$

$$f(\bar{x}) = -0.8414.$$

This is the first problem of Lemarechal (1982). The starting point $x_i^1=1$, $i=1,\ldots,10$, has $f(\bar{x})=5337$. For $\varepsilon_s=10^{-4}$ we obtained $k=51$, $f(x^k)=-0.84136$, $Lf=102$,

$$x^{51} = (-0.1263,-0.0342,-0.0062,0.0269,0.0671,\\ -0.2783, 0.0744, 0.1385,0.0836,0.0383).$$

2.3. Ill -conditioned Linear Programming

The linear programming problem

$$\text{minimize } \langle c,x\rangle \text{ over all } x \in R^N$$

$$\text{satisfying } Ax \le b, \quad x \ge 0,$$

where

$$A_{ij} = 1/(i+j), \quad b_i = \sum_{j=1}^{N} 1/(i+j), \quad i,j=1,\ldots,N, \quad N>2,$$

$$c_i = -1/(1+1) - \sum_{j=1}^{N} 1/(i+j)$$

$$\bar{x} = (1,1,\ldots,1),$$

is ill-conditioned for $N\geq 5$, since A is essentially a section of the Hilbert matrix. The constraint function is

$$F(x) = \max\{\max[(Ax)_i-b_i: i=1,\ldots,N], \max[-x_i: i=1,\ldots,N]\}$$

and $F(\bar{x})=0$.

This problem can be solved by minimizing the exact penalty function

$$f(x) = \langle c,x\rangle + \varrho F(x)_+$$

over all x in R^N, where $\varrho=2N$ is the penalty coefficient. Note that f is polyhedral. We use the feasible starting point $x^1=0$ (with $f(x^1)=0$). Table 2.2 contains results for $\varepsilon_s=10^{-7}$ and $N=5,10,15$, whereas Table 2.3 describes the case $N=15$ for various ε_s.

Table 2.2

N	$f(\bar{x})$	k	$f(x^k)$	Lf
5	-6.26865	14	-6.26865	31
10	-13.1351	23	-13.1351	47
15	-20.0420	32	-20.0420	67

Table 2.3

ε_s	k	$f(x^k)$	Lf
10^{-4}	16	-20.0411	26
10^{-5}	21	-20.0420	41
10^{-6}	25	-20.0420	51
10^{-7}	32	-20.0420	67

The results for the problem

$$\text{minimize } f(x) = \langle c,x \rangle, \quad \text{subject to } F(x) \le 0$$

obtained by the feasible point method (see Section 6.7) are given in Table 2.4.

Table 2.4

N	ε_s	k	$f(x^k)$	Lf	LF
5	10^{-4}	13	-6.26610	30	43
5	10^{-5}	20	-6.26861	52	74
10	10^{-4}	46	-13.1344	100	154
10	10^{-5}	50	-13.1348	110	168
15	10^{-3}	20	-19.9912	56	83
15	10^{-4}	43	-20.0396	117	174
15	10^{-5}	51	-20.0406	131	193

2.4. CRESCENT

$$f(x) = \max\{x_1^2 + (x_2-1)^2 + x_2 - 1,\ -x_1^2 - (x_2-1)^2 + x_2 + 1\},$$

$$\bar{x} = (0,0)\ ,\quad f(\bar{x}) = 0.$$

This objective function has narrow crescent-shaped level sets which force any descent algorithm to make very short steps. The results for $x^1=(-1.5,2)$ $(f(x^1)=4.25)$ are given in Table 2.5.

Table 2.5

ε_s	k	$f(x^k)$	Lf
10^{-6}	16	$8\cdot10^{-6}$	29
10^{-9}	21	$7\cdot10^{-6}$	36
10^{-12}	40	$9\cdot10^{-12}$	62

2.5. SHELL DUAL

The problem is to

$$\text{minimize}\quad 2\sum_{j=1}^{5} d_j y_j^3 + \langle cy,y\rangle - \langle b,z\rangle \quad \text{over all}$$

$$(y,z) \in R^5 \times R^{10}$$

$$\text{satisfying}\ (Ax)_j - 2(Cy)_j - 3d_j y_j^2 - e_j \le 0 \quad \text{for}\quad j=1,\dots,5,$$

$$-y_j \le 0 \quad \text{for}\quad j=1,\dots,5,\quad -z_i \le 0 \quad \text{for}\quad i=1,\dots,10,$$

where $(y,z)=x \in R^{15}$. The problem data are given below.

The optimal point is

$$\bar{x} = (0.3,\ 0.3335,\ 0.4,\ 0.4283,\ 0.224,\ 0,\ 0,\ 5.1741,\ 0,\ 3.0611,\ 11.8396,\ 0,\ 0,\ 0.1039,\ 0)$$

with $f_0(\bar{x})=32.3488$, whereas the starting point

$$x_i^1 = 10^{-4} \quad \text{for}\quad i\neq 7,\quad x_7^1=60$$

has $f_0(x_1)=2400$ and $F(x^1)<0$. Here f_0 denotes the problem objective, while F is the total constraint function $(F=\max(F_i:\ i=1,\dots,20)$, where F_i are the constraints).

matrix A					b
-16	2.	0.	1.	0.	-40
0.	-2	0.	4.	2.	-2.
-3.5	0.	2.	0.	0.	-0.25
0.	-2	0.	-4.	-1.	-4.
0.	-9.	-2.	-1.	-2.8	-4.
2.	0.	-4.	0.	0.	-1.
-1.	-1.	-1.	-1.	-1.	-40.
-1.	-2.	-3.	-2.	-1.	-60.
1.	2.	3.	4.	5.	5.
1.	1.	1.	1.	1.	1.

symmetric matrix C					d	e
30	-20	-10	32	-10	4	-15
-20	39	-6	-31	32	8	-27
-10	-6	10	-6	-10	10	-36
32	-31	-6	39	-20	6	-18
-10	32	-10	-20	30	2	-12

This problem can be solved by minimizing its exact penalty function

$$f(x) = f_0(x) + 500\, F(x)_+$$

over all x in R^{15}. The problem is quite difficult to solve by general-purpose nonsmooth optimization methods (see Lemarechal (1982)). The algorithm stopped by reaching the iteration limit ITMAX=300 with $\varepsilon_s=10^{-4}$. Table 2.6 illustrates its progress.

Table 2.6

k	$f(x^k)$	Lf
100	32.85	259
150	32.54	384
200	32.38	497
250	32.36	626
300	32.35	766

2.6. Electronic Filter Design

$$f(x) = \max\{|e(x,h_i)| : i=1,\dots,41\},$$

$$e(x,h) = H(x,\pi h) - S(h),$$

$$x = (a_1,b_1,c_1,d_1,a_2,b_2,c_2,d_2,A) \in R^9,$$

$$H(x,g)=A\prod_{i=1}^{2}\left(\frac{1+a_i^2+b_i^2+2b_i(2\cos^2 g-1)+2a_i(1+b_i)\cos g}{1+c_i^2+d_i^2+2d_i(2\cos^2 g-1)+2c_i(1+d_i)\cos g}\right)^{1/2},$$

$$g = \pi h,$$

$$S(h) = |1-2h|,$$

$$h_i = (i-1)0.01 \quad \text{for} \quad i=1.,,,.6, \quad h_i=0.07+(i-7) \quad \text{for} \quad i=7,\dots,20,$$

$$h_{21}=0.5,\ h_{22}=0.54,\ h_{23}=0.57,\ h_{24}=0.62,$$

$$h_i=0.63+(i-25)0.03 \quad \text{for} \quad i=25,\dots,35,$$

$$h_i=0.95+(i-36)0.01 \quad \text{for} \quad i=36,\dots,41,$$

$$\bar{x} =(0,0.980039,0,-0.165771,0,-0.735078,0,-0.767228,0.3679),$$

$$x^1=(0,\ 1,\ 0,\ -1.5,\ 0,\ -6.28,\ 0,\ -0.72,\ 0.37),$$

$$f(\bar{x})= 6.1853\cdot 10^{-3}, \quad f(x^1) = 0.6914.$$

This problem originated in the optimal design of electronic filters (Charalambous,1979).

Table 2.7 gives results for various ε_s.

Table 2.7

ε_s	k	$f(x^k)$	Lf
10^{-3}	23	$17\cdot 10^{-3}$	43
10^{-4}	54	$14\cdot 10^{-3}$	105
10^{-5}	201	$6.45\cdot 10^{-3}$	400

2.7. Feedback Controller Design

The following problem arises in the design of robust feedback regulators for linear multivariable control systems (see Gustafson and Desoer (1983) and Kiwiel (1984c) for details). The plant transfer matrix is

$$P(s) = \frac{1}{(s+2)^2(s+3)} \begin{bmatrix} s^2+8s+10 & 3s^2+7s+4 \\ 2s+2 & 3s^2+9s+8 \end{bmatrix}.$$

The compensator transfer matrix $C(s,x)$ depending on the design parameters x R^2 is found in terms of $P(s)$ and a matrix $Q(s,x)$ as

$$C(s,x) = Q(s,x)\ (I-P(s)Q(s,x))^{-1},$$

where

$$Q(s,x) = \frac{1}{2s+4} \begin{bmatrix} \frac{3s^2+9s+8}{d_1(s,x)} & \frac{-3s^2-7s-4}{d_2(s,x)} \\ \frac{-2s-2}{d_1(s,x)} & \frac{s^2-8s+10}{d_2(s,x)} \end{bmatrix},$$

$$d_i(s,x) = (sx_i)^2 + \sqrt{2}sx_i+1\ , \quad i=1,2.$$

Here $1/x_1$ and $1/x_2$ can be interpreted as bandwidths. The size of the maximum singular value $\bar{\sigma}(w,x)$ of the matrix

$$G(w,x) = Q(\sqrt{-1}\ w,x)$$

gives an upper bound on the noise power per hertz in any channel at the plant input. The design requirement

$$\bar{\sigma}(w,x) \le 2.5 \quad \text{for} \quad w \in \Omega = \{1,\ 1.2, \ldots, 2.6,\ 2.8\}$$

can be formulated in terms of the function

$$F(x) := \max\{\langle z, G(w,x)G(w,x)^* z\rangle : \|z\| = 1,\ \ w \in \Omega\} - 6.25$$

as

$$F(x) \le 0.$$

(In view of Lemma 1.2.5., the value and a subgradient of F at x can easily be found by computing the eigenvector of $G(w,x)G(w,x)^*$ corresponding to its maximum eigenvalue.)

The design problem is to choose the compensator parameters $x \in R^2$ such that the bandwidths $1/x_1$ and $1/x_2$ are as large as possible, subject to $F(x) \leq 0$. Thus we want to have small values of both x_1 and x_2, so our design problem is multiobjective (has two criteria). Of course, multicriteria optimization problems should be solved in an interactive mode. We give results for four typical auxiliary subproblems that the designer may wish to solve in order to explore the design possibilities. The subproblems are obtained by choosing different scalarizations of the two objectives and handling the constraint $F(x) \leq 0$ via exact penalties. The subproblem objectives are

Problem 1: $f(x) = 0.8x_1 + 0.2x_2 + 100\, F(x)_+$,

Problem 2: $f(x) = \max\,(x_1, x_2) + 10\, F(x)_+$,

Problem 3: $f(x) = \max\,(x_1, x_2 - 0.5) + 10\, F(x)_+$,

Problem 4: $f(x) = \max\,(x_1 - 0.5, x_2) + 10\, F(x)_+$.

The starting point $x^1 = (1,1)$ has $F(x_1) = 0.420997$. Table 2.8 gives results for $\varepsilon_s = 10^{-5}$.

Table 2.8

Problem	k	Lf	$f(x^k)$	x_1^k	x_2^k	$F(x^k)_+$
1	19	37	0.704194	0.5030	1.5088	0
2	15	45	1.032101	1.0321	1.0321	0
3	21	69	0.740497	0.7405	1.2405	$5 \cdot 10^{-7}$
4	31	78	0.923553	1.4236	0.9236	0

References

Auslender A. (1978). Minimisation de fonctions localement lipschitziennes : Applications à la programmation mi-convexe, mi-différentiable. Nonlinear Programming 3, O.L. Managasarian, P.R. Meyer and S.M. Robinson, eds., Academic Press, New York, pp. 429-460.

Auslender A. (1982). On the differential properties of the support function of the ε-subdifferential of a convex function. Math. Programming, 24, 257-268.

Auslender A. (1985). Numerical methods for nondifferentiable convex minimization. Math. Programming Study (to appear).

Bazaraa M.S. and C.M. Shetty (1979). Nonlinear Programming. Theory and Algorithms. Wiley, New York.

Bertsekas D.P. and S.K. Mitter (1973). A descent numerical method for optimization problems with nondifferentiable cost functionals. SIAM J. Control, 11, 637-652.

Bihain A. (1984). Optimization of upper semidifferentiable functions. J. Optimiz. Theory Appl. 44, 545-568.

Charalambous C. (1979). Acceleration of the least p-th algorithm for minimax optimization with engineering applications. Math. Programming, 17, 270-297.

Cheney E.W. and A.A. Goldstein (1959). Newton's method for convex programming and Chebyshev approximation. Num. Math., 1, 253-268.

Clarke F.H. (1975). Generalized gradients and applications. Trans. Amer. Math. Soc., 205, 247-262.

Clarke F.H. (1976). A new approach to Lagrange multipliers. Math. Oper. Res., 1, 165-174.

Clarke F.H. (1983). Nonsmooth Analysis and Optimization. Wiley - Interscience, New York.

Dantzig G.B. (1963). Linear Programming and Extensions. Princeton University Press, Princeton, New Yersey.

Demyanov V.F. and V.N. Malozemov (1974). Introduction to Minimax. Wiley, New York.

Demyanov V.F. and L.V. Vasilev (1981). Nondifferentiable Optimization. Optimization Software Inc./Springer: New York (to appear, 1985). Russian edition: Nauka, Moscow (1981).

Demyanov V.F., C. Lemaréchal and J. Zowe (1985). Trying to approximate a set-valued mapping. Nondifferentiable Optimization : Theory and Applications, V.F. Demyanov ed., Lecture Notes in Control and Information Sciences, Springer, Berlin (to appear).

Dixon L.C.W. and M. Gaviano (1980). Reflections on nondifferentiable optimization, part 2, convergence. J. Optim. Theory Appl., 32, 259-275.

Eaves B.C. and W.I. Zangwill (1971). Generalized cutting plane algorithms. SIAM J. Control, 9, 529-542.

Fletcher R. (1981). Practical Methods of Optimization, Vol.2, Constrained Optimization. Wiley, New-York.

Fukushima M. (1984). A descent algorithm for nonsmooth convex programming. Math. Programming, 30, 163-175.

Gaudioso M. and M.F. Monaco (1982). A bundle type approach to the unconstrained minimization of convex nonsmooth functions. Math. Programming, 23, 216-226.

Goldstein A.A. (1977). Optimization of Lipschitz continuous functions. Math. Programming, 13, 14-22.

Gupal A.M. (1979). Stochastic Methods for Solving Nonsmooth Extermal Problems. Kiev, Naukova Dumka (in Russian).

Gustafson C.L. and C.A. Desoer (1983). Controller design for linear multivariable feedback systems with stable plants, using optimization with inequality constraints. Int. J. Control, 37, 881-907.

Gwinner J. (1981). Bibliography on nondifferentiable optimization and non-smooth analysis. J. Comput. Appl. Math., 7, 277-285.

Hiriart-Urruty J.B. (1983). The approximate first-order and second-order directional derivatives for a convex function. Proceedings of the Conference on Mathematical Theories of Optimization, Lecture Notes in Mathematics 979, Springer, Berlin.

Huard P. (1967). Resolution of mathematical programming with nonlinear constraints by the method of centers. Nonlinear Programming, J. Abadie, ed., Academic Press, New York.

Kelley J.E. (1960). The cutting plane method for solving convex programs: J. SIAM, 8, 703-712.

Kiwiel K.C. (1981a). A globally convergent quadratic approximation algorithm for inequality constrained minimax problems. CP-81-9, International Institute for Applied Systems Analysis, Laxenburg, Austria. (revised version: A phase I - phase II method for inequality constrained minimax problems. Control Cyb., 12 (1983) 55-75).

Kiwiel K.C. (1981b). A variable metric method of centers for nonsmooth minimization. CP-81-27, International Institute for Applied Systems Analysis, Laxenburg, Austria.

Kiwiel K.C. (1983). An aggregate subgradient method for nonsmooth convex minimization. Math. Programming, 27, 320-341.

Kiwiel K.C. (1984a). A linearization algorithm for constrained nonsmooth minimization. System Modelling and Optimization, P. Thoft-Christensen, ed., Lecture Notes in Control and Information Sciences 59, Springer, Berlin, pp. 311-320.

Kiwiel K.C. (1984b). A quadratic approximation method for minimizing a class of quasidifferentiable functions. Numer. Mathematik, 45, 411-430.

Kiwiel K.C. (1984c). An algorithm for optimization problems with singular values of control systems. Proc. IFAC 9th World Congress, J. Gertler and L. Keviczky, eds., Pergamon Press, Oxford (to appear).

Kiwiel K.C. (1985a). An exact penalty function algorithm for nonsmooth constrained convex minimization problems. IMA J. Num. Anal. (to appear).

Kiwiel K.C. (1985b). A method for minimizing the sum of a convex function and a continuously differentiable function. J. Optim. Theory Appl. (to appear).

Kiwiel K.C. (1985c). A linearization algorithm for nonsmooth minimization. Math. Oper. Res. (to appear).

Kiwiel K.C. (1985d). An algorithm for linearly constrained convex nondifferentiable minimization problems. J. Math. Anal. Appl. (to appear).

Kiwiel K.C. (1985e). A descent method for nonsmooth convex multiobjective minimization. Large Scale Systems (to appear).

Kiwiel K.C. (1985f). A decomposition method of descent for minimizing a sum of convex nonsmooth functions. J. Optim. Theory Appl. (to appear).

Kiwiel K.C. (1985g). An algorithm for nonsmooth convex minimization with errors. Math. Comput. (to appear).

Kiwiel K.C. (1985h). A method of linearizations for minimizing certain quasidifferentiable functions. Math. Programming Study (to appear).

Kiwiel K.C. (1985i). Descent methods for nonsmooth convex constrained minimization. Nondifferentiable Optimization: Theory and Applications, V.F.Demyanov, ed., Lecture Notes in Control and Information Sciences (to appear).

Lasdon L.S. (1970). Optimization Theory for Large Systems. Macmillan, London.

Lemarechal C. (1975). An extension of Davidon methods to nondifferentiable problems. Nondifferentiable Optimization, M.L. Baliński and P.Wolfe, eds., Mathematical Programming Study 3, North-Holland, Amsterdam, pp. 95-109.

Lemarechal C. (1976). Combining Kelley's and conjugate gradient methods. Abstract, IX International Symposium on Mathematical Programming, Budapest.

Lemarechal C. (1978). Nonsmooth optimization and descent methods. RR-78-4, International Institute for Applied Systems Analysis, Laxenburg, Austria.

Lemarechal C. (1978b). Bundle methods in nonsmooth optimization. Nonsmooth Optimization, C.Lemarechal and R.Mifflin, eds., Pergamon Press, Oxford, pp. 79-102.

Lemarechal C. (1980). Extensions diverses des methods de gradient et applications. These d´Etat, Universite de Paris IX.

Lemarechal C. (1981). A view of line searches. Optimization and Optimal Control, W.Oettli and J.Stoer, eds., Lecture Notes in Control and Information Sciences 30, Springer, Berlin, pp. 59-78.

Lemarechal C. (1982). Numerical experiments in nonsmooth optimization. Progress in Nonsmooth Optimization, E.Nurminski, ed., CP-82-S8, International Institute for Applied Systems Analysis, Laxenburg, Austria, pp. 61-84.

Lemarechal C. and R.Mifflin, eds. (1978). Nonsmooth Optimization. Pergamon Press, Oxford.

Lemarechal C. and R.Mifflin (1982). Global and superlinear convergence of an algorithm for one-dimensional minimization of convex functions. Math. Programming, 24, 241-256.

Lemarechal C., J.-J.Strodiot and A.Bihain (1981). On a bundle algorithm for nonsmooth optimization. Nonlinear Programming 4, O.L.Mangasarian, R.R.Mayer and S.M.Robinson, eds., Academic Press, New York, pp. 245-281.

Lemarechal C. and J.-J.Strodiot (1985). Bundle methods, cutting plane algorithms and ϵ-Newton directions. Nondifferentiable Optimization: Theory and Applications, V.F.Demyanov, ed. Lecture Notes in Control and Information Sciences (to appear).

Lemarechal C. and J.Zowe (1983). Some remarks on the construction of higher order algorithms for convex optimization. Appl. Math. Optim., 10, 51-68.

Madsen K. and H.Schjaer-Jackobsen (1978). Linearly constrained minimax optimization. Math. Programming, 14, 208-223.

Mifflin R. (1977a). Semismooth and semiconvex functions in constrained optimization. SIAM J. Control Optim., 15, 959-972.

Mifflin R. (1977b). An algorithm for constrained optimization with semismooth functions. Math. Oper. Res., 2, 191-207.

Mifflin R. (1978). A feasible descent algorithm for linearly constrained least squares problems. Nonsmooth Optimization, C. Lemarechal and R.Mifflin, eds., Pergamon Press, Oxford, pp. 103-126.

Mifflin R. (1982). A modification and an extension of Lemarechal´s algorithm for nonsmooth minimization. Nondifferential and Variational Techniques in Optimization, D.C.Sorensen and R.J.-B. Wets, eds., Mathematical Programming Study 17, pp. 77-90.

Mifflin R. (1983). A superlinearly convergent algorithm for one-dimensional constrained minimization with convex functions. Math. Oper. Res., 8, 185-195.

Mifflin R. (1984). Better than linear convergence and safeguarding in nonsmooth minimization. System Modelling and Optimization, P. Thoft-Christensen, ed., Lecture Notes in Control and Information Sciences 59, Springer, Berlin, pp. 321-330.

Mifflin R. (1985). A nested optimization application. Nondifferentiable Optimization: Theory and Applications, V.F.Demyanov, ed. Lecture Notes in Control and Information Sciences, Springer, Berlin (to appear).

Nurminski E.A. (1979). Numerical Methods for Solving Deterministic and Stochastic Minimax Problems. Kiev, Naukova Dumka (in Russian).

Nurminski E.A. (1981). On a decomposition of structured problems. WP-81-32, International Institute for Applied Systems Analysis, Laxenburg, Austria.

Nurminski E.A. (1982). Bibliography on nondifferentiable optimization. Progress in Nondifferentiable Optimization, E.Nurminski, ed., CP-82-S8, International Institute for Applied Systems Analysis, Laxenburg, Austria.

Pironneau O. and E.Polak (1972). On the rate of convergence of certain methods of centers. Math. Programming, 2, 230-257.

Pironneau O. and E.Polak (1973). Rate of convergence of a class of methods of feasible directions. SIAM J. Num. Anal., 10, 161-174.

Polak E. (1970). Computational Methods in Optimization. A Unified Approach. Academic Press, New York.

Polak E. and D.Q.Mayne (1981). A robust secant method for optimization problems with inequality constraints. J. Optim. Theory Appl., 33, 463-477.

Polak E., D.Q.Mayne and Y.Wardi (1983). On the extension of constrained optimization algorithms from differentiable to nondifferentiable problems. SIAM J. Control Optim., 21, 179-203.

Polak E., R.Trahan and D.Q.Mayne (1979). Combined phase I - phase II methods of feasible directions. Math. Programming, 17, 61-73.

Powell M.J.D. (1978). Algorithms for nonlinear constraints that use Lagrangian functions. Math. Programming, 14, 224-248.

Pshenichny B.N. (1980). Convex Analysis and Extremal Problems. Nauka, Moscow (in Russian).

Pshenichny B.N. and Yu.M.Danilin (1975). Numerical Methods for Extremal Problems. Nauka, Moscow (English translation, Mir, Moscow, 1978).

Rockafellar R.T. (1970). Convex Analysis. Princeton University Press, Princeton, New Yersey.

Rockafellar R.T. (1978). The theory of subgradients and its applications to problems of optimization. Lecture Notes, University of Montreal.

Rockafellar R.T. (1981). The theory of subgradients and its applications to problems of optimization. Convex and nonconvex functions. Research Notes in Mathematics 1, K.H.Hoffman and R.Wille, eds., Heldermann, Berlin.

Rzewski S.V. (1981). ε-Subgradient method for solving the convex programming problem. Zurn. Vyc. Mat. Mat. Fiz., 25, 1126-1132 (in Russian).

Shor N.Z. (1979). Methods for minimizing nondifferentiable functions and their applications. Kiev, Naukova Dumka (in Russian). (English translation: Minimization methods for nondifferentiable functions, Springer-Verlag, Berlin, 1985).

Strodiot J.-J., V.H.Nguyen and N.Heukemes (1983). ε-Optimal solutions in nondifferentiable convex programming and some related questions. Math. Programming, 25, 307-328.

Topkis D.M. (1970a). Cutting - plane methods without nested constraint sets. Oper. Res., 18, 404-413.

Topkis D.M. (1970b). A note on the cutting - plane method without nested constraint sets. Oper. Res., 18, 1216-1220.

Topkis D.M. (1982). A cutting - plane algorithm with linear and geometric rates of convergence. J. Optim. Theory Appl., 36, 1-22.

Wierzbicki A.P. (1978a). A quadratic approximation method based on augmented Lagrangian functions for nonconvex nonlinear programming problems. WP-78-61, International Institute for Applied Systems Analysis, Laxenburg, Austria.

Wierzbicki A.P. (1978b). Lagrangian functions and nondifferentiable optimization. WP-78-63, International Institute for Applied Systems Analysis, Laxenburg, Austria.

Wierzbicki A.P. (1982). Lagrangian functions and nondifferentiable optimization. Progress in Nondifferentiable Optimization, E.Nurminski, ed., CP-82-S8, International Institute for Applied Systems Analysis,Laxenburg, Austria, pp. 173-213.

Wolfe P. (1975). A method of conjugate subgradients for minimizing nondifferentiable convex functions. Nondifferentiable Optimization, M.L.Balinski and P.Wolfe, eds., Mathematical Programming Study 3, North-Holland, Amsterdam, pp. 145-173.

Wolfe P. (1976). Finding the nearest point in a polytope. Math. Programming, 11, 128-149.

Wolfe P. (1978). Sufficient minimization of piecewise-linear univariate functions. Nondifferentiable Optimization, C.Lemarechal and R.Mifflin, eds,, Pergamon Press, Oxford, pp. 127-130.

Zowe J. (1985). Nondifferentiable optimization - a motivation and a short introduction into the subgradient- and the bundle concept. ASI Proceedings on Computational Mathematical Programming (to appear).

INDEX